Fundamentals of Cavitation

FLUID MECHANICS AND ITS APPLICATIONS
Volume 76

Series Editor: R. MOREAU
MADYLAM
Ecole Nationale Supérieure d'Hydraulique de Grenoble
Boîte Postale 95
38402 Saint Martin d'Hères Cedex, France

Aims and Scope of the Series

The purpose of this series is to focus on subjects in which fluid mechanics plays a fundamental role.

As well as the more traditional applications of aeronautics, hydraulics, heat and mass transfer etc., books will be published dealing with topics which are currently in a state of rapid development, such as turbulence, suspensions and multiphase fluids, super and hypersonic flows and numerical modelling techniques.

It is a widely held view that it is the interdisciplinary subjects that will receive intense scientific attention, bringing them to the forefront of technological advancement. Fluids have the ability to transport matter and its properties as well as transmit force, therefore fluid mechanics is a subject that is particulary open to cross fertilisation with other sciences and disciplines of engineering. The subject of fluid mechanics will be highly relevant in domains such as chemical, metallurgical, biological and ecological engineering. This series is particularly open to such new multidisciplinary domains.

The median level of presentation is the first year graduate student. Some texts are monographs defining the current state of a field; others are accessible to final year undergraduates; but essentially the emphasis is on readability and clarity.

For a list of related mechanics titles, see final pages.

Fundamentals of Cavitation

by

JEAN-PIERRE FRANC

Research Director (CNRS), Turbomachinery and Cavitation Research Group, Laboratory of Geophysical and Industrial Fluid Flows (LEGI) of the Grenoble University (Institut National Polytechnique de Grenoble (INPG) & Université Joseph Fourier (UJF), France

and

JEAN-MARIE MICHEL

Presently retired, was Research Director (CNRS) and Head of the Cavitation Research Group in the Laboratory of Geophysical and Industrial Fluid Flows (LEGO) of the Grenoble University (Institut National Polytechnique de Grenoble (INPG) & Université Joseph Fourier (UJF), France

KLUWER ACADEMIC PUBLISHERS

DORDRECHT / BOSTON / LONDON

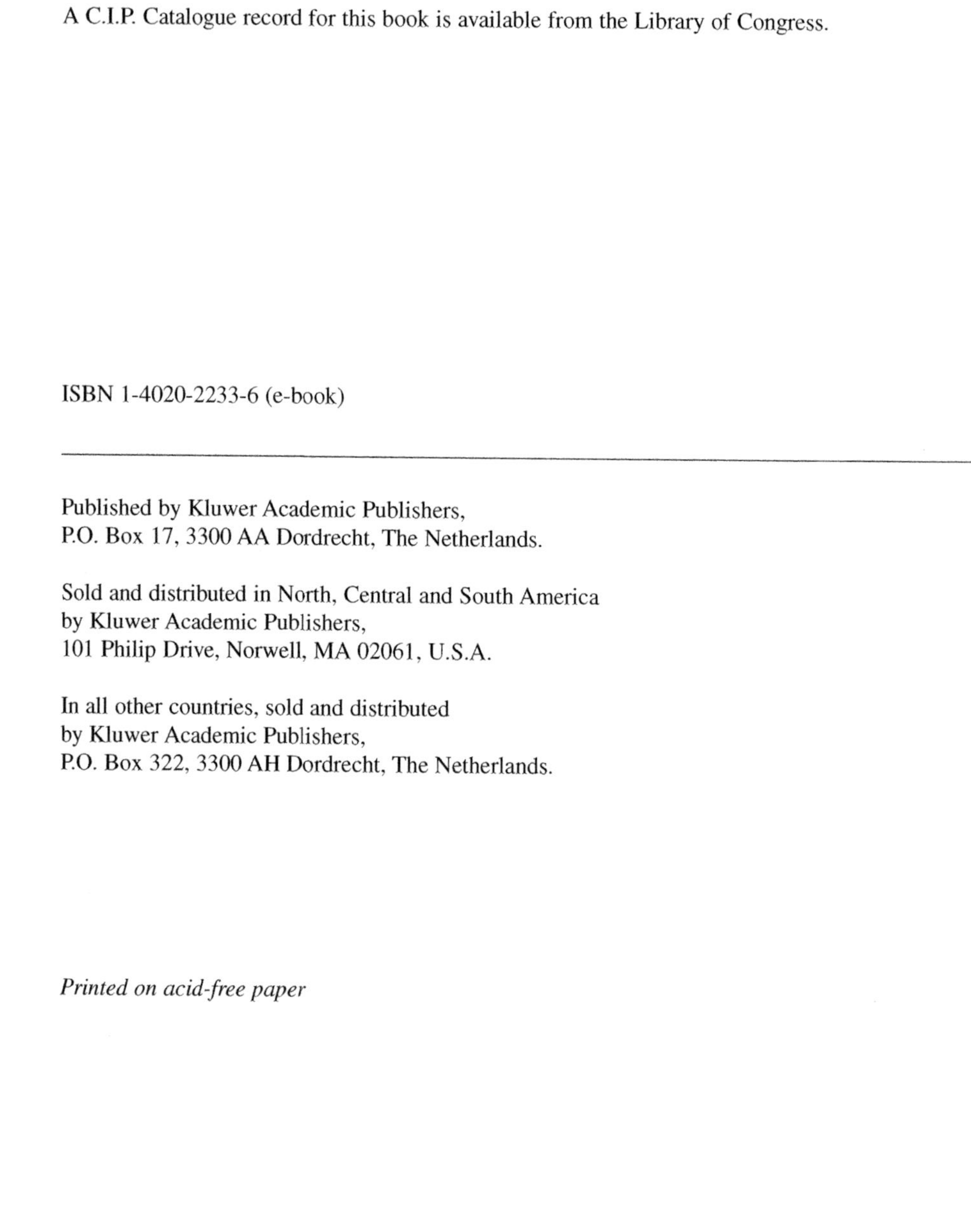

A C.I.P. Catalogue record for this book is available from the Library of Congress.

ISBN 1-4020-2233-6 (e-book)

Published by Kluwer Academic Publishers,
P.O. Box 17, 3300 AA Dordrecht, The Netherlands.

Sold and distributed in North, Central and South America
by Kluwer Academic Publishers,
101 Philip Drive, Norwell, MA 02061, U.S.A.

In all other countries, sold and distributed
by Kluwer Academic Publishers,
P.O. Box 322, 3300 AH Dordrecht, The Netherlands.

Printed on acid-free paper

Grenoble Sciences

"Grenoble Sciences", directed by Prof. Jean Bornarel, was created ten years ago by the Joseph Fourier University of Grenoble, France (Science, Technology and Medecine) to select and publish original projects. Anonymous referees choose the best projects and then a Reading Committee interacts with the authors as long as necessary to improve the quality of the manuscripts.

(Contact: Tel: (33) 4 76 51 46 95 - E-mail: Grenoble.Sciences@ujf-grenoble.fr)

The "Fundamentals of Cavitation" Reading Committee included the following members:

- **Professeur Hiroharu Kato**, Faculty of Engineering, Tokyo University, Japan
- **Professeur Kirill V. Rozhdestvensky**, Saint-Petersburg State Marine Technical University, Russia
- **Professeur Dr.-Ing. Bernd Stoffel**, Darmstadt University of Technology, Germany

Front Cover Photo:
Photograph Bassin d'Essais des Carènes, Délégation Générale à l'Armement (DGA), Val de Reuil, France.

TABLE OF CONTENTS

FOREWORD

This book treats cavitation, which is a unique phenomenon in the field of hydrodynamics, although it can occur in any hydraulic machinery such as pumps, propellers, artificial hearts, and so forth. Cavitation is generated not only in water, but also in any kind of fluid, such as liquid hydrogen. The generation of cavitation can cause severe damage in hydraulic machinery. Therefore, the prevention of cavitation is an important concern for designers of hydraulic machinery. On the contrary, there is great potential to utilize cavitation in various important applications, such as environmental protection.

There have been several books published on cavitation, including one by the same authors. This book differs from those previous ones, in that it is both more physical and more theoretical. Any theoretical explanation of the cavitation phenomenon is rather difficult, but the authors have succeeded in explaining it very well, and a reader can follow the equations easily. It is an advantage in reading this book to have some understanding of the physics of cavitation. Therefore, this book is not an introductory text, but a book for more advanced study.

However, this does not mean that this book is too difficult for a beginner, because it explains the cavitation phenomenon using many figures. Therefore, even a beginner on cavitation can read and can understand what cavitation is. If the student studies through this book (with patience), he or she can become an expert on the physics of cavitation.

In conclusion, this book is very comprehensive and instructive for advanced students, scientists, and engineers, who want to understand the true nature of cavitation.

The authors, Dr. Jean-Marie MICHEL and Dr. Jean-Pierre FRANC, are professors at the University of Grenoble, although Dr. MICHEL retired recently. They have much experience in the teaching and study of cavitation. Dr. MICHEL and Dr. FRANC have presented many important papers in internationally recognized academic journals such as the *Journal of Fluid Mechanics*, which are referenced in this book.

Dr. MICHEL and Dr. FRANC are the most suitable persons to write a book on cavitation such as this. I take great pleasure in being the first to congratulate them on their most recent contribution to this very unique and fascinating field.

March, 2003, Tokyo, Japan
Dr. Hiroharu KATO,
Professor Emeritus,
University of Tokyo

PREFACE

The present book is aimed at providing a comprehensive presentation of the phenomena involved in cavitation. It is focused on hydrodynamic cavitation, i.e. the kind of cavitation which occurs in flowing liquids, contrary to acoustic cavitation which is induced by an oscillating pressure field in a liquid almost at rest. Nevertheless, the principles which govern the hydrodynamic bubble and the acoustic bubble are basically the same.

Briefly, cavitation is the occurrence of vapor cavities inside a liquid. It is well known that in static conditions a liquid changes to vapor if its pressure is lowered below the so-called vapor pressure. In liquid flows, this phase change is generally due to local high velocities which induce low pressures. The liquid medium is then "broken" at one or several points and "voids" appear, whose shape depends strongly on the structure of the flow.

This book deals with all types of cavitation which develop in real liquid flows. This includes bubble cavitation (spherical bubbles in the simplest case), sheet cavitation, supercavitation and superventilation, cavitation in shear and vortex flows and some other patterns. It covers the field of cavitation inception as well as developed cavitation, which is encountered in advanced hydraulics at high speed.

It is intended for graduate students, research workers and engineers facing cavitation problems, particularly in the industrial fields of hydraulic machinery and marine propulsion. A special effort has been made to explain the physics of cavitation in connection with various phenomena such as surface tension, heat and mass transfer, viscosity and boundary layers, compressibility, nuclei content, turbulence, etc... In addition to the physical foundations of the phenomenon, various methods of investigation, either experimental or computational, are presented and discussed so that the reader can deal with original problems.

The book results from about 40 years of research carried out at Grenoble University in various fields of cavitation science, with the financial support of several firms and institutions, particularly the French Navy. Initially, two main influences converged to stimulate the creation by Pr. J. DODU of the cavitation research group: the strong hydraulic experience of private Companies in Grenoble and the advice of renowned foreign scientists (M.S. PLESSET, M.P. TULIN, B.R. PARKIN, A.J. ACOSTA... and so many others) who delivered a detailed account of the *state of the art* to Pr. J. DODU. Many of those initial scientific and industrial relationships have remained active over the years. Here we particularly wish to acknowledge the very fine contribution we received from Mr Y. LECOFFRE, either in the design of experimental rigs or the initiation of new research programs. We must also remember the name of Pr. A. ROWE, whose acute insight into hydrodynamics and pioneering work on numerical modeling of cavitating flows are still present in our minds.

The book is made of rather short chapters, each designed to correspond to one or two lectures. It is the result of our teaching program given over many years including a number of seminars. The lists of references (at the end of each chapter) are limited to major contributions, while the very abundant literature devoted to cavitation can be found from the quoted review papers.

After an introductory chapter, classic results relative to liquid breakdown are recalled in *chapter 2*. It is shown that a liquid can actually sustain absolute pressures smaller than the vapor pressure (and even tensions) without cavitating. This leads to the idea of nuclei (i.e. points of weakness in the liquid continuum) which is a fundamental concept in cavitation. The physics of the microbubble as a nucleus is presented in detail with a special emphasis on stability. *Chapter 2* ends with the definition of the quality of a liquid sample in terms of nuclei content, a key concept for the prediction of cavitation patterns in real liquid flows.

Chapters 3 to *5* are concerned with the isolated bubble. In *chapter 3*, basic results on the dynamics of the spherical bubble are presented and the famous RAYLEIGH-PLESSET equation is derived. Throughout the book, this fundamental equation is used to throw light on essential questions, such as scale effects. The evolution of a bubble in a non-symmetrical environment will result in deviations from sphericity which are discussed in *chapter 4*, together with the problem of the path of a bubble within a liquid flow. The effects of liquid compressibility and thermal diffusion are presented in *chapter 5*.

Chapters 6 to *9* address sheet cavitation, which appears on blades of propellers, foils of boats, or behind axisymmetric bodies such as torpedoes. *Chapters 6* and *9* are devoted to the neighbouring problems of supercavitation and superventilation respectively. A special effort has been made to present the analytical approach derived by our colleagues from Russia and Ukraine on the basis of the so-called *"Logvinovich independence principle of cavity expansion"* in the case of axisymmetric cavities. We are especially grateful to Pr. V.V. SEREBRYAKOV, Pr. Y.N. SAVCHENKO and their colleagues from the Kiev University. Through them, we became aware of the very significant research which was carried out in those countries over the years. One example (*chapter 9*) is the theoretical modeling of ventilated cavity pulsations by Pr. E.V. PARISHEV of Moscow University.

Partial sheet cavitation is addressed in *chapter 7* with special attention given to cloud cavitation, re-entrant jets and more generally to cavitation instabilities. Because of their practical importance, those subjects have been studied in a number of laboratories in the recent past, in Europe, Japan and the USA. We would like to thank especially Pr. Y. TSUJIMOTO (Osaka University) for numerous and fruitful discussions on cavitation instabilities. The interaction between traveling bubbles and attached sheet cavities is addressed in *chapter 8*. It is shown that the boundary layer on the wall together with the water nuclei content strongly influence the type of cavitation that can occur. Basic principles for the prediction of cavitation patterns on hydrofoils or pump blades are proposed.

Chapters 10 and *11* are devoted to vortex cavitation, tip vortex cavitation and shear cavitation respectively. Several results presented in *chapter 10* were obtained in the framework of a joined program supported by the French Navy. We are particularly grateful to Pr. D.H. FRUMAN who directed that research program, and to the colleagues of the associated laboratories : Bassin d'Essais des Carènes (Val de Reuil, France), École Navale (Brest, France), Institut de Machines Hydrauliques (Lausanne, Switzerland). The difficult subject of shear cavitation is approached in *chapter 11*, with a special emphasis on the physical analysis derived by Pr. R.E.A. ARNDT (University of Minnesota).

The main effects of cavitation on hydraulic equipments are also examined in this book. An overview on cavitation erosion is given in *chapter 12*, whereas the reader will find information on cavitation noise in several chapters.

We are very indebted to H. KATO (formerly Professor at the Tokyo University), Pr. K.V. ROZHDESTVENSKY (Saint Petersburg State Marine Technical University) and Pr. B. STOFFEL (Darmstadt University of Technology) who kindly accepted to review our manuscripts and made precious comments and suggestions. We are also grateful to our colleague Pr. F. MCCLUSKEY who accepted the difficult charge of correcting the English and spent a considerable time improving our manuscript, far beyond matters of pure form.

Thanks also to our colleagues in the Grenoble Cavitation research group (J.C. JAY, M. MARCHADIER, M. RIONDET and J.F. VERDYS) whose technical competence allowed us to obtain a significant number of the experimental results presented in this book.

Our editors, Kluwer and Grenoble Sciences, deserve special acknowledgements. We are particularly grateful to Pr René MOREAU, Scientific Editor of the series *Fluid mechanics and its applications* and to the local team of Grenoble Sciences : Pr Jean BORNAREL (Grenoble University) who supported this project since its very beginning, Nicole SAUVAL who ensured the administrative work, Sylvie BORDAGE, Julie RIDARD and Thierry MORTURIER for their constant and exceptional care in the production of the manuscript.

Finally, we would like to thank once more our colleague and friend Hiro KATO whose scientific road crossed our own one so many times and who accepted to write the foreword for this book.

J.P. FRANC & J.M. MICHEL
November, 2003

List of Symbols

> *Numbers in square brackets refer to the corresponding chapters.*

Symbol	Definition	Unit	Dimension
a	Constant in the VAN DER WAALS state law [1]		ML^5T^{-2}
	Thermal diffusivity [2, 7]	m^2/s	L^2T^{-1}
	Viscous core radius [11]	m	L
$\mathbf{a_n}$	Amplitude of spherical harmonics [3]	m	L
b	Constant in the VAN DER WAALS state law [1]	m^3	L^3
B	STEPANOV factor [5, 7]		
c	Chord length [6, 7]	m	L
	Speed of sound [5, 7]	m/s	LT^{-1}
$\mathbf{c_v, c_p}$	Heat capacities at constant volume (resp. constant pressure) [2, 5]	J/kg/°K	$L^2T^{-2}\theta^{-1}$
$\mathbf{C_s}$	Concentration of a gas dissoved in a liquid [2]	kg/m^3	ML^{-3}
$\mathbf{C_p}$	Pressure coefficient [1, 10, 11]		
$\mathbf{C_D}$	Drag coefficient [4, 6]		
$\mathbf{C_L}$	Lift coefficient [4, 6]		
$\mathbf{C_Q}$	Flowrate coefficient [5, 7, 9]		
D	Coefficient of mass diffusion [2]	m^2/s	L^2T^{-1}
	Diameter [6, 9]	m	L
	Drag force [6]	N	LMT^{-2}
e	Cavity thickness [6, 7]	m	L
$\mathbf{e_{ij}}$	Deformation rate [11]	s^{-1}	T^{-1}
f	Frequency [3, 7, 9]	s^{-1}	T^{-1}
Fr	FROUDE number [6, 9]		
h	Enthalpy [5]	J/kg	L^2T^{-2}
	Heat convective transfer coefficient [7]	$W/m^2/°K$	$MT^{-3}\theta^{-1}$
H	HENRY constant [2]	s^{-1}/m^2	
k	Polytropic exponent [5]		
l, dl	Curvilinear distance, length element	m	L
ℓ	Cavity length [6, 7, 9]	m	L

L	Latent heat of vaporization	J/kg	L^2T^{-2}
	Lift force[6]	N	MLT^{-2}
$\dot{m}$	Mass flowrate through a unit surface area[1]	$kg/m^2/s$	$ML^{-2}T^{-1}$
	Mass loss rate[12]	kg/s	MT^{-1}
M	Virtual mass of an immersed body[4]	kg	M
n(R)	Density distribution of nuclei size[2]	$/cm^3/\Delta R$	L^{-4}
N	Density concentration of nuclei[2]	$/cm^3$	L^{-3}
p	Absolute pressure[1]	Pa	$ML^{-1}T^{-2}$
p_c	Cavity pressure[1, 6, 7, 9]	Pa	$ML^{-1}T^{-2}$
p_g	Partial gas pressure inside a cavity[1, 9]	Pa	$ML^{-1}T^{-2}$
p_r	Pressure at the reference point[1]	Pa	$ML^{-1}T^{-2}$
$p_v(T)$	Vapor pressure at temperature T[1]	Pa	$ML^{-1}T^{-2}$
q	Mass flowrate through a unit surface area[7]	$kg/m^2/s$	$ML^{-2}C$
Q	Heat transfer[1]	J	ML^2T^{-2}
Q_m	Mass flowrate of air[9]	kg/s	MT^{-1}
r	Radial coordinate[3, 6, 9]	m	L
R	Spherical bubble radius[3]	m	L
	Radius of an axisymmetric cavity[6, 9]	m	L
$\dot{R}$	Bubble interface velocity[3]	m/s	LT^{-1}
Re	REYNOLDS number		
$\vec{R}_p$	Force exerted by the liquid on an immersed body[4]	N	MLT^{-2}
s	Curvilinear distance[3, 8]	m	L
S	Cavity cross-sectional area[6, 9]	m^2	L^2
	STROUHAL number[7, 9, 11]		
	Surface tension of the liquid[2, 3]		
t	Time	s	T
t_{rr}	Radial stress[3]	Pa	$ML^{-1}T^{-2}$
T	Absolute temperature[1, 7]	°K	θ
	Period[3, 9]	s	T
u	Radial component of the velocity[3, 5]	m/s	LT^{-1}
V	Velocity[4]	m/s	LT^{-1}
$\mathcal{V}$	Bubble or cavity volume[5, 9]	m^3	L^3

Symbol	Description	Unit	Dimension
W	Relative velocity [4]	m/s	LT^{-1}
We	WEBER number [3]		
x, y, z	Cartesian coordinates [3, 6, 10]	m	L

Greek characters

Symbol	Description	Unit	Dimension
α	Angle of attack [6]		
α_ℓ	Liquid thermal diffusivity [5]	m^2/s	L^2T^{-1}
δ	Boundary layer thickness [8]	m	L
Δ	Increment operator		
ε	Small parameter		
	Strain [12]		
φ	Non-dimensional frequency [9]		
	Velocity potential [4, 6]	m^2/s	L^2T^{-1}
γ	Ratio of heat capacities ($\gamma = c_p / c_v$) [2, 3, 9]		
Γ	Circulation [10, 11]	m^2/s	L^2T^{-1}
λ	Thermal conductivity [2, 5]	W/m/°K	$MLT^{-3}\theta^{-1}$
	Wavelength [9, 11]	m	L
μ	Dynamic viscosity [3, 11]	kg/m/s	$ML^{-1}T^{-1}$
ν	Kinematic viscosity [3, 11]	m^2/s	L^2T^{-1}
ρ	Density [1, 2, 3...]	kg/m^3	ML^{-3}
σ	Normal stress [12]	Pa	$ML^{-1}T^{-2}$
σ_v	Cavitation number [1]		
σ_{vi}, σ_{vd}	Incipient (resp. desinent) cavitation number [1, 2, 8, 10, 11]		
σ_a	Relative pressure of air inside a ventilated cavity [9]		
σ_c	Relative underpressure of a developed cavity [1, 6, 7, 9]		
Σ	BRENNEN thermodynamic parameter [5]	$m/s^{3/2}$	$LT^{-3/2}$
τ	Characteristic time [2, 3, 6, 9]	s	T
	RAYLEIGH time for the bubble collapse [3]	s	T
ω	Rotation rate [11]	/s	T^{-1}
Ω	Vorticity [10, 11]	/s	T^{-1}

Subscripts

c	Cavity
min	Minimum value
g	Gas
v	Vapor
ℓ	Liquid
0	Initial value. Mean value
r	Reference point

1. INTRODUCTION

THE MAIN FEATURES OF CAVITATING FLOWS

1.1. THE PHYSICAL PHENOMENON

1.1.1. DEFINITION

Cavitation, i.e. the appearance of vapor cavities inside an initially homogeneous liquid medium, occurs in very different situations. According to the flow configuration and the physical properties of the liquid, it can present various features.

Cavitation can be defined as the breakdown of a liquid medium under very low pressures. This makes cavitation relevant to the field of continuum mechanics and it applies to cases in which the liquid is either static or in motion.

This book is particularly concerned with hydrodynamic cavitation, i.e. cavitation in flowing liquids. This includes flows through Venturi nozzles, in narrow passages (e.g. hydraulic valves) or around wings or propeller blades.

However, cavitation can also occur in a static or nearly static liquid. When an oscillating pressure field is applied over the free surface of a liquid contained in a reservoir, cavitation bubbles may appear within the liquid bulk if the oscillation amplitude is large enough. This type of cavitation is known as acoustic cavitation.

A further example of cavitation in a liquid almost at rest is the sudden and rapid acceleration of a solid body with sharp edges (such as a disk) in still water. Bubbles can appear close to these edges almost instantaneously, whereas the velocity of the liquid itself remains negligible.

The above definition of cavitation introduces the concept of a pressure threshold, beneath which liquid cohesion is no longer ensured. Ideally, the threshold would be determined from physical considerations on a microscopic scale. Taking into account the actual state of scientific knowledge, together with the need for practical solutions to often complicated industrial systems, it is more useful to refer only to macroscopic fluid properties.

A simple everyday example is that of a syringe. Correct filling requires the piston motion to be relatively slow, otherwise, the liquid column breaks and the filling stops. Because of head losses inside the needle, the pressure within the syringe drops below the atmospheric value. The pressure difference increases with piston velocity. Furthermore, at the syringe inlet, where the flow is a submerged liquid jet, additional turbulent pressure fluctuations occur. Both mechanisms contribute to a reduction in local pressure, possibly to a value below the vapor pressure of the liquid, thus producing vapor. A similar phenomenon can be found in volumetric pumps for fuel injection in engines. Head losses and rapid acceleration of the liquid column can result in low pressures, causing cavitation and consequent partial filling of the chamber.

1.1.2. *VAPOR PRESSURE*

The concept of vapor pressure is best considered from the viewpoint of classical thermodynamics. In the phase diagram for, say, water (fig. 1.1), the curve from the triple point T_r to the critical point C separates the liquid and vapor domains. Crossing that curve is representative of a reversible transformation under static (or equilibrium) conditions, i.e. evaporation or condensation of the fluid at pressure p_v, known as the vapor pressure. This is a function of the temperature T.

Following from this, cavitation in a liquid can be made occur by lowering the pressure at an approximately constant temperature, as often happens locally in real flows. Cavitation thus appears similar to boiling, except that the driving mechanism is not a temperature change but a pressure change, generally controlled by the flow dynamics.

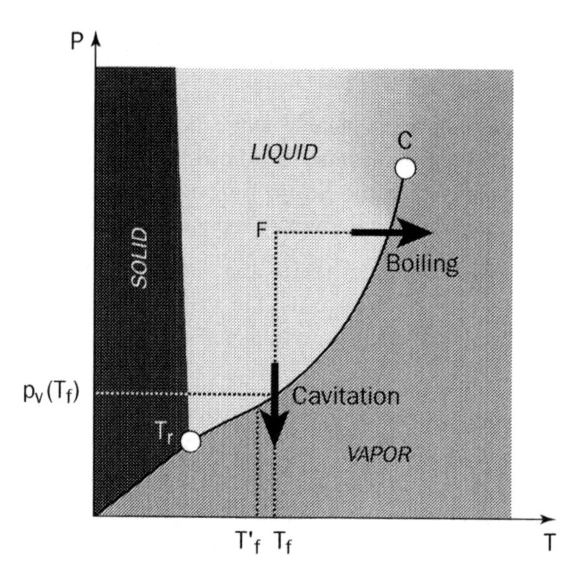

1.1
Phase diagram

In most cases (with cold water, in particular), only a relatively small amount of heat is required for the formation of a significant volume of vapor. The surrounding liquid (the heat source for vaporization) therefore shows only a very minor temperature change. The path in the phase diagram is practically isothermal (see fig. 1.1).

However, in some cases, the heat transfer needed for the vaporization is such that phase change occurs at a temperature T' lower than the ambient liquid temperature T. The temperature difference $T - T'$ is called thermal delay in cavitation.

It is greater when the ambient temperature is closer to the critical temperature of the fluid. This phenomenon may become important e.g. when pumping cryogenic liquids in rocket engines. It will be considered in chapters 5 and 8.

From a purely theoretical point of view, several steps can be distinguished during the first instants of cavitation:

- breakdown or void creation,
- filling of this void with vapor, and
- eventual saturation with vapor.

In reality, those phases are effectively simultaneous with the second step being so rapid that instantaneous saturation of the void with vapor can be justifiably assumed.

It must be kept in mind that the curve $p_v(T)$ is not an absolute boundary between liquid and vapor states. Deviations from this curve can exist in the case of rapid phase change.

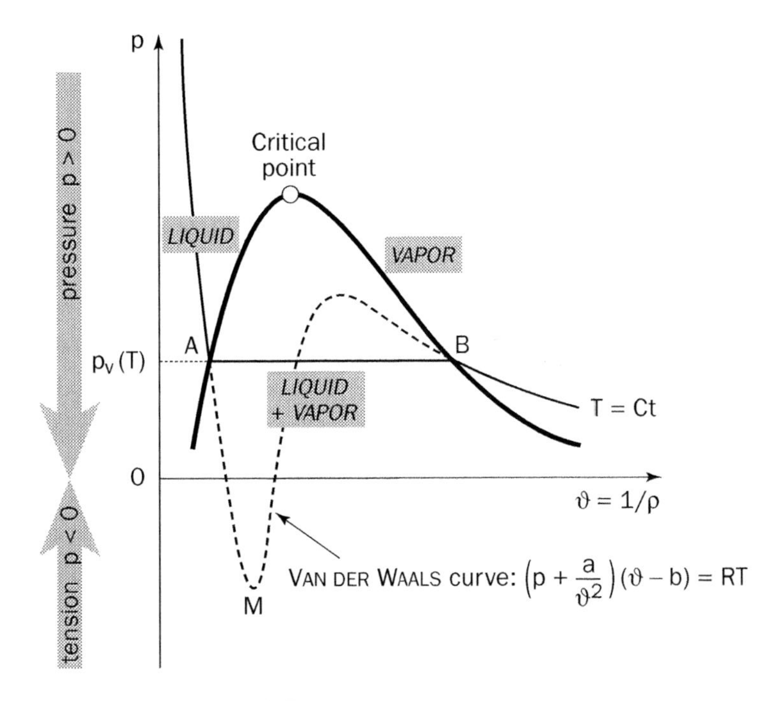

1.2 - Andrews-isotherms

Even in almost static conditions, a phase change may occur at a pressure lower than p_v. For example, consider the so-called ANDREWS-isotherms in the $p-\vartheta$ diagram, where $\vartheta = 1/\rho$ is the specific volume and ρ the density (fig. 1.2). Such curves can be approximated in the liquid and vapor domains by the VAN DER WAALS equation of state. The transformation from liquid to vapor along the path AM can be avoided, provided special care is taken in setting up such an experiment. Along this path, the liquid is in metastable equilibrium and even can withstand negative absolute pressures, i.e., tensions, without any phase change.

In conclusion, the condition that the local absolute pressure be equal to the vapor pressure at the global system temperature does not ensure in all cases that cavitation actually occurs. The difference between the vapor pressure and the actual pressure at cavitation inception is called static delay. In some cases, there is also a dynamic delay, which is due to inertial phenomena associated with the time necessary for vapor cavities to be observable.

1.1.3. *THE MAIN FORMS OF VAPOR CAVITIES*

Cavitation can take different forms as it develops from inception. Initially, it is strongly dependent on the basic non-cavitating flow structure. However, as it develops, the vapor structures tend to disturb and modify the basic flow. Cavitation patterns can be divided into three groups. These are:

- Transient isolated bubbles. These appear in the region of low pressure as a result of the rapid growth of very small air nuclei present in the liquid (see chap. 2 to 5 and chap. 8). They are carried along by the main flow and subsequently disappear when they enter areas of high enough pressure.
- Attached or sheet cavities. Such cavities are often attached to the leading edge of a body, e.g. on the low-pressure side of blades and foils (see chap. 6 to 8).
- Cavitating vortices. Cavitation can appear in the low-pressure core of vortices in turbulent wakes or, as a regular pattern in tip vortices of 3-D wings or propeller blades (chap. 10 and 11).

Some patterns do not fall easily in these classes. For example, on the low-pressure surfaces of foils or propeller blades, vapor structures with a very short lifetime can appear. They have the form of attached cavities but are transported similarly to traveling bubbles.[1]

---

1. If needed, it would be possible to give an objective foundation to the classification by considering the relative velocity of the liquid at the interface with respect to the mass center of the vapor figure. If the velocity is approximately normal to the interface, the figure should be considered as a bubble; then attention is mainly paid to the volume variation. On the other hand, if the relative velocity is more tangential to the interface, the figure should be considered as a cavity, in the absence of circulation, or as a cavitating vortex, in the case of circulation.

1.2. CAVITATION IN REAL LIQUID FLOWS

1.2.1. CAVITATION REGIMES

For practical purposes, it is useful to consider two distinct steps in cavitation development:

- cavitation inception, i.e. the limiting regime between the non-cavitating and the cavitating flow;
- developed cavitation, which implies a certain permanency and extent of the cavitation or a significant fall in performance of machines.

The distinction is important in the context of acceptance or otherwise of cavitation in industrial situations. In the case of undeveloped cavitation, inception or desinence thresholds are of interest. For developed cavitation, the manufacturer must focus on the consequences of cavitation on the operation of the hydraulic system (see § 1.2.3).

In the case of attached cavities, a further distinction may be useful: partial cavities, which close on the wall, and supercavities, which close away from the boundary (typically a foil).

1.2.2. TYPICAL SITUATIONS FAVORABLE TO CAVITATION

In this section, typical situations in which cavitation can appear and develop within a flow are described briefly.

- Wall geometry may give rise to sharp local velocity increases and resulting pressure drops within a globally steady flow. This happens in the case of a restriction in the cross-sectional area of liquid ducts (Venturi nozzles), or due to curvature imposed on flow streamlines by the local geometry (bends in pipe flow, upper sides of blades in propellers and pumps).
- Cavitation can also occur in shear flows due to large turbulent pressure fluctuations (see jets, wakes, etc.).
- The basic unsteady nature of some flows (e.g. water hammer in hydraulic control circuits, or ducts of hydraulic power plants, or in the fuel feed lines of Diesel engines) can result in strong fluid acceleration and consequently in the instantaneous production of low pressures at some points in the flow leading to cavitation.
- The local roughness of the walls (e.g. the concrete walls of dam spillways) produces local wakes in which small attached cavities may develop.
- As a consequence of the vibratory motion of the walls (e.g. liquid cooling of Diesel engines, standard A.S.T.M.E. erosion device) oscillating pressure fields are created and superimposed on an otherwise uniform pressure field. If the oscillation amplitude is large enough, cavitation can appear when the negative oscillation occurs.

- Finally, attention has to be drawn to the case of solid bodies that are suddenly accelerated by a shock in a quiescent liquid, particularly if they have sharp edges. The liquid acceleration needed to get round these edges produces low pressures even if the velocities are relatively small immediately after the shock.

1.2.3. *THE MAIN EFFECTS OF CAVITATION IN HYDRAULICS*

If a hydraulic system is designed to operate with a homogeneous liquid, additional vapor structures due to cavitation can be interpreted, by analogy with the case of mechanical systems, as mechanical clearances. The vapor structures are often unstable, and when they reach a region of increased pressure, they often violently collapse since the internal pressure hardly varies and remains close to the vapor pressure. The collapse can be considered analogous to shocks in mechanical systems by which clearances between neighboring pieces disappear. Following this, a number of consequences can be expected:

- alteration of the performance of the system (reduction in lift and increase in drag of a foil, fall in turbomachinery efficiency, reduced capacity to evacuate water in spillways, energy dissipation, etc.);
- the appearance of additional forces on the solid structures;
- production of noise and vibrations;
- wall erosion, in the case of developed cavitation if the velocity difference between the liquid and the solid wall is high enough.

Thus, at first glance, cavitation appears as a harmful phenomenon that must be avoided. In many cases, the free cavitation condition is the most severe condition with which the designer is faced. To avoid the excessive financial charges that would be associated with this, a certain degree of cavitation development may be allowed. Of course, this can be done only if the effects of developed cavitation are controlled.

The negative effects of cavitation are often stressed. However, cavitation is also used in some industrial processes to concentrate energy on small surfaces and produce high pressure peaks. For this purpose, ultrasonic devices are often used. Examples of such positive applications include:

- the cleaning of surfaces by ultrasonics or with cavitating jets,
- the dispersion of particles in a liquid medium,
- the production of emulsions,
- electrolytic deposition (the ion layers that cover electrodes are broken down by cavitation, accelerating the deposition process),
- therapeutic massage and bacteria destruction in the field of medical engineering,
- the limitation of flowrates in confined flows due to the development of supercavities.

1.3. *SPECIFIC FEATURES OF CAVITATING FLOW*

1.3.1. *PRESSURE AND PRESSURE GRADIENT*

In non-cavitating flows, the reference pressure level has no effect on flow dynamics and attention is paid only to the pressure gradient. On the other hand, cavitating flows are primarily dependent on this level, since by simply lowering the reference pressure, cavitation can appear and develop. Thus, it is essential to consider the absolute value of the pressure, and not simply its gradient.

To predict cavitation inception by theoretical or numerical analysis, one has to compare the calculated value of the pressure in a critical region of the flow to a threshold value, typically the vapor pressure. The method of calculation depends on the flow configuration.

- In the case of one-dimensional, steady flows in pipes, the use of the BERNOULLI equation, taking into account head losses, is sufficient to identify the region of minimum pressure together with the value of this minimum.
- Steady flows without significant shear, such as flows around wings and propeller blades, can be considered as potential flows. Classical methods require that the kinematic problem is solved first, with the pressure again calculated using BERNOULLI's equation. In these cases, the minimum pressure is generally located on the boundary of the flow, a conclusion usually supported by experimental evidence.
- The case of turbulent shear flow is among the most complicated. Consequently, until recently it has been treated experimentally and/or empirically. Progress in computational fluid dynamics has made it possible to predict cavitation inception, at least for the simplest configurations. Some encouraging results have recently been obtained in this field (see chap. 11).
- In the case of tip vortices, it is possible to use simple vortex models, such as those of RANKINE or BURGERS. Effectively, the problem reduces to the estimation of two parameters –the circulation around the vortex and the size of its viscous core.

Pressure also plays an important role in the case of developed cavitation and is the source of additional complexity in the modeling of cavitating flows.

- For example, the modeling of cavities attached to foils or blades requires a condition of constant pressure along the cavity boundary. This modifies the nature of the mathematical problem to be solved. From the physical point of view, the change in the pressure distribution causes a change in the pressure gradient and therefore a change in boundary layer behavior.
- When a large number of bubbles explode on the low-pressure side of a foil, the initial, non-cavitating, pressure distribution can be significantly modified and the interaction between the basic, non-cavitating flow and the bubbly flow must be taken into account.

- Finally, the evolution of turbulent, cavitating vortices in a wake cannot be predicted via the usual fluid mechanics equations governing the conservation of mass and circulation of the vortex filaments. When the core of a vortex filament cavitates and becomes vapor laden, it is then dependent on the local pressure field. In other words, cavitation breaks the link between the elongation rate $\delta\ell$ of a rotating filament and its vorticity ω, expressed by the classical relation $\omega / \delta\ell = \text{Constant}$.

On the experimental side, two main difficulties appear with respect to the measurement of the pressure. First, pressure transducers, which must be flush mounted on the walls to avoid any disturbance to the flow itself, do not necessarily give valuable information on the pressure within a turbulent shear flow [LESIEUR 1998].

The second difficulty is rather technical. It will become clear, from subsequent discussion that phenomena connected to bubble collapse have a small characteristic size (less than about 0.1 mm) and a short duration (of the order of a microsecond). This means that transducers must have a very small spatial resolution and a very short rise time. These conditions are not easily met technically. Moreover, in the case of erosion studies, pressures of the order of hundreds of Megapascals, or even Gigapascals, have to be measured. Obviously, the mechanical resistance of pressure transducers becomes a central problem for the experimentalist.

In general, experimental studies on cavitation inception and the physics of developed cavitation employ specially configured equipment that has been progressively developed over the years. The most common set-up is the cavitation tunnel, a vertical closed loop with the test section at the top and a circulation pump in the lower part. This avoids cavitation problems within the pump, owing to the gravitational force. The absolute pressure in the test section is adjusted to below atmospheric pressure, using a vacuum pump. In some cases, a compressor is necessary to increase the pressure above the atmospheric level.

1.3.2. LIQUID-VAPOR INTERFACES

Cavitating flows, like other two-phase liquid-gas flows, are characterized by the presence of numerous interfaces. However, their response to external perturbations, for example a pressure rise, can be very different from the case of liquid-gas flows.

Two-phase flows containing gas bubbles are not usually subject to rapid changes in mean density (except in the case of shock waves). This is because the non-condensable nature of the gas confers a kind of global stability to the flow.

In cavitating flows, however, the interfaces are subjected on one side to a constant pressure, practically equal to the vapor pressure. Thus, they cannot sustain an increase or decrease in external pressure without rapidly evolving in both shape and size. They are extremely unstable.

It is almost impossible to use intrusive probes to take measurements within a cavitating flow due mainly to the cavitation the probe itself generates. However, if the liquid is transparent, it is possible to visualize the interfaces, as they reflect light very effectively. Interfaces can generally be considered as material surfaces and their observation from one-shot photographs (short flash durations of the order of a microsecond) or from high-speed photography or video (at a typical rate of ten thousand frames per second) gives an idea of the flow dynamics.

Concerning the exchange of liquid and vapor across an interface, the mass flowrate $\dot{m}$ (per unit surface area) across the interface is proportional to the normal velocities of either the liquid or the vapor relative to the interface. Mass conservation across the interface gives (see fig. 1.3):

$$\dot{m} = \rho_\ell \left[v_{\ell n} - \frac{dn}{dt} \right] = \rho_v \left[v_{vn} - \frac{dn}{dt} \right] \tag{1.1}$$

In this equation, the indices ℓ and v refer to the liquid and vapor phases respectively, and the index n to the normal component of the velocities. The symbol dn/dt is the normal velocity of the interface.[2]

1.3
The liquid/vapor interface

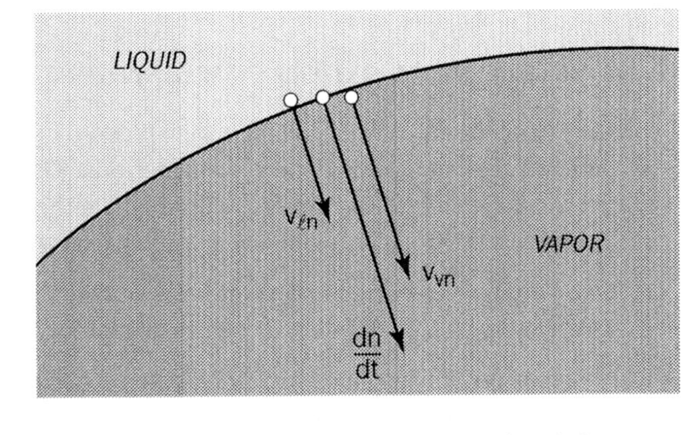

If the flowrate through the interface is negligible, i.e. $\dot{m} = 0$ (which is actually assumed in most cases), the three normal velocities are equal and the interface is a material surface, i.e. a surface made of the same fluid particles at different instants.

Two cases are of particular interest:

- For a spherical bubble whose radius $R(t)$ is a function of time, the normal velocity of the interface is dR/dt. In the case of negligible flowrate: $v_{\ell n} = v_{vn} = dR/dt$.
- For a steady cavity attached to a wall surrounded by a flowing liquid, dn/dt is zero. If it is assumed that the mass flowrate through the interface is negligible,

2. If the equation of the interface is $f(x, y, z, t) = 0$, the normal velocity of the interface is given by $\frac{dn}{dt} = -\frac{1}{\left\| \vec{\text{grad}}\, f \right\|} \frac{\partial f}{\partial t}$.

the normal velocities of the liquid and the vapor at the interface are also zero. Thus the liquid velocity of the outside flow at the interface is tangential to the cavity wall.

1.3.3. THERMAL EFFECTS

There are two principal effects on cavitation phenomena due to a temperature variation to be expected.

First, at constant ambient pressure, an increase in fluid temperature results in a greater aptitude to cavitate. Smaller pressure variations are necessary to reach the phase change curve, due to the increase in vapor pressure (see fig. 1.1).

Second, as vaporization requires heat transfer from the liquid bulk to the liquid/vapor interface, the thermal delay $T - T'$ (see fig. 1.1) tends to increase with temperature. To solve this problem where the temperature inside the cavity T' is unknown, a thermal equilibrium expression has to be set out. Heat transfer via conduction is expected in the case of bubbles. Convection is generally predominant in the case of attached cavities.

1.3.4. SOME TYPICAL ORDERS OF MAGNITUDE

Instabilities of the interfaces can lead to explosion or collapse of cavities, with large variations in size and velocity over short periods. This makes their scaling, together with their experimental or numerical analysis, rather difficult. Some typical values encountered in the field of cavitation are given below.

- The duration of collapse of a 1 cm radius spherical vapor bubble in water under an external pressure of one bar, is approximately one millisecond.
- The duration of the final stage of bubble or cavitating vortex collapse, which is important in the erosion process, is of the order of one microsecond.
- The normal velocity of an interface generally lies between some meters per second and some hundreds of meters per second.
- The overpressures due to the implosion of vapor structures (bubbles and vortices) can reach several thousand bars.

1.4. NON-DIMENSIONAL PARAMETERS

1.4.1. CAVITATION NUMBER σ_V

In a hydraulic system liable to cavitate, such as a turbine, a pump, a gate or a foil in a hydrodynamic tunnel, let us define p_r as the pressure at a conventional reference point r where it is easily measurable. Usually, r is chosen in a region close to that where cavitation inception is expected. If T is the operating temperature of the liquid and Δp a pressure difference that characterizes the system, the cavitation number (also called cavitation parameter, or THOMA cavitation number) is defined by:

$$\sigma_v = \frac{p_r - p_v(T)}{\Delta p} \tag{1.2}$$

For example:

- in the case of a gate:

$$\sigma_v = \frac{p_{downstream} - p_v(T)}{p_{upstream} - p_{downstream}} \tag{1.3}$$

- for a foil placed at a submersion depth h in a horizontal free surface channel where the pressure on the surface is p_0 and the flow velocity U:

$$\sigma_v = \frac{p_0 + \rho g h - p_v(T)}{\frac{1}{2}\rho U^2} \tag{1.4}$$

- for a pump:

$$\sigma_v = \frac{p_{inlet} - p_v(T)}{\rho V_p^{\,2}} \tag{1.5}$$

  where V_p is the velocity at the periphery of the runner.

It should be noted that the cavitation number is defined using dynamical parameters and not geometrical ones.

In a non-cavitating flow, this non-dimensional parameter cannot be considered as a scaling parameter. The difference $p_r - p_v$ has no physical significance for a single phase flow as it cannot be obtained by integration of the pressure gradient along a real path. The cavitation parameter becomes a similarity parameter only at cavitation inception.

1.4.2. *CAVITATION NUMBER AT INCEPTION, σ_{VI}*

The number σ_{vi} is the value of the parameter σ_v corresponding to cavitation inception at any point of the flow system. Cavitation appears because of either a decrease in pressure at the reference point (i.e., the ambient pressure) or an increase in the Δp-value. For experimental convenience (in particular for improved repeatability), the number σ_{vd}, corresponding to cavitation disappearance from an initial regime of developed cavitation, may also be used.

Operation in non-cavitating conditions requires that :

$$\sigma_v > \sigma_{vi} \tag{1.6}$$

The threshold σ_{vi} depends on all the usual factors considered in fluid mechanics such as flow geometry, viscosity, gravity, surface tension, turbulence levels, thermal parameters, wall roughness and the gas content of the liquid in terms of dissolved and free gases (i.e., gas nuclei, see chap. 2).

In general, the smaller the value of σ_{vi} for a given system, the better behaved is the flow. For example, for the flow around a 10 mm diameter circular cylinder, the cavitation inception number is about 1.5, whereas for elliptical cylinders at zero incidence, with a chord length of 80 mm and axis ratios of 1/4 and 1/8, the σ_{vi}-values are 0.45 and 0.20, respectively.

When σ_v becomes smaller than σ_{vi}, cavitation usually becomes increasingly developed. Very exceptionally, it may happen that after initial development, cavitation finally disappears as the consequence of a further lowering of σ_v (see chap. 6 and 8).

In many circumstances, particularly for the numerical modeling of cavitating flows, the following estimate is taken for σ_{vi}:

$$\sigma_{vi} = -Cp_{min} \tag{1.7}$$

where Cp_{min} is the minimum pressure coefficient, which is normally negative. The pressure coefficient Cp at a point M is defined by the relation:

$$Cp = \frac{p_M - p_r}{\Delta p} \tag{1.8}$$

In this expression, p_r is the absolute pressure at the reference point, as in relation (1.2). Two assumptions lie behind equation (1.7). First, cavitation occurs at the point of minimum pressure and second, the pressure threshold value is that of the vapor pressure. Clearly, these assumptions may be over-restrictive. Thus, the estimate (1.7) must be considered cautiously.

1.4.3. RELATIVE UNDERPRESSURE OF A CAVITY, σ_C

If a developed cavity is attached to the low-pressure surface of a blade, or if a large number of bubbles are present, the pressure in the region covered by the cavity is uniform. Referring to this pressure as p_c, a non-dimensional parameter known as the relative underpressure of the cavity is defined as:

$$\sigma_c = \frac{p_r - p_c}{\Delta p} \tag{1.9}$$

It is a true scaling parameter, as the numerator expresses an actual pressure difference inside the flow domain. This number is extensively used in the numerical modeling of flows with developed cavities and plays an important role in cavity dynamics. It is often, but improperly, referred to as the cavitation number.

Usually, the pressure p_c in the cavity is the sum of two components: the vapor pressure p_v and a partial pressure p_g due to the presence of non-condensable gas inside the cavity. If this last term is negligible, the relative underpressure of the cavity σ_c becomes equal to the cavitation parameter σ_v, which probably explains the confusion referred to above.

1.5. *SOME HISTORICAL ASPECTS*

The word "cavitation" appeared in English scientific literature at the end of the nineteenth century. It seems that the problem of cavitation in rotating machinery handling liquids was identified by TORRICELLI, and later by EULER and NEWTON. In the middle of the nineteenth century, DONNY and BERTHELOT measured the cohesion of liquids. The negative effect of cavitation on the performance of a ship propeller was first noted by PARSONS (1893), who built the first cavitation tunnel. The cavitation number was introduced by THOMA and LEROUX around the years 1923-1925.

Subsequently, many experiments were carried out to study the physical aspects of the phenomenon and to examine its effects on industrial systems. Theoretical and numerical approaches were widely used. There were two main fields of research.

The first focused on bubble dynamics [RAYLEIGH 1917, LAMB 1923, COLE 1948, BLAKE 1949, PLESSET 1949]. The simplicity of the spherical shape made their studies (either theoretical or experimental) relatively easy. A large amount of work has been published on bubble dynamics.

The second field was related to developed cavities or supercavities and was based on the old wake theory [HELMHOLTZ 1868, KIRCHHOFF 1869, LEVI-CIVITA 1907, VILLAT 1913, RIABOUCHINSKI 1920 [3]]. This theory considers wakes as regions of uniform pressure, limited by surfaces on which the tangential velocity is not continuous. It is more suited to cavitating wakes than to single phase wakes. Later, TULIN (1953) and WU (1956) made use of linearization procedures to adapt the theory to the case of slender bodies such as wings and blades.

Vortical cavitation was only considered more recently, in particular by GENOUX and CHAHINE (1983) and by LIGNEUL (1989), who studied the cavitating torus and tip vortices, respectively.

3. Those references can be found in JACOB's book: *Introduction mathématique à la mécanique des fluides.*

REFERENCES

BIRKHOFF G. & ZARANTONELLO E.H. –1957– Jets, wakes and cavities. *Academic Press Inc.*

BLAKE F.G. –1949– The tensile strength of liquids: a review of the literature. *Harvard Acoustics Res. Lab.* TM 9, June.

COLE R.H. –1948– Underwater explosions. *Princeton University Press.*

GENOUX P. & CHAHINE G.L. –1983– Équilibre statique et dynamique d'un tore de vapeur tourbillonnaire. *J. Méc. Théor. Appl.* **2**(5), 829-857.

JACOB C. –1959– Introduction mathématique à la mécanique des fluides. *Gauthier-Villars Ed.*

KNAPP R.T., DAILY J.W. & HAMMITT F.G. –1970– Cavitation. *McGraw-Hill Book Company Ed.*, 578 p.

LAMB H. –1923– The early stages of a submarine explosion. *Phil. Mag.* **45**, 257 *sq.*

LESIEUR M. –1998– Vorticity and pressure distributions in numerical simulations of turbulent shear flows. *Proc. 3rd Int. Symp. on Cavitation*, vol. 1, Grenoble (France), April 7-9, 9-18.

LEVKOVSKY Y.L. –1978– Structure of cavitating flows. *Sudostroenie Publishing House*, Leningrad (Russia), 222 p.

LIGNEUL P. –1989– Theoretical tip vortex cavitation inception threshold. *Eur. J. Mech. B/Fluids* **8**, 495-521.

PERNIK A.D. –1966– Problems of cavitation. *Sudostroenie Publishing House*, Leningrad (Russia), 439 p.

PLESSET M.S. –1949– The dynamics of cavitation bubbles. *J. Appl. Mech.* **16**, 277 *sq.*

RAYLEIGH (Lord) –1917– The pressure developed in a liquid during the collapse of a spherical cavity. *Phil. Mag.* **34**, 94 *sq.*

ROZHDESTVENSKY V.V. –1977– Cavitation. *Sudostroenie Publishing House*, Leningrad (Russia), 247 p.

TULIN M.P. –1953– Steady two-dimensional cavity flows about slender bodies. *DTMB*, Rpt 834.

WU T.Y.T. –1956– A free streamline theory for two-dimensional fully cavitated hydrofoils. *J. Math. Phys.* **35**, 236-265.

2. NUCLEI AND CAVITATION

2.1. INTRODUCTION

2.1.1. LIQUID TENSION

In chapter 1, possible differences between the actual value of the cavitation threshold and the vapor pressure were discussed. In real flows as in laboratory flows, liquids can actually sustain absolute pressures lower than the vapor pressure at the operating temperature and even negative pressures, i.e. tensions.

To explain these discrepancies, one must first refer to the classical data relating to liquid breakdown. In the nineteenth century [DONNY 1846, BERTHELOT 1850, REYNOLDS 1882], experiments demonstrated that a liquid at rest could sustain negative pressures without vaporization occurring. For water, the values were of the order of several tens of bars. More recent experiments [TEMPERLEY 1946, BRIGGS 1950, REES & TREVENA 1966] have shown that the experimental values are rather scattered (for example, BRIGGS obtained 277 bars). They depend on the experimental procedure, the preliminary treatment of the liquid (for example, degassing or pressurization over a long period) and the degree of cleanliness of the container wall. It is often not clear whether the limit corresponds to a loss of cohesion in the bulk liquid or a loss of adhesion of the liquid to the walls.

These experimental values are lower than the estimates calculated from theoretical models. For example, considering a microscopic bubble with a diameter of the order of the intermolecular distance d_0, the expression $2S/d_0$ (S is the liquid surface tension) gives a value of 7,000 bars (with $d_0 \approx 0.1$ nm, S = 0.072 N/m for water). This estimate is obviously open to criticism, as such a very small size is not compatible with the assumption of a continuum. If the molecular nature of the liquid is then taken into account, together with oscillations due to temperature fluctuations, the limiting tension is reduced by about one order of magnitude.

Another estimate can be derived from the VAN DER WAALS equation (see § 1.1.2, fig. 1.2). The ordinate of the minimum M on the isotherm can be considered a limiting value of the tension sustained by a liquid. In the case of water, it is about 500 bars at room temperature.

2.1.2. CAVITATION NUCLEI

The differences observed with respect to the vapor pressure $p_v(T)$ in typical experiments on cavitation are much smaller than the experimental results and theoretical estimates mentioned above. They don't usually exceed a few bars at most for tap water. Thus, for liquids currently used in industry, the existence of points of weakness in the liquid continuum is to be expected. Those points are formed by small gas and vapor inclusions and operate as starting points for the liquid breakdown. They are known as cavitation nuclei. Numerous experiments show that those nuclei actually exist. Their size is between a few micrometers and some hundreds of micrometers. They remain spherical at this scale due to surface tension. They can be referred to as microbubbles.

The assumption of heterogeneities inside a homogeneous medium in order to explain phase changes is common in thermophysics, for example in boiling, condensation, and solidification. Nuclei also proved to be the origin of great differences in cavitation inception found in the past when tests on similar bodies were made in different facilities.

Various questions arise concerning nuclei: How do they appear? Are they stable? What is their effect on liquid cohesion and then on conditions of cavitation inception? How can they be measured? How can the nucleus content of a liquid be characterized?

Nuclei are present either on walls or in the liquid bulk. Surface nuclei consist of gas trapped in small wall crevices that are not filled with the liquid (see § 2.3.2). The wetting capacity of the liquid is therefore of great importance. It is possible that bulk nuclei are produced by cosmic rays, i.e., by a mechanism of energy deposition similar to the one used in bubble chambers for the experimental study of atomic particles. Another example of micronic bubble production by energy deposition is found in the breakdown of insulating liquids subjected to high voltage [AITKEN *et al.* 1996, JOMNI *et al.* 1999]. However, the most efficient way to produce nuclei is the reduction in pressure of a saturated liquid. Regions downstream of developed cavities may also be an abundant source of nuclei, as it will be seen in chapter 7.

Once present, nuclei evolve under two main influences. First, free nuclei (i.e., not attached to a wall) rise due to gravity. Second, all nuclei exchange gas via diffusion with the dissolved gas present in the surrounding liquid. In general, as the mass diffusion coefficients are very small, the diffusion process is slow and typical diffusion times are long, of the order of a second (see § 2.3.2). This is a large value in comparison with the time necessary for bubble collapse, which typically takes milliseconds (see chap. 3). Thus, in the following section, mechanical equilibrium of a spherical nucleus is assumed and the mass of enclosed gas is supposed constant. The problem of gas diffusion will be considered at the end of the chapter as it concerns the stability of gas nuclei over a long period.

The void fraction resulting from the presence of free nuclei is extremely small. For example, for a concentration of one hundred nuclei per cubic centimeter (this value is actually rather high) with a diameter of 0.1 mm, the void fraction is 0.52×10^{-4}. Thus, the liquid density remains practically unchanged. The same conclusion holds for the velocity of sound in the liquid.

2.2. EQUILIBRIUM OF A NUCLEUS

2.2.1. EQUILIBRIUM CONDITION [BLAKE 1949]

Consider a spherical microbubble, containing gas and vapor, in equilibrium within a liquid at rest. The liquid is supposed capable of withstanding pressures below the vapor pressure p_v and even tensions. Its metastable state can be represented by a point on the branch AM of the VAN DER WAALS curve in figure 1.2. The microbubble is the point of weakness from which the breakdown of the liquid medium can begin.

2.1
Microbubble in a liquid

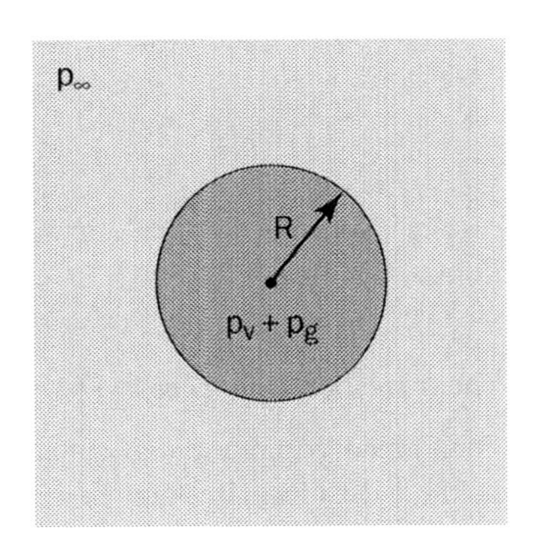

The bubble radius R is considered sufficiently small for the difference in hydrostatic pressure $2\rho gR$ to be negligible compared to the pressure difference corresponding to the surface tension $2S/R$. This condition requires that R be smaller than the limiting value $\sqrt{(S/\rho g)}$, namely 2.7 mm for water. It is fulfilled in this case since micro-microbubbles of diameter smaller than about 0.5 mm only are considered. The pressure can then be considered uniform in the bulk of the surrounding liquid (expressed as p_∞) and the microbubble is actually spherical.

The equilibrium of the interface requires the following condition to be satisfied:

$$p_\infty = p_g + p_v - \frac{2S}{R} \tag{2.1}$$

in which p_g is the partial pressure of the gas inside the bubble, S the surface tension and R the radius.

It is assumed that the pressure changes slowly so that mechanical equilibrium is still satisfied and that heat transfer between gas and liquid is possible. However, the change in pressure must be rapid enough to ensure that gas diffusion at the interface is negligible. In other words, the transformation is assumed isothermal and the mass of gas constant.

For the initial state, denoted by subscript 0, equation (2.1) is written:

$$p_{\infty 0} = p_{g0} + p_v - \frac{2S}{R_0} \tag{2.2}$$

As the gas pressure is inversely proportional to the volume in an isothermal transformation, then from equation (2.1) one obtains:

$$p_\infty = p_{g0}\left[\frac{R_0}{R}\right]^3 + p_v - \frac{2S}{R} \tag{2.3}$$

The curve $p_\infty(R)$ is shown in figure 2.2. The two mechanisms considered in equation (2.1), i.e.,

- the internal pressure effect, which tends to increase the bubble size, and
- the surface tension effect, which acts in the opposite direction, result in the existence of a minimum given by:

$$\begin{cases} R_c = R_0\sqrt{\dfrac{3p_{g0}}{2S/R_0}} \\ p_c = p_v - \dfrac{4S}{3R_c} \end{cases} \tag{2.4}$$

The corresponding locus is also represented in figure 2.2.

For a given liquid and a given gas at a fixed temperature, a nucleus is characterized by the mass of gas it contains, which is proportional to the quantity $p_{g0}\,R_0^3$. To define a nucleus with a constant mass of gas, one can use either one of the doublets (R_0, p_∞), (R_0, p_{g0}), etc. or preferably one of the quantities R_c or p_c.

2.2.2. *STABILITY AND CRITICAL PRESSURE OF A NUCLEUS*

Consider another nucleus containing more gas and having a radius $R'_0 > R_0$ under pressure $p_{\infty 0}$. The equilibrium curve of this nucleus is shown in figure 2.2. The corresponding gas pressure p'_{g0} is smaller than p_{g0}, as seen in equation (2.2).

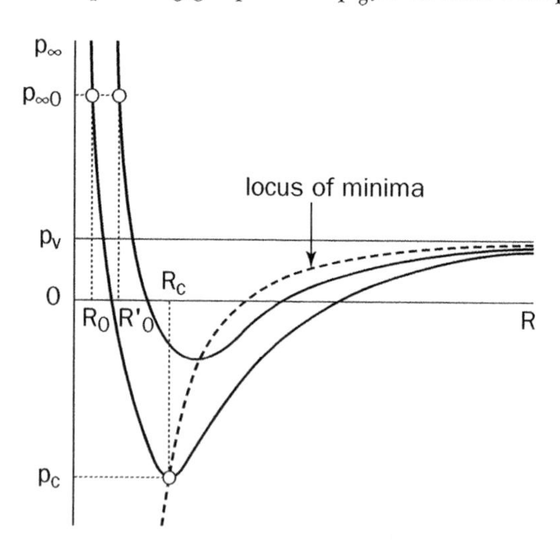

2.2
Equilibrium of a spherical nucleus

Suppose a virtual transformation is carried out resulting in the radius of the first nucleus becoming R'_0 so that the surface tension terms are identical. In its new state, the first nucleus has a gas pressure given by $p_{g0}(R_0/R'_0)^3$. This is lower than p'_{g0} as can be seen from a comparison of the mass of gas enclosed in each nucleus. This is proportional to $p_{g0} R_0^3$. Thus, the balance of forces tends to return the first nucleus to its initial state. Hence, the mechanical equilibrium of the spherical nucleus is stable on the branch of the curve that has a negative slope. It is straightforward to check that the equilibrium is unstable on the other branch.

The minimum pressure, p_c, that the nucleus can withstand under stable conditions is a limiting value referred to as the critical pressure. The difference $p_v - p_c$ is the static delay to cavitation. Expression (2.4) suggests that the smaller the nucleus, the greater the delay.

Figure 2.3 gives another representation of the equilibrium curves, limited to the stable domain, in the case of air nuclei in water. It is based on the following equation:

$$p_\infty = \frac{4}{27} \frac{(2S/R)^3}{(p_v - p_c)^2} + p_v - \frac{2S}{R}$$

obtained by combination of equations (2.3) and (2.4). The evolution of a given nucleus, characterized by its critical pressure, under varying external pressures, can be followed on a vertical line. On this diagram, the larger nuclei, which are more likely to provoke a breakdown of the liquid medium, have the smaller static delay.

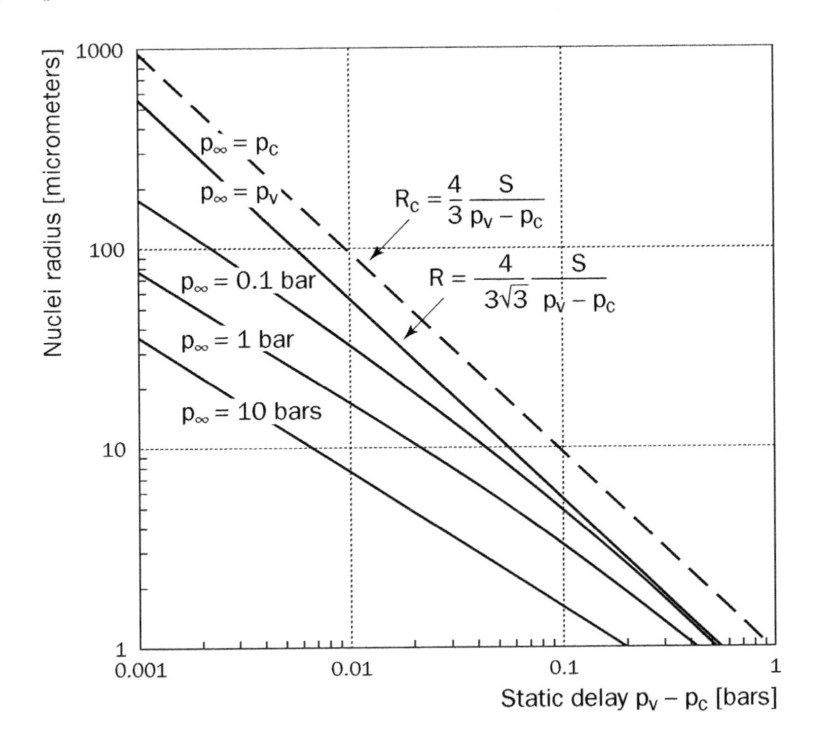

2.3 - Radius of equilibrium of air nuclei in water for various external pressures p_∞

2.2.3. *NUCLEUS EVOLUTION IN A LOW PRESSURE REGION*

Via BLAKE's model the evolution of an isolated nucleus passing through a low-pressure region of limited extent is more easily understood. Consider, for example, the evolution of a nucleus in a Venturi (fig. 2.4). The local pressure at a distance s in this approximately one-dimensional flow is denoted $p(s)$. For the nucleus, this local pressure $p(s)$ plays the role of p_∞, as introduced in section 2.2.1.

Two cases must be considered according to the value of the minimum pressure p_{min} with regard to the critical pressure p_c of the nucleus.

- If $p_{min} > p_c$, the nucleus grows slightly and then returns to its initial state as it passes through the throat.
- If $p_{min} < p_c$, it becomes unstable and much larger than it was initially. Figure 2.4 shows its evolution in the (R, p_∞) diagram, as deduced from dynamic calculations (see chap. 3). Due to inertial forces, its maximum size is reached at point D, a little downstream from throat C. At D, it is far from equilibrium, so that the increasing external pressure compels it to collapse violently.

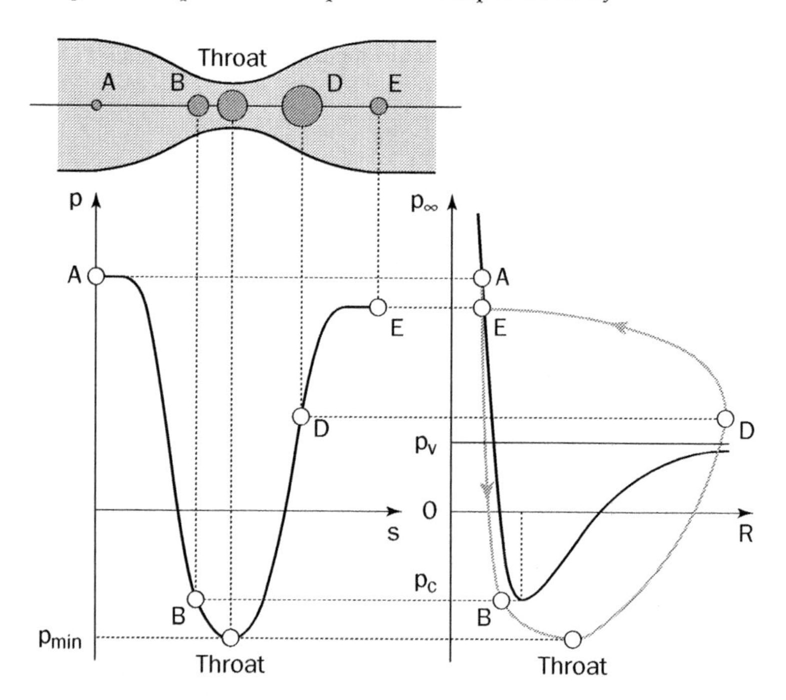

2.4 - Typical nucleus evolution in a Venturi. Unstable case: $p_{min} < p_c$

This type of bubble evolution can be applied to many practical situations in which the time spent in the low-pressure region is long compared to other characteristic times, particularly the collapse time (see chap. 3 for an estimation of this time). Note that, on the stable branch of the equilibrium curve (fig. 2.2), small oscillations around equilibrium can be expected. Thus, if the time required for the pressure drop is of the same order –or even shorter– than the natural period of oscillation of the nucleus, the bubble size will not grow monotonically but will undergo oscillations.[1]

2.3. *HEAT AND MASS DIFFUSION*

In section 2.2, it was assumed that the nucleus undergoes an isothermal transformation with a constant mass of gas. The conditions of validity of both assumptions and the consequences of possible transfers of gas with the surrounding liquid are now examined.

2.3.1. *THE THERMAL BEHAVIOR OF THE GAS CONTENT*

If, contrary to the assumption made in section 2.2.1, the change in pressure around the nucleus occurs in such a short time that heat transfer between the gas and the liquid cannot be achieved, the transformation of the gas is effectively adiabatic. In this case, equations (2.4) for the critical radius and critical pressure have to be replaced by the following:

$$\begin{aligned} R'_c &= R_0 \left[\frac{3\gamma p_{g0}}{2S/R_0} \right]^{1/(3\gamma-1)} \\ p'_c &= p_v - \left[1 - \frac{1}{3\gamma} \right] \frac{2S}{R'_c} \end{aligned} \qquad (2.5)$$

where γ is the ratio of the gas heat capacities. This model is not entirely consistent since the vapor pressure is still taken at a constant temperature.

To decide on the correct gas behavior in each particular case, the heat transfer must be analysed. A rather simple way is to proceed as follows [PLESSET & HSIEH 1960].

Consider a nucleus which undergoes a pressure drop during a time Δt, such that its radius increases from R to $R + \Delta R$. For simplicity, the bubble is supposed to contain only non-condensable gas, and no phase change is considered. During this transformation, the gas temperature inside the bubble is assumed to change from T to $T + \Delta T$. The temperature variation ΔT can be determined from the energy balance of the gas:

$$\left[\frac{4}{3} \pi R^3 \right] \rho_g c_{vg} \Delta T = \Delta Q - p [4\pi R^2 \Delta R] \qquad (2.6)$$

---

1. The question of resonant excitation of nuclei by turbulent fluctuations in the liquid flow was considered by CRIGHTON and FFOWCS WILLIAMS (1969). They showed that such a mechanism is not plausible because the high resonant frequencies of nuclei are associated with sizes of turbulent fluctuations that are very small in comparison with nuclei sizes (see § 3.4.1).

The left-hand side of this equation is the variation of the internal energy of the gas (ρ_g and c_{vg} stand for the density and specific heat at constant volume), whereas the two terms on the right-hand side are the heat received by the bubble and the work of the surface pressure forces respectively.

Because of the temperature variation ΔT, a thermal boundary layer develops inside the liquid (see fig. 2.5). The temperature gradient induces a heat flux by conduction which determines the heat ΔQ transferred from the liquid to the bubble. If a_ℓ is the thermal diffusivity of the liquid, the boundary layer thickness is of the order of $\sqrt{a_\ell \Delta t}$, so that FOURIER's law leads to the following estimate:

$$\Delta Q \approx -\lambda_\ell \frac{\Delta T}{\sqrt{a_\ell \Delta t}} 4\pi R^2 \Delta t = -\sqrt{\lambda_\ell \rho_\ell c_{p\ell}} \sqrt{\Delta t}\, 4\pi R^2 \Delta T \tag{2.7}$$

where λ_ℓ and $c_{p\ell}$ are the conductivity and heat capacity of the liquid respectively.

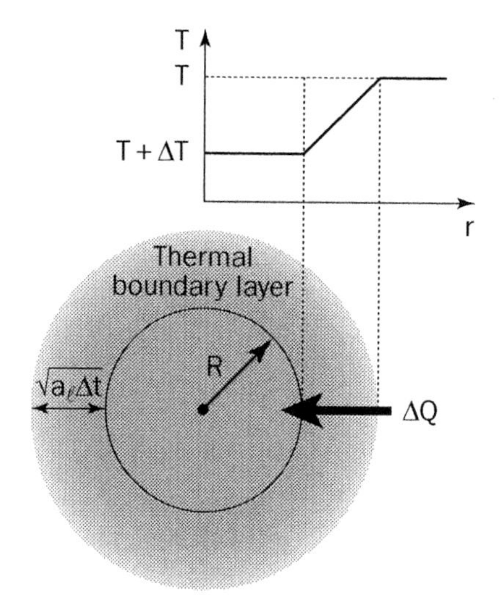

2.5
Thermal boundary layer and schematic temperature distribution in the liquid around the bubble

The energy balance (2.6) allows an estimation of the temperature variation of the gas in the bubble:

$$\Delta T \approx \frac{\Delta T_{ad.}}{1 + \sqrt{\dfrac{\Delta t}{\Delta t_r}}} \tag{2.8}$$

where $\Delta T_{ad.}$ is the temperature variation in the adiabatic case $\Delta Q = 0$:

$$\Delta T_{ad.} = -\frac{3p\,\Delta R}{R\,\rho_g\,c_{vg}} \tag{2.9}$$

and Δt_r is a characteristic time defined by:

$$\Delta t_r = \frac{(\rho_g\,c_{vg}\,R)^2}{9\lambda_\ell\,\rho_\ell c_{p\ell}} \tag{2.10}$$

The transformation can be considered adiabatic if the exchanged heat is much lower than the internal energy variation or, in terms of the characteristic times, if the transit time Δt is much smaller than the characteristic time Δt_r, which can be interpreted as the time required for heat transfer.

Inversely, if $\Delta t_r << \Delta t$, equation (2.8) gives $\Delta T = 0$ and the transformation can be considered as isothermal. In other words, the time available for heat transfer is so large that heat transfer can proceed until thermal equilibrium is reached.

In the case of the water/air system at ambient temperature ($\lambda_\ell = 0.6\ W/m/K$, $\rho_\ell = 1{,}000\ kg/m^3$, $c_{p\ell} = 4{,}180\ J/kg/K$, $c_{vg} = 715\ J/kg/K$), $\Delta t_r = 0.023\ (\rho_g R)^2$. For example, with $\rho_g = 3\ kg/m^3$ and $R = 0.5\ mm$ [2] the required time is $\Delta t_r = 0.051\ \mu s$, a very small value.

For the transit time, we take the case of a nucleus that passes through a low-pressure region 1 cm in length at a velocity of 50 m/s.[3] The corresponding transit time $\Delta t = 0.2$ ms is about four thousand times greater than the time Δt_r required for heat transfer. Thus, it can be inferred that, in many practical water-air systems, the transformation can be considered isothermal.

As the characteristic time Δt_r is proportional to the radius squared, the question must be re-examined if the nucleus explodes and becomes a macroscopic bubble of very large radius (see chap. 3).

2.3.2. *GAS DIFFUSION AND NUCLEUS STABILITY*

As stated in the previous sections, gas can be present in a liquid in two forms, dissolved, or trapped in free nuclei. The question of any exchange between the forms is examined here on the basis of the physical laws concerning the saturation of the liquid and the transfer of gas in non-equilibrium conditions.

Dissolved gas and nuclei in a static liquid

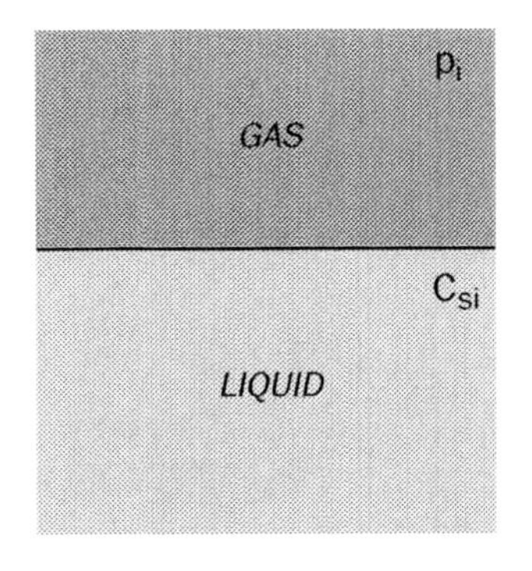

2.6
Diffusive equilibrium between a liquid and the atmosphere above it

The diffusion equilibrium between a liquid and the atmosphere above it is described by HENRY's law:

$$C_{si} = H_i(T)\, p_i \tag{2.11}$$

in which p_i is the partial pressure of gas i in the atmosphere (fig. 2.6), C_{si} is its concentration at saturation in the liquid and H_i the corresponding HENRY's constant. C_{si} is expressed in kg/m^3, and H_i in $(m/s)^{-2}$.

2. These values are chosen large enough to produce adiabatic behavior.
3. These values are rather extreme and again favorable to adiabatic transformation.

If diffusion equilibrium is not achieved, a concentration gradient **C** exists, resulting in a mass flux given by FICK's law:

$$\vec{q}_i = -D_i \, \vec{grad}\, C_i \tag{2.12}$$

In equation (2.12), D_i is the diffusivity coefficient of element i. The balance of mass transfer for a small domain gives the classic diffusion equation:

$$\frac{\partial C_i}{\partial t} = D_i \, \Delta C_i \tag{2.13}$$

In the case of air dissolved in water, concentrations are often expressed in parts per million, a concentration of 1 ppm being equivalent to $10^{-3}\ kg/m^3$. The concentrations at saturation in water are 19 ppm for pure nitrogen and 43 ppm for pure oxygen, under an external pressure of 1 bar. Taking into account the partial pressure of each element in the atmosphere, 0.79 bar and 0.21 bar respectively, one obtains 15 ppm for nitrogen and 9 ppm for oxygen, i.e., 24 ppm in total under atmospheric pressure. Then, the HENRY constant is $0.24 \times 10^{-6}\ (s/m)^2$ while the diffusivity coefficient is $D = 2 \times 10^{-9}\ m^2/s$ for air in water.

Let us now consider the equilibrium of a spherical nucleus in a static liquid (fig. 2.7). Mechanical equilibrium requires that relation (2.1) be satisfied:

$$p_\infty = p_g + p_v - \frac{2S}{R} \tag{2.14}$$

Its diffusive equilibrium is expressed, from (2.11), by:

$$C_s = H p_g \tag{2.15}$$

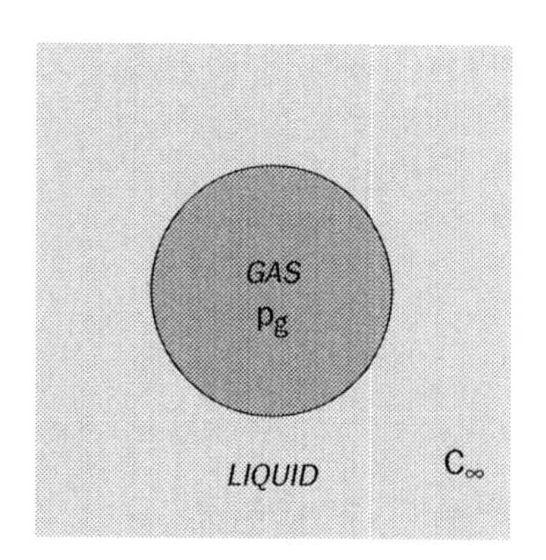

2.7
Nucleus in a static liquid

Two cases must be considered:

- If the concentration far from the nucleus, C_∞, is smaller than C_s, the gas tends to migrate from the nucleus to the liquid. The radius decreases, the surface tension term increases. With a constant pressure p_∞ far from the nucleus, the term p_g grows and increases the diffusive imbalance and the nucleus tends to be resorbed.

- If C_∞ is larger than C_s, the opposite phenomenon occurs: the nucleus volume increases and the diffusive imbalance continues to increase, as in the previous case.

In conclusion, the double mechanical and diffusive equilibrium of the nucleus is always unstable. The conclusion is the same if surface tension is neglected, but the difference between the concentrations remains constant and the instability is weaker.

EPSTEIN and PLESSET (1950) give estimates of the characteristic time for air diffusion in the absence of surface tension. The integration of equation (2.13) leads to the following equation for the evolution of the radius R of the nucleus:

$$\frac{dR}{dt} = -D\frac{C_s - C_\infty}{\rho_g}\left[\frac{1}{\sqrt{\pi D t}} + \frac{1}{R}\right] \tag{2.16}$$

where ρ_g is the gas density. If the first term is ignored, i.e. if the nucleus radius is assumed small with respect to the thickness of the concentration boundary layer, the total resorption time is:

$$\tau_{res} = \frac{R_0^{\ 2}}{2D}\frac{\rho_g}{C_s - C_\infty} \tag{2.17}$$

Equation (2.17) gives 10,420 seconds, for the air/water system at atmospheric pressure, with the following values: $R_0 = 1\ \text{mm}$, $C_s = 0.024\ \text{kg}/\text{m}^3$, $C_\infty = 0$, $\rho_g = 1\ \text{kg}/\text{m}^3$. The table below gives the values of τ for a nucleus of initial radius 0.01 mm and $C_s/\rho_g = 0.02$. The parameter f is the ratio C_∞/C_s. The calculated time is the total time required for nucleus resorption if $f < 1$. If $f > 1$, it is the time needed by the nucleus to grow to a radius ten times larger than its initial value.

f	0	0.25	0.50	0.75	1	1.25	1.50	1.75	2.0	5.0
τ(s)	1.05	1.44	2.21	4.58	∞	466	228	149	110	46

If surface tension is taken into account, a resorption time of about 6 seconds is found for $f = 1$. In most practical situations, the characteristic diffusion time appears to be at least of the order of one second.

Nucleus stability with respect to gas diffusion

According to the previous analysis, nuclei cannot exist for long periods of time in static liquids. Other mechanisms must be invoked to explain their commonly observed long lifetimes.

It was previously assumed that the nucleus is surrounded by an infinite liquid medium that constitutes a source (or sink) capable of giving (or receiving) an unlimited amount of gas. Actually, the gas transfer involves a limited domain of liquid close to the nucleus [MORI 1977, CHA 1981, TSAI 1990, ACHARD & CANOT 1992]. The equilibrium can thus be stable since each nucleus has a limited capacity of gas transfer. However, such a mechanism requires all nuclei to be the same size –which is not the case in reality– since the gas concentration and the pressure are uniformly distributed inside the domain. This is a weakness of this model.

The stability of nuclei can also be explained by assuming that they are enveloped by a thin organic film that prevents gas transfer [FOX & HERZFELD 1954]. Such microbubbles actually exist in the sea, but their response to cavitation has not been systematically studied.

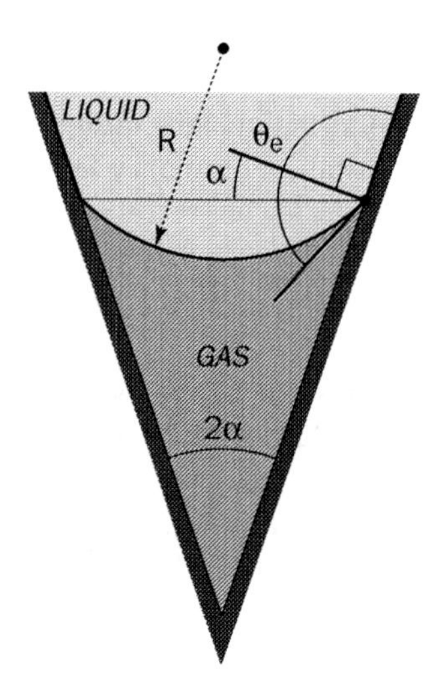

2.8
Model of a crevice as a nucleus
[from KNAPP et al., 1970]

Finally, small crevices in solid walls or in solid particles carried by the liquid flow can shelter stable nuclei [HARVEY *et al.* 1950]. Consider, for example, a conical crevice (fig. 2.8) with an angle 2α, which is partly filled by a liquid. If the contact angle between the liquid and the wall θ_e is larger than $\pi/2+\alpha$, which is the case in figure 2.8, the interface curvature is reversed and the surface tension term in equation (2.14) changes sign. If the inequality $C_\infty < C_s$ holds, the gas migrates towards the liquid and the volume of the trapped gas decreases. Consequently, both the radius R and the gas pressure p_g decrease while the external pressure remains constant. The difference $C_\infty - C_s$ diminishes and the situation tends to a stable equilibrium.

Liquid convection and rectified diffusion

In the previous paragraphs, it was assumed that the nucleus did not move with respect to the surrounding liquid. If there is a relative motion as is the case for liquids flowing past wall crevices, the liquid convection continuously renews the gas source at a small distance from the nucleus and gas transfer is considerably increased [PARKIN & KERMEEN 1962].

Another situation deserves attention. If a nucleus undergoes pressure oscillations due, for example, to an ultrasonic oscillation field, and if the excitation frequency is properly adjusted, the size of the nucleus will vary with the imposed pressure. If the diffusive equilibrium corresponding to the mean pressure is reached, then, during half a period of growth phase, the internal pressure of the nucleus becomes lower than the mean pressure and the gas migrates towards it. The reverse gas flux tends to take place during the half period when the size of the nucleus becomes smaller. On the whole, as the exchange area is larger in the first half period, there is a net flux of gas to the advantage of the nucleus. This phenomenon, called rectified diffusion, was studied by PLESSET and HSIEH (1960) who give the following expression for the time required for doubling the nucleus radius:

$$\tau = \frac{9}{4} \frac{R_0{}^2 \rho_g}{\varepsilon^2 D C_\infty} \tag{2.18}$$

where ε is the relative amplitude of the pressure oscillations. For $R_0 = 0.1\ \mathrm{mm}$, $\rho_g = 1\ \mathrm{kg/m^3}$, $\varepsilon = 0.5$ (a rather large value), in the case of air-water one finds $\tau = 2{,}300\ \mathrm{s}$. Thus, for most industrial situations where pressure fluctuations are due to turbulence, this mechanism of nucleus growth by rectified diffusion is not very efficient.

2.4. *NUCLEUS POPULATION*

2.4.1. *MEASUREMENT METHODS*

In the case of real liquids, nuclei of various sizes are present. From the previous static analysis, it is expected that the response of the liquid to the pressure variations imposed by the flow will strongly depend upon the concentration of cavitation nuclei. The measurement of the nucleus population is therefore essential. Two main kinds of measuring techniques are used:

- optical methods, which give the equilibrium radius R_0 of the nuclei under an external pressure $p_{\infty 0}$;
- dynamic methods, which give the critical pressure p_c of the nuclei [OLDENZIEL 1982].

In addition, it is necessary to determine the volume of the liquid sample analyzed. For optical methods such as holography and DOPPLER phase anemometry, the nucleus population is characterized by a histogram which gives, per unit volume, the number of nuclei of radius R_0 in a bandwidth ΔR_0. It is expressed in terms of nuclei$/\mathrm{cm}^3/\Delta R_0$ (fig. 2.9-a).

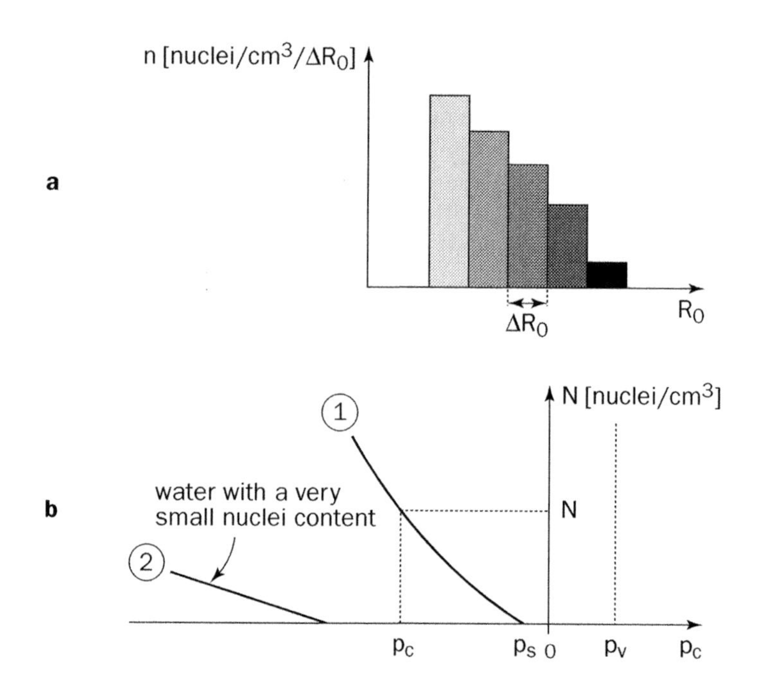

2.9 - Characterization of the nucleus population by (a) a size-based histogram obtained by optical methods or (b) a cumulative histogram of critical pressure obtained by dynamic methods

Histogram 2 corresponds to a much smaller nuclei content than histogram 1.

In the case of dynamic methods, the nuclei go through a low-pressure region, such as a Venturi. All nuclei with a critical pressure higher than the minimum are destabilized. They explode and are counted subsequently by an ultrasonic transducer sensitive to the noise they emit during collapse. A cumulative histogram[4] $N(p_c)$ is obtained (fig. 2.9-a and 2.10-a), with N given in nuclei / cm^3.

Comparison between optical and dynamic methods requires a conversion between the critical pressure and the equilibrium size R_0 by means of the equilibrium relation (2.2).

Dynamic methods enable us to count nuclei with a diameter of less than 10 μm, which is the usual limit of the holographic technique. Generally, optical methods give higher concentrations (by up to more than one order of magnitude) than

4. Note that a population characterized by a variable x can be described by either a distribution function $n(x)$ or a cumulative histogram $N(x)$ where N is the number of elements whose size is larger than x. N and n are connected by the equation: $N(x) = \int_x^\infty n(u)\,du$.

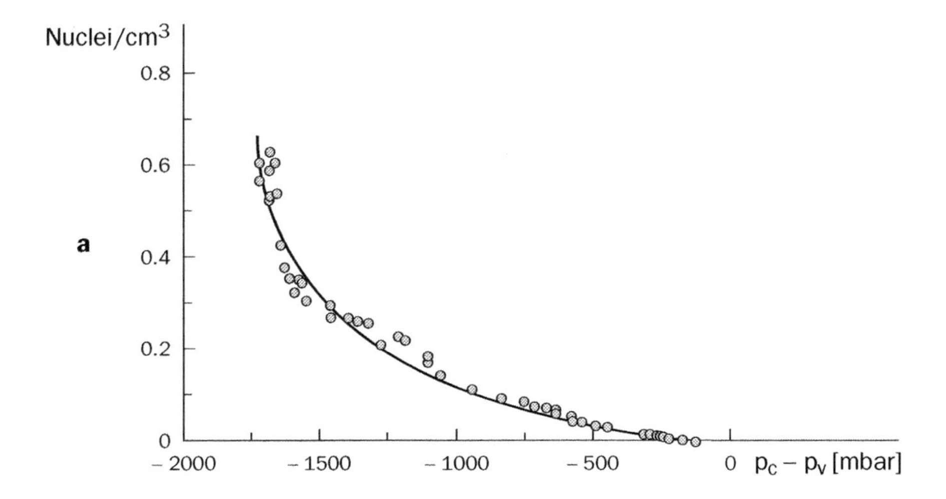

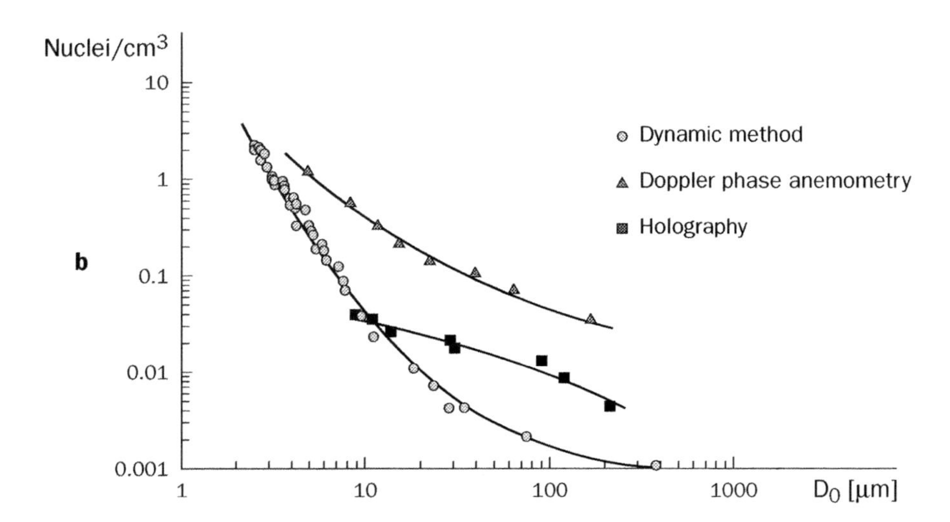

2.10 - (a) Typical histogram of nuclei population as a function of critical pressure (dynamic method of measurement) and (b) comparison of cumulative histograms in size for three different measuring techniques
[from Bassin d'Essais des Carènes, 1992]

dynamic methods. This could be due either to the difficulty in estimating the volume of the liquid sample in holography, or to the fact that all objects counted are not necessarily cavitation nuclei, but may be solid particles or microbubbles with an organic envelope. The difference between dynamic and holographic methods is lower for carefully filtered water.

In recent years, the use of Venturi devices has become common in test facilities such as cavitation tunnels. They offer several advantages in comparison with optical techniques.

- Nuclei are characterized by their critical pressure, which is the relevant quantity in cavitation. In particular, it has been shown that concentration of nuclei measured with a Venturi closely agrees with the density of bubbles that explode on the upper surface of a 2-D foil [BRIANÇON-MARJOLLET *et al.* 1990].
- Very significant volumes of liquid, typically 0.1 m^3, are used for counting.
- The measurement can be carried out in a short time, approximately ten minutes for a complete histogram.
- The relative uncertainties regarding the size of the nuclei are fewer, at least for small nuclei.

Details of the design of Venturi devices can be found in PHAM *et al.* (1997).

It must be noted that detection of collapse by piezoelectric transducers can lead to an overestimate of the bubble concentration for two main reasons:
- big bubbles may be broken into smaller ones during their collapse;
- in some circumstances, the same bubble can collapse, then rebound and collapse again.

In both cases, multipeaking is observed in the noise signal, leading to possible errors in the measurement of the bubble population.

Modern cavitation tunnels are also equipped with systems of nucleus seeding. The concentration of nuclei which can explode in the vicinity of a body tested in a laboratory water flow in which nuclei are injected is typically of the order of one nucleus per cubic centimeter. Very large cavitation effects can be obtained with such a concentration.

2.4.2. *CONDITIONS FOR INCEPTION OF BUBBLE CAVITATION*

The largest critical pressure p_s in a nucleus population, which is the critical pressure of the biggest nucleus, is called the susceptibility pressure of the liquid (see fig. 2.9-b). It is an important element in the onset of bubble cavitation.

If thermal and dynamic delays in cavitation inception are assumed negligible, cavitation inception occurs as soon as the biggest nucleus is destabilized, i.e., as soon as the minimum pressure is lower than the susceptibility pressure (fig. 2.11). Hence, the condition for bubble cavitation inception is:

$$p_s = p_{min} \tag{2.19}$$

In figure 2.11, the pressure distribution 1 corresponds to cavitation inception, while a larger number of nuclei are destabilized in the case of pressure distribution 2.

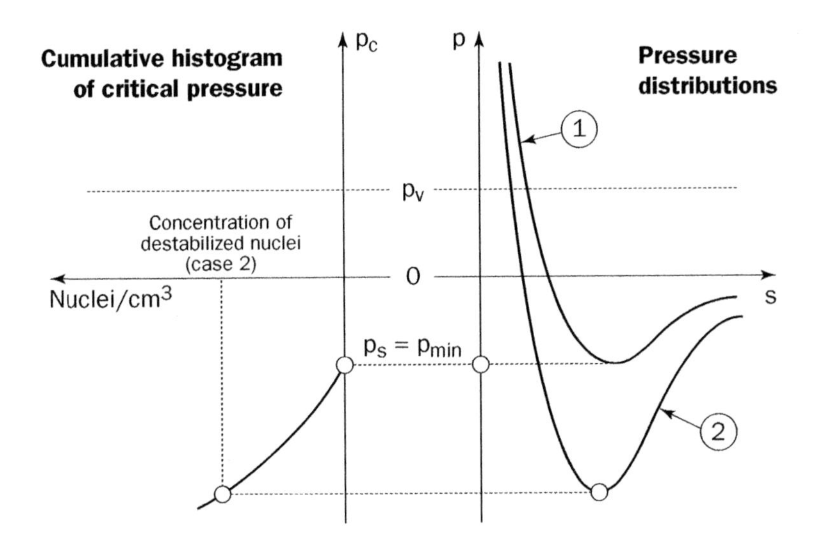

2.11 - Nucleus population, pressure distribution and bubble cavitation

For a given nucleus population indicated on the left-hand side, pressure distribution 1 corresponds to the inception of bubble cavitation whereas pressure distribution 2 corresponds to developed bubble cavitation.

Taking into account the expressions for the cavitation parameter σ_v and the pressure coefficient Cp given in sections 1.4.1 and 1.4.2, the condition for cavitation inception (2.19) takes the following non-dimensional form:

$$\sigma_{vi} = -Cp_{min} - \Delta\sigma_s \tag{2.20}$$

where $\Delta\sigma_s$ stands for the relative susceptibility underpressure:

$$\Delta\sigma_s = \frac{p_v - p_s}{\Delta p} \tag{2.21}$$

The shift $\Delta\sigma_s$ increases when the biggest nuclei are eliminated, subsequently increasing the resistance of the liquid to cavitation. For identical concentrations of nuclei, it diminishes when the pressure difference Δp increases, which is the case for an increase in liquid velocity.

It may be difficult to experimentally determine an exact value for the threshold σ_{vi}. Consideration must be given not only to the size of detectable nuclei (typically 1 mm in diameter) but also to the frequency of such cavitation events.

The response of different concentrations of nuclei to the same pressure distribution can be very different. Dynamic calculations show that large nuclei can grow smoothly to a size of one millimeter, while small nuclei tend to explode as soon as they are destabilized. Then, if the liquid does not contain any large nuclei, the inception of cavitation can take the form of a sudden explosion of nuclei to a size of one centimeter or more.

REFERENCES

ACHARD J.L. & CANOT E. –1992– *Bubble evolution in liquid-gas solutions, viewed as an elementary catastrophe.* Instabilities in Multiphase Flows. *Plenum Press Ed.*, 37-51.

AITKEN F., MC CLUSKEY F. & DENAT A. –1996– An energy model for artificially generated bubbles in liquids. *J. Fluid Mech.* **327**, 373-392.

BERTHELOT M. –1850– Sur quelques phénomènes de dilatation forcée de liquides. *Ann. Chim. Phys.* **30**(3), 232-237.

BLAKE F.G. –1949– The onset of cavitation in liquids. *Harvard Acoustics Res. Lab.* TM 9.

BRIANÇON-MARJOLLET L., FRANC J.P. & MICHEL J.M. –1990– Transient bubbles interacting with an attached cavity and the boundary layer. *J. Fluid Mech.* **218**, 355-376.

BRIGGS L.J. –1950– Limiting negative pressure of water. *J. Appl. Phys.* **21**, July, 721-722.

CHA Y.S. –1981– On the equilibrium of cavitation nuclei in liquid-gas solution. *ASME J. Fluids Eng.* **103**, 425 *sq.*

CRIGHTON D.G. & FFOWCS WILLIAMS J.E. –1969– Sound generation by turbulent two-phase flow. *J. Fluid Mech.* **36**, part 3, 585-603.

DONNY F.M.L. –1846– *Ann. Chim. Phys.* **16**(3), 167-190.

EPSTEIN P.S. & PLESSET M.S. –1950– On the stability of gas bubbles in liquid-gas solutions. *J. Chem. Phys.* **18**, 1505-1509.

FOX F.E. & HERZFELD K.F. –1954– Gas bubbles with organic skin as cavitation nuclei. *J. Acoust. Soc. Am.* **26**, 984-989.

HABERMAN W.L. & MORTON R.K. –1953– An experimental investigation of the drag and shape of air bubbles rising in various liquids. *DTMB*, Rpt 802, September.

HARVEY E.N., MC ELROY W.D. & WHITELEY A.H. –1947– On cavity formation in water. *J. Appl. Phys.* **18**, 162-172.

JOMNI F., AITKEN F. & DENAT A. –1999– Dynamics of microscopic bubbles generated by a corona discharge in insulating liquids: influence of pressure. *J. Electrostatics* **47**, 49-59.

KNAPP R.T., DAILY J.W. & HAMMITT F.G. –1970– Cavitation. *McGraw-Hill Book Company Ed.*

MEDWIN H. –1977a– Acoustical determination of bubble-size spectra. *J. Acoust. Soc. Am.* **62**(4), 1041-1044.

MEDWIN H. –1977b– Counting bubbles acoustically: a review. *Ultrasonics* **15**(1), 7-13.

MORI Y., HIJIKATA K. & NAGATANI T. –1977– Fundamental study of bubble dissolution in liquid. *Int. J. Heat Mass Transfer* **20**, 41 *sq*.

OLDENZIEL D.M. –1982– New instruments in cavitation research: the cavitation susceptibility meter. *J. Fluids Eng.* **104**(2), 136-142.

PARKIN B.R. & KERMEEN R.W. –1962– The roles of convective air diffusion and liquid tensiles during cavitation inception. *Proc. IAHR Symp. on Cavitation and Hydraulic Machinery*, Sendai (Japan), September.

PHAM T.M., MICHEL J.M. & LECOFFRE Y. –1997– Dynamical nuclei measurement: on the design and performance evaluation of an optimized center-body meter. *J. Fluids Eng.* **119**(4), 744-751.

PLESSET M.S. –1957– Physical effects in cavitation and boiling. *Proc. 1st Int. Symp. on Naval Hydrodynamics*, Washington DC (USA), 297-323.

PLESSET M.S. & HSIEH D.H. –1960– Theory of gas bubble dynamics in oscillating pressure fields. *Phys. Fluids* **3**, 882-892.

REES E.P. & TREVENA D.H. –1966– Cavitation thresholds in liquids under static conditions. *ASME Cavitation Forum*, 12 *sq*.

REYNOLDS O. –1882– On the internal cohesion of liquids and the suspension of a column of mercury to a height of more than double that of a barometer. *Mem. Manchester Lit. Phil. Soc.* **7**, 1-19.

TEMPERLEY H.N.V. –1946– The behavior of water under hydrostatic tension. *Proc. Phys. Soc. London* **58**, 436-443.

TSAI J.F. & CHEN Y.N. –1990– A generalized approach on equilibrium theory of cavitation nuclei in liquid-gas solutions. *J. Fluids Eng.* **112**, 487-491.

3. THE DYNAMICS OF SPHERICAL BUBBLES

3.1. BASIC EQUATIONS

3.1.1. INTRODUCTION

In this chapter we consider the dynamic evolution of a spherical bubble with a fixed center, which undergoes uniform pressure variations at infinity. This simple model demonstrates the main features of many practical cases such as bubble collapse, bubble formation from a nucleus, bubble oscillations, etc. Experience shows that more complicated situations, involving the motion of the bubble center for example, can be approximately dealt with using this model.

From an historical viewpoint, the liquid motion induced by a spherical cavity in an infinite medium under uniform pressure at infinity seems to have been first considered by BESANT in 1859. It was solved for a non-viscous liquid by RAYLEIGH (1917) to interpret the phenomenon of cavitation erosion. In 1948, COLE used the model of a spherical bubble containing a non-condensable gas and applied it to submarine explosions. PLESSET (1954) considered the general case of bubble evolution for a viscous and non-compressible liquid.

3.1.2. ASSUMPTIONS

The main assumptions are the following:

- the liquid is incompressible and either Newtonian or inviscid;
- gravity is neglected;
- the air content of the bubble is constant, its inertia is neglected as is any exchange of heat with the surroundings. This adiabatic assumption is valid when considering rather large bubbles;
- the bubble is saturated with vapor whose partial pressure is the vapor pressure at the liquid bulk temperature.

The functions to be determined, in the liquid domain $r \geq R(t)$, are the velocity $u(r,t)$ and the pressure $p(r,t)$ induced by the evolution of the bubble (fig. 3.1).

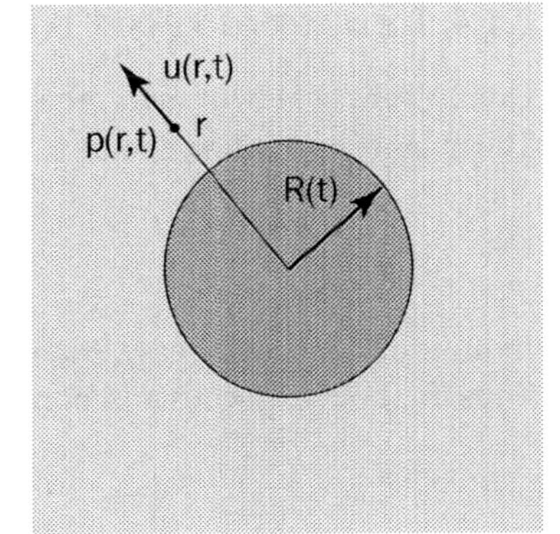

3.1

3.1.3. BOUNDARY AND INITIAL CONDITIONS

The mass transfer through the interface is neglected (see § 1.3.2), so that the liquid velocity at the interface $u(R,t)$ is equal to the interface velocity $\dot{R} = dR/dt$.

In the case of a viscous fluid of kinematic viscosity μ, the normal stress at the interface is:

$$t_{rr}(R,t) = -p(R,t) + 2\mu \left.\frac{\partial u}{\partial r}\right|_{r=R} \tag{3.1}$$

The balance of normal forces is given by:

$$-t_{rr}(R,t) = p_v + p_g(t) - \frac{2S}{R} \tag{3.2}$$

where p_g stands for the partial pressure of the gas inside the bubble. Assuming an adiabatic transformation of the gas, the instantaneous gas pressure is related to the initial pressure p_{g0} by the following expression:

$$p_g(t) = p_{g0}\left[\frac{R_0}{R(t)}\right]^{3\gamma} \tag{3.3}$$

where γ is the ratio of heat gas capacities c_{pg} and c_{vg}.

Thus, the pressure on the cavity interface is given by:

$$p(R,t) = p_v + p_{g0}\left(\frac{R_0}{R}\right)^{3\gamma} - \frac{2S}{R} + 2\mu \left.\frac{\partial u}{\partial r}\right|_{r=R} \tag{3.4}$$

Far from the bubble, the liquid is assumed at rest so that $u(\infty,t) \to 0$ and the pressure $p(\infty,t)$ also denoted $p_\infty(t)$ is assumed given.

For the initial conditions (subscript 0), the bubble is assumed to be in equilibrium, i.e., $\dot{R}(0) = 0$, so that equation (2.2) is satisfied:

$$p_{\infty 0} = p_{g0} + p_v - \frac{2S}{R_0} \tag{3.5}$$

3.1.4. RAYLEIGH-PLESSET EQUATION

Due to spherical symmetry, the flow is of source (or sink) type and so irrotational. The mass conservation equation for an incompressible fluid $\operatorname{div}\vec{V} = 0$ gives:

$$u(r,t) = \dot{R}\frac{R^2}{r^2} \tag{3.6}$$

In this very particular case, the viscous term of the NAVIER-STOKES equation is zero. Thus, for both a viscous and non-viscous fluid, the momentum equation is:

$$\frac{\partial u}{\partial t} + u\frac{\partial u}{\partial r} = -\frac{1}{\rho}\frac{\partial p}{\partial r} \tag{3.7}$$

from which one infers, taking equation (3.6) into account:

$$\ddot{R}\frac{R^2}{r^2}+2\dot{R}^2\left[\frac{R}{r^2}-\frac{R^4}{r^5}\right]=-\frac{1}{\rho}\frac{\partial p}{\partial r} \tag{3.8}$$

Integrating with respect to r and considering the conditions at infinity, one obtains:

$$\frac{p(r,t)-p_\infty(t)}{\rho}=\ddot{R}\frac{R^2}{r}+2\dot{R}^2\left[\frac{R}{r}-\frac{R^4}{4r^4}\right] \tag{3.9}$$

This equation is equivalent to the BERNOULLI equation for a variable unsteady flow of non-viscous liquid. On the interface $r=R$, equation (3.9) gives:

$$\frac{p(R,t)-p_\infty(t)}{\rho}=R\ddot{R}+\frac{3}{2}\dot{R}^2 \tag{3.10}$$

Finally, with expression (3.4) for the pressure at the interface, and noting that:

$$\left.\frac{\partial u}{\partial r}\right|_{r=R}=-\frac{2\dot{R}}{R} \tag{3.11}$$

equation (3.10) becomes:

$$\rho\left[R\ddot{R}+\frac{3}{2}\dot{R}^2\right]=p_v-p_\infty(t)+p_{g0}\left(\frac{R_0}{R}\right)^{3\gamma}-\frac{2S}{R}-4\mu\frac{\dot{R}}{R} \tag{3.12}$$

This equation, known as the RAYLEIGH-PLESSET equation, allows us to determine the temporal evolution of the radius R and consequently the pressure field in the liquid when the law $p_\infty(t)$ is given. For a non-viscous liquid, the last term on the right-hand side vanishes. The corresponding equation is known as the RAYLEIGH equation.

Both equations are differential and highly non-linear, due to the inertial terms. This results in various specific features which are presented in this chapter.

In the two following sections, the RAYLEIGH equation will be used to solve the problem of bubble collapse (BESANT problem) and bubble explosion. In most cases, the inertial forces are dominant and viscosity does not play a significant role. The role of surface tension is often secondary in the case of bubble collapse.

3.1.5. INTERPRETATION OF THE RAYLEIGH-PLESSET EQUATION IN TERMS OF ENERGY BALANCE

Noting that:

$$R\ddot{R}+\frac{3}{2}\dot{R}^2=\frac{1}{2\dot{R}R^2}\frac{d}{dt}\left[\dot{R}^2R^3\right] \tag{3.13}$$

the RAYLEIGH-PLESSET equation (3.12) can be written as follows:

$$\frac{d}{dt}\left(2\pi\rho\dot{R}^2R^3\right)=\left[p_v+p_{g0}\left(\frac{R_0}{R}\right)^{3\gamma}-p_\infty(t)\right]4\pi R^2\dot{R}-8\pi SR\dot{R}+16\pi\mu R\dot{R}^2 \qquad (3.14)$$

The term on the left-hand side represents the variation in kinetic energy of the liquid body. The first term on the right-hand side represents pressure forces acting on the liquid, while the surface tension forces are represented by the second term. The dissipation rate due to viscosity is expressed as $\iiint 2\mu\, e_{ij}\, e_{ij}\, d\tau$, where e_{ij} stands for the deformation rate tensor and the integral is taken over the entire liquid volume. This gives the last term.

3.2. THE COLLAPSE OF A VAPOR BUBBLE

3.2.1. ASSUMPTIONS

In the present section, the effects of viscosity, non-condensable gas and surface tension are all ignored.

Before the initial time, the bubble is supposed to be in equilibrium under pressure $p_{\infty 0}$, which is equal to p_v, according to equation (3.5). From the instant $t=0$, a constant pressure p_∞, higher than p_v, is applied to the liquid. It results in the collapse of the bubble in a characteristic time τ called the RAYLEIGH time.

This simple model allows us to describe the global features of the first bubble collapse for an almost inviscid liquid such as water. However, it does not provide an account of the successive rebounds and collapses actually observed in various physical situations. It should be noted that, if surface tension were not ignored, the collapse would be only slightly accelerated.

3.2.2. THE INTERFACE VELOCITY

With the previous assumptions, the RAYLEIGH-PLESSET equation (3.12) can be integrated using relation (3.13) to give:

$$\rho\dot{R}^2R^3=-\frac{2}{3}\left(p_\infty-p_v\right)\left(R^3-R_0{}^3\right) \qquad (3.15)$$

As $\dot{R}$ is negative during collapse, one obtains:

$$\frac{dR}{dt}=-\sqrt{\frac{2}{3}\frac{p_\infty-p_v}{\rho}\left[\frac{R_0{}^3}{R^3}-1\right]} \qquad (3.16)$$

The radius tends to 0 and the radial inwards motion accelerates without limit. The numerical integration of this equation allows the calculation of the radius $R(t)$ as a function of time. The characteristic collapse time or RAYLEIGH time is:

$$\tau = \sqrt{\frac{3}{2}\frac{\rho}{p_\infty - p_v}} \int_0^{R_0} \frac{dR}{\sqrt{\frac{R_0^3}{R^3} - 1}} \cong 0.915\, R_0 \sqrt{\frac{\rho}{p_\infty - p_v}} \tag{3.17}$$

The constant 0.915 is the approximate value of $\sqrt{\frac{\pi}{6}}\frac{\Gamma(5/6)}{\Gamma(4/3)}$ where Γ is the factorial gamma function.

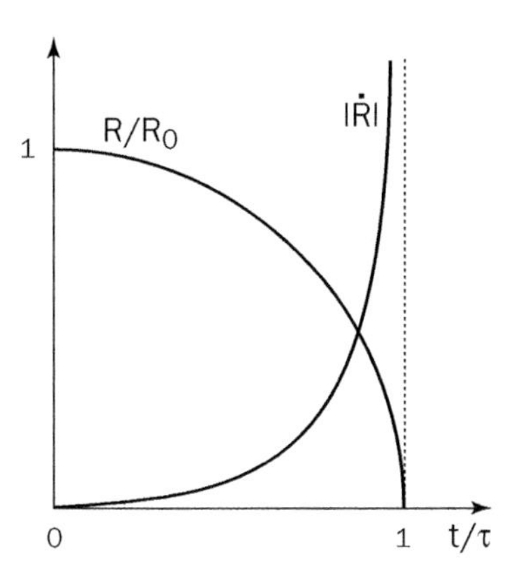

3.2
Evolution of R(t) and Ṙ(t) during bubble collapse

The value of τ is in good agreement with the experimental values for a large range of initial values of the bubble diameter from about one micrometer to one meter. As an example, in the case of water, a bubble with an initial radius of 1 cm collapses in about one millisecond under an external pressure of 1 bar.

The behavior of R(t) and $\dot{R}(t)$ are shown in figure 3.2. While the mean value of the collapse velocity is R_0/τ, $\dot{R}$ tends to infinity at the end of collapse. For R approaching 0, the interface velocity has the following strong singularity:

$$\left|\dot{R}\right| \cong \sqrt{\frac{2}{3}\frac{p_\infty - p_v}{\rho}} \left[\frac{R_0}{R}\right]^{3/2} \cong 0.747\,\frac{R_0}{\tau}\left[\frac{R_0}{R}\right]^{3/2} \tag{3.18}$$

At the end of the collapse, the radius evolves according to the law:

$$\frac{R}{R_0} \cong 1.87 \left[\frac{\tau - t}{\tau}\right]^{2/5} \tag{3.19}$$

With the previous numerical values, it is found that $\left|\dot{R}\right| \approx 720$ m/s for $R/R_0 = 1/20$. Such high values of velocity, of the order of half of the velocity of sound in water, lead us to believe that liquid compressibility must be taken into account in the final stages of collapse.

It must be kept in mind that some other physical aspects, such as the presence of non-condensable gas or the finite rate of vapor condensation, will modify bubble behavior. However, the RAYLEIGH model exhibits the main features of bubble collapse, particularly its short duration and the rapid change in its time scale.

3.2.3. *THE PRESSURE FIELD*

The pressure field $p(r,t)$ can be determined from equation (3.9) in which $\dot{R}$ is known from equation (3.16) and $\ddot{R}$ can be deduced by derivation, which gives:

$$\ddot{R} = -\frac{p_\infty - p_v}{\rho} \frac{R_0^{\;3}}{R^4} \tag{3.20}$$

The result of the calculation is:

$$\Pi(r,t) = \frac{p(r,t) - p_\infty}{p_\infty - p_v} = \frac{R}{3r}\left[\frac{R_0^{\;3}}{R^3} - 4\right] - \frac{R^4}{3r^4}\left[\frac{R_0^{\;3}}{R^3} - 1\right] \tag{3.21}$$

The behavior of non-dimensional pressure Π at several instants is shown in figure 3.3. It exhibits a maximum within the liquid as soon as the bubble radius becomes smaller than $\left(1/\sqrt[3]{4}\right)R_0 \cong 0.63\,R_0$. The maximum pressure is:

$$\Pi_{max} = \frac{p_{max} - p_\infty}{p_\infty - p_v} = \frac{\left[\dfrac{R_0^{\;3}}{4R^3} - 1\right]^{4/3}}{\left[\dfrac{R_0^{\;3}}{R^3} - 1\right]^{1/3}} \tag{3.22}$$

and it occurs at distance r_{max} from the bubble center given by:

$$\frac{r_{max}}{R} = \left[\frac{\dfrac{R_0^{\;3}}{R^3} - 1}{\dfrac{R_0^{\;3}}{4R^3} - 1}\right]^{1/3} \tag{3.23}$$

When R/R_0 becomes small, the two previous relations give approximately:

$$\Pi_{max} \approx \frac{1}{4^{4/3}}\left[\frac{R_0}{R}\right]^3 \cong 0.157\left[\frac{R_0}{R}\right]^3 \tag{3.24}$$

$$\frac{r_{max}}{R} \approx \sqrt[3]{4} \cong 1.59 \tag{3.25}$$

Very high pressures close to the bubble interface are reached. For example, for $R/R_0 = 1/20$, $p_{max} = 1{,}260$ bars if $p_\infty - p_v$ is one bar.

Attention must be paid to the kind of pressure wave that appears in figure 3.3 during the collapse of the bubble. As only pressure and inertia forces are taken into account in the present model, this pressure wave propagating inward must be considered as the effect of inertia forces only. More complicated models exhibit a similar behavior.

From a physical viewpoint, the violent behavior of bubble collapse results from two main facts:

- the pressure inside the bubble is constant and does not offer any resistance to liquid motion;
- the conservation of the liquid volume, through spherical symmetry (eq. 3.6), tends to concentrate liquid motion to a smaller and smaller region.

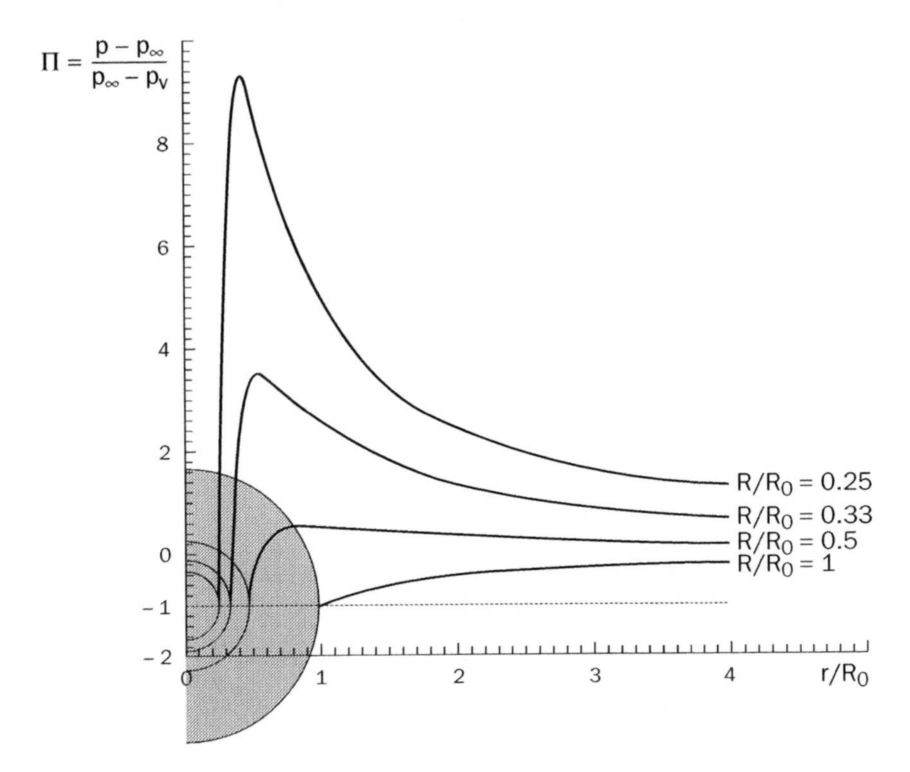

3.3 - Evolution of the pressure field during bubble collapse

3.2.4. *Remark on the effect of surface tension*

If surface tension is taken into account, equation (3.16) becomes:

$$\frac{dR}{dt} = -\sqrt{\frac{2}{3}\frac{p_\infty - p_v}{\rho}\left[\frac{R_0^{\,3}}{R^3} - 1\right] + \frac{2S}{\rho R_0}\frac{R_0^{\,3}}{R^3}\left[1 - \frac{R^2}{R_0^{\,2}}\right]} \tag{3.26}$$

The accelerating effect of surface tension becomes significant if:

$$\frac{2S}{\rho R_0} > \frac{2}{3}\frac{p_\infty - p_v}{\rho} \tag{3.27}$$

i.e.

$$R_0 < \frac{3S}{p_\infty - p_v} \tag{3.28}$$

For water, with $S = 0.072\ N/m$, and assuming a pressure difference of 1 bar, one finds a value of 2.2 µm. Such a small value is rather exceptional in current hydraulics. It results from the fact that pressure influences the collapse through volume variations, contrary to the surface effect of the S-term.

3.3. THE EXPLOSION OF A NUCLEUS

3.3.1. THE INTERFACE VELOCITY

In this section, we consider a nucleus originally in equilibrium under pressure $p_{\infty 0}$, subjected at time $t = 0$ to a lower pressure $p_\infty < p_{\infty 0}$, which may be higher or lower than the vapor pressure p_v (fig. 3.4). The sudden decrease in pressure makes it grow due to the gas it contains. Viscous effects are still ignored but surface tension is taken into account.

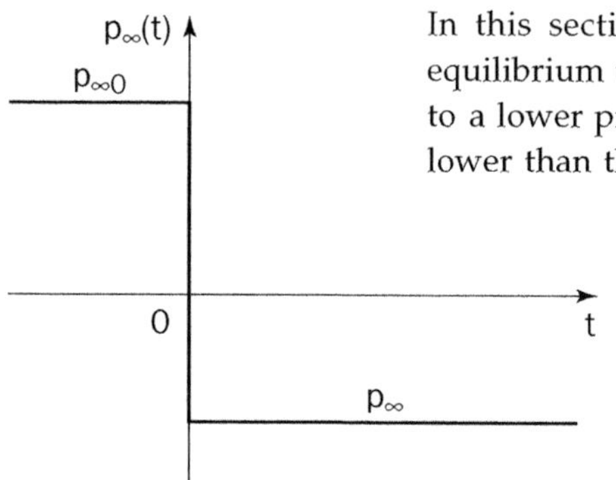

3.4

As the pressure p_∞ is assumed constant, the RAYLEIGH-PLESSET equation (3.12) can be integrated analytically, using relation (3.13) to obtain:

$$\dot{R}^2 = \frac{2}{3}\frac{p_v - p_\infty}{\rho}\left[1 - \frac{R_0^3}{R^3}\right] + \frac{2}{3(\gamma - 1)}\frac{p_{g0}}{\rho}\frac{R_0^3}{R^3}\left[1 - \left(\frac{R_0}{R}\right)^{3(\gamma-1)}\right] - \frac{2S}{\rho R}\left[1 - \frac{R_0^2}{R^2}\right] \tag{3.29}$$

As in chapter 2, p_{g0} is the initial gas pressure. As the nucleus is expected to develop into a large bubble, an adiabatic transformation of the gas is considered, in accordance with the considerations of section 2.3.1. In the case of an isothermal transformation for the gas trapped in the bubble, the second term on the right-hand side of equation (3.29) would become:

$$2\frac{p_{g0}}{\rho}\frac{R_0^3}{R^3}\ln\frac{R}{R_0} \tag{3.30}$$

In fact, the conclusions of the following discussion are not substantially modified for an isothermal gas transformation.

Equation (3.29) is of the classical type $\dot{R}^2 = f(R)$ and is frequently encountered in rational mechanics of solid bodies. The basis of the analysis is the existence of roots of $f(R)$, which determine the limits of the domain of variation of the radius R, since motion is possible only if $f(R)$ is positive.

3.3.2. *THE EQUILIBRIUM CASE* ($p_\infty = p_{\infty 0}$)

The equilibrium for $R = R_0$ requires the function $f(R)$ to have a double root in R_0, so that $\dot{R}_0 = 0$ and $\ddot{R}_0 = 0$. Hence:

- $f(R_0) = 0$: this condition is automatically satisfied by the choice of the initial conditions;
- $\dot{f}(R_0) = 0$: this condition is equivalent to the usual equilibrium condition (2.2):

$$p_\infty = p_{g0} + p_v - \frac{2S}{R_0} \tag{3.31}$$

The stability of the equilibrium state in the vicinity of R_0 requires the additional condition $\ddot{f}(R_0) < 0$. This condition is equivalent to the stability criterion mentioned in section 2.2.2 according to which the equilibrium is stable only on the descending branch of the curve $p_\infty(R)$.

3.3.3. *THE CASE OF NUCLEUS GROWTH* ($p_\infty < p_{\infty 0}$)

Two main cases have to be considered (see fig. 3.5).

- If $f(R)$ has a root R_1 larger than R_0 (which is possible either for $p_\infty < p_v$ or $p_\infty > p_v$), the sign of $\dot{R}$ changes both at the limits $R = R_0$ and $R = R_1$ and the radius oscillates periodically between the two values. As no dissipation is considered here, no damping occurs during oscillation.

  The motion is highly dependent upon the value of the ratio R_1 / R_0. If the ratio is large (case b, fig. 3.5), the role of the gas pressure is small when R is large and close to R_1, but it becomes dominant for R close to R_0. The motion is strongly non-linear. Bubble regrowth is made possible by the elastic behavior of the enclosed gas. The oscillations look like a succession of explosions and collapses.

  On the contrary, if the ratio R_1 / R_0 is close to unity (case a, fig. 3.5), the motion becomes harmonic. The period is given by the classical expression:

$$T_0 = \frac{2\pi\sqrt{2}}{\sqrt{-\ddot{f}(R_0)}} \tag{3.32}$$

so that the bubble behaves like a linear oscillator of frequency:

$$f_0 = \frac{1}{2\pi R_0}\sqrt{\frac{1}{\rho}\left[3\gamma p_{g0} - \frac{2S}{R_0}\right]} \tag{3.33}$$

- If the function $f(R)$ has no root greater than R_0, the sign of the interface velocity $\dot{R}$ does not change and the bubble grows indefinitely (case d, fig. 3.5). Equation (3.29) shows that the interface velocity $\dot{R}$ reaches a limit $\dot{R}_\infty$ given by:

$$\dot{R}_\infty = \sqrt{\frac{2}{3}\frac{p_v - p_\infty}{\rho}} \tag{3.34}$$

This value agrees fairly well with experiment.

3.3.4. DYNAMIC CRITERION

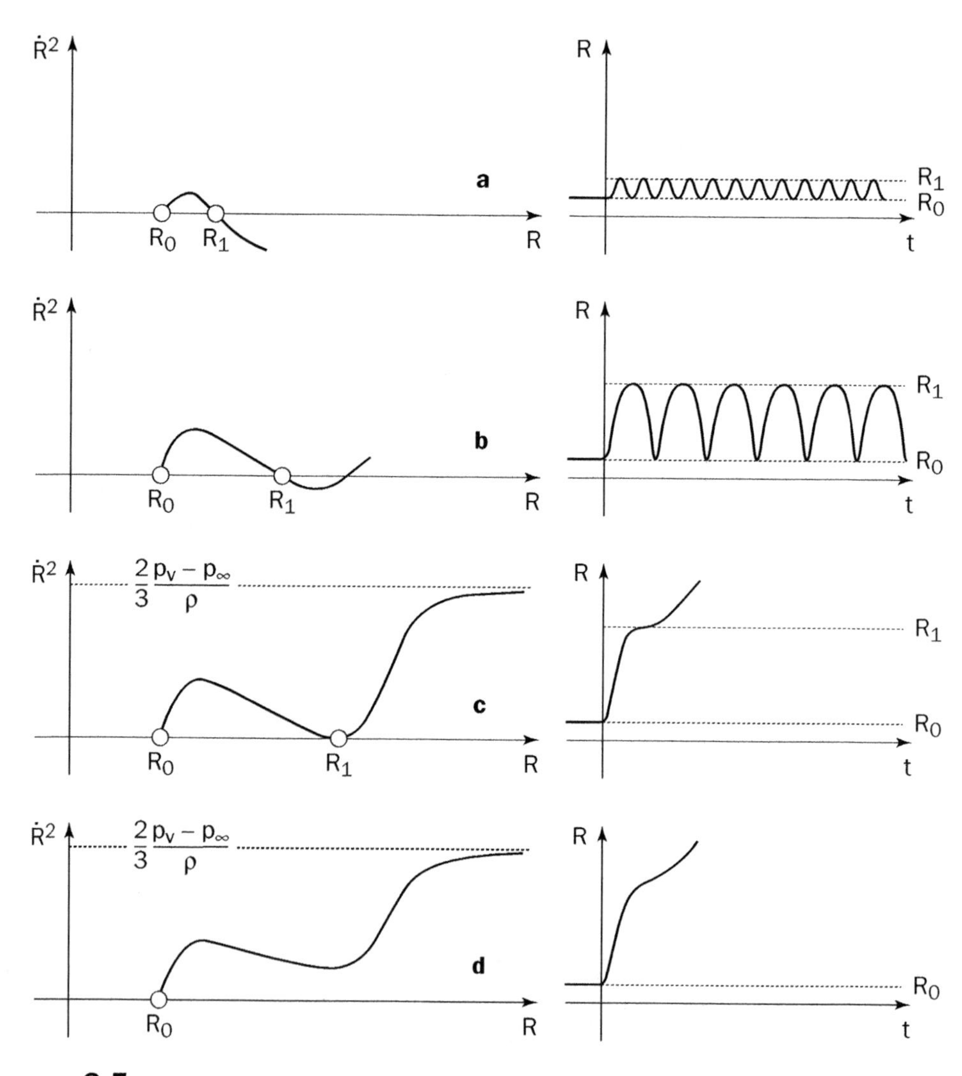

3.5 - **Typical evolutions of a bubble: (a) linear oscillations - (b) non-linear oscillations - (c) illustration of the dynamic criterion - (d) unlimited growth**

The intermediate case obtained when the function f has a double root in R_1 can be considered as a kind of dynamic criterion for the stability of the nucleus (case c, fig. 3.5). It corresponds to the limiting case between oscillations and unlimited growth for a nucleus exposed to a step change in pressure.

The dynamic critical pressure is generally not too far from the normal static value. For example, for a nucleus of radius 5 μm, under a pressure of one bar, the dynamic criterion gives about −4,500 Pa while the static criterion gives −3,400 Pa.

However, it should be kept in mind that here the change in external pressure is made instantaneously, which is not completely realistic in practice. Thus, this dynamic criterion, which obviously depends on the form of the function $p_\infty(t)$ has a limited practical significance.

3.3.5. *REMARK ON TWO PARTICULAR CASES*

It is assumed here that the nucleus contains no gas ($p_{g0} = 0$) and surface tension and viscosity are not taken into account ($S = 0, \mu = 0$).

- If the bubble is exposed to a constant pressure equal to the vapor pressure, i.e., $p_\infty(t) = p_v$, the right-hand side of equation (3.12) is zero and the bubble radius evolves according to the following law:

$$\frac{R}{R_0} = \left[1 + \frac{5}{2}\frac{\dot{R}_0 t}{R_0}\right]^{2/5} \tag{3.35}$$

  A positive value of $\dot{R}_0$ results in infinite growth with a decreasing velocity. A negative value of $\dot{R}_0$ leads to bubble collapse in a time $-2R_0/5\dot{R}_0$. The velocity increases with the singularity $R^{-3/2}$, referred to in section 3.2.2.

  This particular case demonstrates the instability of a nucleus even when pressures inside the bubble and at infinity are equal.

- Let us assume now that, from the initial time $t = 0$, the pressure decreases linearly with time according to the law:

$$P_\infty(t) - p_v = -m\frac{t}{\tau} \tag{3.36}$$

  where τ stands for a characteristic time and m is a positive constant. R is found to have an asymptotic behavior $R/R_0 = kt^n$ given by:

$$R = \sqrt{\frac{8}{33}\frac{m\tau^2}{\rho}}\left(\frac{t}{\tau}\right)^{3/2} \tag{3.37}$$

3.4. THE EFFECT OF VISCOSITY

3.4.1. LINEAR OSCILLATIONS OF A BUBBLE

In the presence of a slightly oscillating pressure field applied far from the bubble:

$$p_\infty(t) = p_{\infty 0}(1 + \varepsilon \sin \omega t) \tag{3.38}$$

then, with $\varepsilon << 1$, the radius will oscillate as follows:

$$R = R_0[1 + \kappa(t)] \tag{3.39}$$

where κ will remain small. The linearization of the RAYLEIGH-PLESSET equation (3.12) gives:

$$\ddot{\kappa} + \frac{4\nu}{R_0^2}\dot{\kappa} + \frac{3\gamma p_{g0} - 2S/R_0}{\rho R_0^2}\kappa = -\frac{1}{\rho R_0^2} p_{\infty 0}\, \varepsilon \sin \omega t \tag{3.40}$$

From this equation, the natural frequency f_0 already given in equation (3.33) is found. The values of the natural frequency are spread over a wide range according to the size of the bubble and the pressure at infinity. As an example, a bubble of 10 μm radius in equilibrium in water ($S = 0.072\ N/m$) under atmospheric pressure ($p_{\infty 0} = 10^5\ Pa$) has a partial pressure of gas inside equal to $p_{g0} = p_{\infty 0} - p_v + 2S/R \cong 112{,}000\ Pa$ and a natural frequency $f_0 \cong 340\ kHz$. For similar conditions, a micro-bubble of 1 μm radius would have a natural frequency of 4.7 MHz.

In addition, it appears that viscosity has a damping effect with the damping rate being given by $4\nu/R_0^2$.

3.4.2. EFFECT OF VISCOSITY ON EXPLOSION OR COLLAPSE OF BUBBLES

The problem was considered as early as 1952 by PORITSKY for a bubble containing no gas and later by ZABABAKHIN (1960) and SHIMA and FUJIWARA (1980). PORITSKY introduced the non-dimensional numbers:

$$\mu' = \frac{4\mu}{R_0\sqrt{\varepsilon \rho (p_\infty - p_v)}} \tag{3.41}$$

$$\sigma' = \frac{S}{\varepsilon R_0 (p_\infty - p_v)} \tag{3.42}$$

with $\varepsilon = 1$ for a collapse and $\varepsilon = -1$ for an explosion.

Both explosion and collapse are slowed down by viscosity. PORITSKY found that if μ' is larger than about 0.46, the collapse takes an infinite time. The existence of a critical value of μ' was demonstrated by SHU (1952). With $R_0 = 1\ mm$ and a pressure difference equal to one bar, the value 0.46 corresponds to a viscosity about 1,200 times higher than the viscosity of water. Thus, in the case of water, the slowdown of collapse or explosion by viscosity is very weak.

For other applications, for example small bubbles created by electric discharge in an insulating, strongly viscous, organic liquid, collapse in an infinite time was observed [JOMNI 1997, JOMNI *et al.* 1999]. In that case, the radius was of the order of 4 µm; liquid density $\rho = 835\ kg/m^3$; kinematic viscosity $\nu = 100.10^{-6}\ m^2/s$; and the pressure difference was close to 1.5 bar. The value of the non-dimensional parameter μ' was close to 1.1.

3.5. *NON-LINEAR OSCILLATIONS OF A BUBBLE*

In equation (3.40) the nucleus is considered as a simple linear oscillator submitted to a small oscillating external force. In fact, in many circumstances, for example in the case of vibratory erosion test devices, ε cannot be considered small with respect to 1 and the response of the nucleus is far from linear.

In such cases, the bubble can exhibit a response curve with subharmonics. Moreover, the response can be non-periodic, as shown by figure 3.6, which presents two examples of numerical results by BOROTNIKOVA and SOLOUKIN (1964) for the transient motion of a gas bubble in an oscillating pressure field. In each case, the periodicity is far from being achieved. In the second case, corresponding to a strong excitation, subharmonics appear.

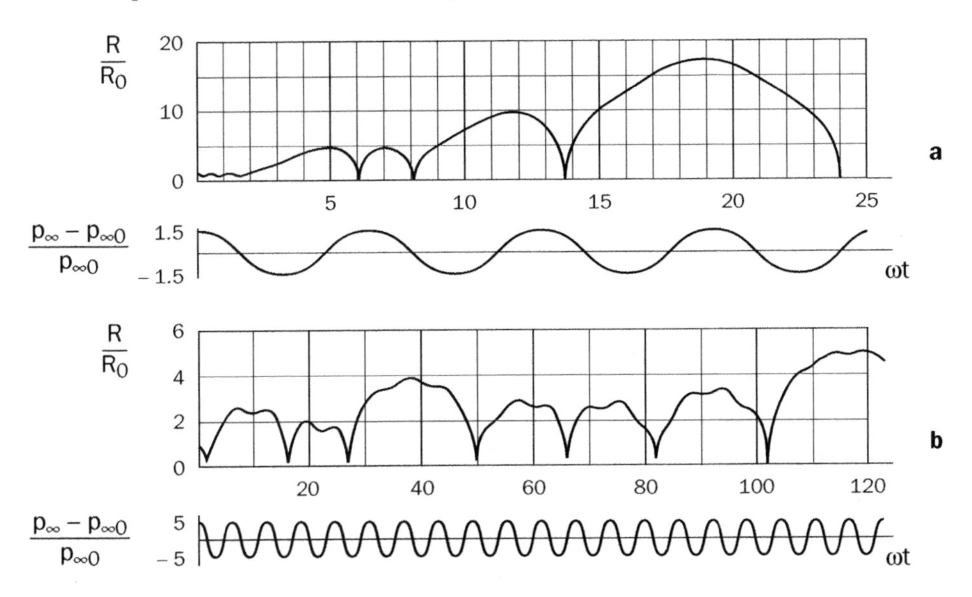

3.6 - Two examples of numerical results for the transient motion of a gas bubble in an oscillating pressure field $p_\infty(t) = p_{\infty 0}(1 + \varepsilon \cos \omega t)$: (a) $\varepsilon = 1.5$, $\omega / \omega_0 = 0.0154$ - (b) $\varepsilon = 5$, $\omega / \omega_0 = 1.54$

ω_0 is defined as $\omega_0 = 2\pi f_0$ where f_0 is the natural frequency of the bubble [from PLESSET & PROSPERETTI, 1977] – Reprinted, with permission, from the Annual Review of Fluid Mechanics, Volume 9, ©1977 by Annual Review (www.annualreviews.org).

Non-linearities of the RAYLEIGH-PLESSET equation can play an important role in the behavior of bubbles under periodic excitation. The reader interested in the non-linearities of the RAYLEIGH-PLESSET equation as seen from the viewpoint of bifurcations and chaos physics will find useful information in KELLER and MIKSIS (1980), SMEREKA *et al.* (1987), PARLITZ *et al.* (1990).

3.6. *SCALING CONSIDERATIONS*

3.6.1. *NON-DIMENSIONAL FORM OF THE RAYLEIGH-PLESSET EQUATION*

Let a, τ and P be characteristic scales of radius R, time t and pressure p respectively, so that the following non-dimensional variables are defined:

$$\begin{cases} \overline{R} = \dfrac{R}{a} \\ \overline{t} = \dfrac{t}{\tau} \\ \overline{p} = \dfrac{p}{P} \end{cases} \tag{3.43}$$

The RAYLEIGH-PLESSET equation (3.12) takes the non-dimensional form:

$$\overline{R}\,\ddot{\overline{R}} + \frac{3}{2}\dot{\overline{R}}^2 = -\mathrm{Th}\,\frac{p_\infty(t) - p_v}{P} + \frac{\overline{P}_0}{\overline{R}^{3\gamma}} - \frac{\mathrm{We}}{\overline{R}} - \frac{1}{\mathrm{Re}}\frac{\dot{\overline{R}}}{\overline{R}} \tag{3.44}$$

with the following set of non-dimensional numbers:

- REYNOLDS number $\mathrm{Re} = \dfrac{a^2}{4\nu\tau}$,
- WEBER number $\mathrm{We} = \dfrac{2S\tau^2}{\rho a^3}$,
- the pressure numbers $\mathrm{Th} = \dfrac{P\tau^2}{\rho a^2}$ and $\dfrac{p_{g0}\,\overline{R}_0^{\,3\gamma}\tau^2}{\rho a^2}$, the first (Th) being similar to the THOMA cavitation number.

The scaling is well suited to the problem if the values of $\overline{R}$ and $\overline{t}$ are of the order of unity. For non-irregular behavior of the physical phenomenon, the quantities $\dot{\overline{R}}$ and $\ddot{\overline{R}}$ are consequently also of the order of unity, as usually assumed in classical dimensional analysis. Then, the above non-dimensional numbers allow us to compare the importance of the different terms in equation (3.44): pressure, non-condensable gas content, surface tension, and viscosity.

The choice of the reference length and time scales deserves caution as the radius and the velocities can change by several orders of magnitude during the bubble evolution. Thus, in general, the scaling is well adapted to a limited phase of the phenomenon only.

Despite this difficulty, in general the following scales are to be taken:

- for the length scale a, the initial radius R_0 (then $\overline{R}_0 = 1$) or the characteristic size of a body close to the bubble, such as its chord length c (see § 3.6.4);
- for P, a characteristic pressure difference of the system under consideration, for example $p_{\infty 0} - p_v$;
- for τ, one of the characteristic time scales described in the following section.

3.6.2. *CHARACTERISTIC TIME SCALES OF THE RAYLEIGH-PLESSET EQUATION*

The following characteristic time scales can be considered:

- the pressure time τ_P, defined as:

$$\tau_P = a\sqrt{\frac{\rho}{P}} \tag{3.45}$$

This is analogous to the RAYLEIGH time for the bubble collapse;

- the viscous time τ_v:

$$\tau_v = \frac{a^2}{4\nu} \tag{3.46}$$

which describes the slowing down of the bubble wall motion due to the effect of viscosity;

- the surface tension time τ_S:

$$\tau_S = a\sqrt{\frac{\rho a}{2S}} \tag{3.47}$$

which is a measure of the collapse time under the effect of surface tension only.

The previous non-dimensional parameters can be interpreted as ratios of the characteristic time scales as follows:

$$\begin{cases} Th = \dfrac{\tau^2}{\tau_P^{\ 2}} \\ We = \dfrac{\tau^2}{\tau_S^{\ 2}} \\ Re = \dfrac{\tau_v}{\tau} \end{cases} \tag{3.48}$$

These time scales are mainly useful in the numerical solution of the RAYLEIGH-PLESSET equation. In practice, the computational time step needs to be smaller than all the above characteristic time scales, in which a is equal to the actual value of the radius.

Besides, one must pay attention to the natural period of oscillation $T_0 = 1/f_0$ of the bubble. If the time required for the pressure variation is of the same order as the natural period T_0, bubble oscillations are expected. Hence, the time interval must also be smaller than T_0 in order to describe those oscillations precisely.

3.6.3. QUALITATIVE DISCUSSION OF THE RAYLEIGH-PLESSET EQUATION

The characteristic times of the RAYLEIGH-PLESSET equation are also useful for a qualitative discussion of bubble behavior. For example, in the case of a collapsing bubble containing no gas (i.e. $p_{g0} = 0$) the relevant scales are chosen as:

- length scale $a = R_0$,
- time scale, the collapse time τ,
- pressure scale $P = p_{\infty 0} - p_v$.

Several cases can occur according to the order of magnitude of the different terms on the right-hand side of equation (3.44).

If the pressure term is dominant, then Th is much larger than both We and Re^{-1}. Moreover, Th is of the order of unity, as the left-hand term of equation (3.44) is itself of the order of unity for such a well suited choice of reference scales. Because of relations (3.48), it is equivalent to say that the collapse time is scaled by τ_p and that τ_p is much smaller than τ_v or τ_S.

The same can be said if either Re^{-1} or We are predominant in comparison with the two other non-dimensional numbers. If so, the correct time scales are respectively τ_v or τ_S and in each case these times are smaller than the other two.

In other circumstances, two mechanisms may have equivalent influences, which require the corresponding characteristic times to have the same order of magnitude. For example:

- $\tau_p = \tau_v$ if the initial radius is $R_0 = R_1 = 4\nu \sqrt{\dfrac{\rho}{p_{\infty 0} - p_v}}$,
- $\tau_v = \tau_S$ if the initial radius is $R_0 = R_2 = \dfrac{8\rho\nu^2}{S}$,
- $\tau_S = \tau_p$ if $R_0 = R_3 = \dfrac{2S}{p_{\infty 0} - p_v}$.

The radius R_2 depends only on the liquid characteristics. For water at room temperature, one finds $R_2 = 0.11\ \mu m$, while for the organic liquid considered in section 3.4.2, with $S = 27.10^{-3} N/m$, the R_2-value is 2.47 mm.

Finally, there is the particular of the three mechanisms being equally important in equation (3.44), which requires that the three characteristic times be equal. In this case, the following additional condition must be fulfilled:

$$P = p_{\infty 0} - p_v = \frac{S^2}{4\rho\nu^2} \tag{3.49}$$

For water, this gives $P = 13$ bars, and for the organic liquid considered earlier $P = 22$ Pa.

3.6.4. CASE OF A TRANSIENT BUBBLE NEAR A FOIL

In many cases, information on the behavior of nuclei that explode and then collapse on a foil or a blade can easily be obtained as follows.

It is assumed that the bubble keeps its spherical shape and follows a streamline at the local speed $U(s)$ where s denotes the curvilinear distance along the streamline. It is usual to consider the streamline that coincides with the body surface. Those points will be further discussed in chapter 8.

The local pressure $p(s)$ on the body is taken to be the pressure $p_\infty(t)$ used in the RAYLEIGH-PLESSET equation. The transformation:

$$t = \int_{s_0}^{s} \frac{ds}{U(s)} \tag{3.50}$$

is used for a change of variables. The relevant scales are the following:

$$\begin{cases} a = R_0 \\ P = \dfrac{1}{2}\rho U_\infty^{\ 2} \\ \tau = \dfrac{R_0}{U_\infty} \end{cases} \tag{3.51}$$

and the corresponding non-dimensional numbers are:

$$\begin{cases} Th = \dfrac{1}{2} \\ We = \dfrac{2S}{\rho U_\infty^{\ 2} R_0} \\ Re = \dfrac{U_\infty R_0}{4\nu} \end{cases} \tag{3.52}$$

The RAYLEIGH-PLESSET equation takes the following non-dimensional form:

$$\overline{R}\,\ddot{\overline{R}} + \frac{3}{2}\dot{\overline{R}}^2 = -\frac{1}{2}\left[-Cp(t) - \sigma_v\right] + \frac{\overline{P_0}}{\overline{R}^{3\gamma}} - \frac{\overline{We}}{\overline{R}} - \frac{1}{Re}\frac{\dot{\overline{R}}}{\overline{R}} \tag{3.53}$$

where σ_v is the cavitation number defined by:

$$\sigma_v = \frac{p_\infty - p_v}{\frac{1}{2}\rho U_\infty^{\ 2}} \tag{3.54}$$

and Cp the local pressure coefficient:

$$Cp(s) = \frac{p(s) - p_\infty}{\frac{1}{2}\rho U_\infty^{\ 2}} \tag{3.55}$$

U_∞ is the reference flow velocity at infinity. Within the framework of steady potential flow theory, the local flow velocity U(s) and the local pressure coefficient Cp(s) are connected by the BERNOULLI equation, which gives:

$$U(s) = U_\infty \sqrt{1 - Cp(s)} \tag{3.56}$$

Concerning the initial conditions, it is convenient to inject the nucleus at the abscissa s_0 where the pressure coefficient is zero i.e. where the pressure is equal to the upstream reference pressure. In addition, $\dot{R}_0$ is usually chosen to be zero at this location.

The low-pressure surface of a hydrofoil exhibits a minimum in the pressure distribution. The dynamic behavior of a nucleus that slides down the low-pressure surface is described by equation (3.53). It depends primarily upon its critical pressure p_c compared with the minimum pressure p_{min}.

If the minimum pressure is lower than the critical pressure, the nucleus is destabilized and begins to grow. Whereas the biggest nuclei grow slowly, the smallest ones explode violently as soon as their critical pressure is reached. By then, they are larger than their original size by several orders of magnitude. On the whole, when nuclei of different critical pressures are destabilized, their size evolution R(t) is practically independent of their initial size R_0.

A particular case of practical interest, as it corresponds approximately to the case of blades working near their design point, is obtained when the pressure on the low-pressure side of a foil is practically uniform. If surface tension, viscosity and gas effects are neglected, the radial velocity $\dot{R}$ tends to the limit value given by expression (3.34), or, in non-dimensional form, by:

$$\frac{dR/dt}{U_\infty} \cong \sqrt{\frac{1}{3}(-Cp - \sigma_v)} \tag{3.57}$$

As the local liquid velocity is connected to the pressure coefficient by the BERNOULLI equation (3.56), then, in this particular case of a constant pressure coefficient:

$$\frac{dR}{ds} \cong \sqrt{\frac{-Cp - \sigma_v}{3(1 - Cp)}} \tag{3.58}$$

so that the bubble radius increases proportionally to the curvilinear distance s:

$$R(s) \cong \sqrt{\frac{-Cp_{min} - \sigma_v}{3(1 - Cp_{min})}}\, s \tag{3.59}$$

This result agrees fairly well with the experimental behavior of bubbles exploding on a foil.

Finally, it should be noted that, from the discussion given in chapter 2 (§ 2.3.1), the transit time Δt should be compared with the thermal time Δt_r required for heat transfer. Generally, Δt is far larger than Δt_r so that the gas transformation can be considered isothermal and γ must be made equal to unity in the RAYLEIGH-PLESSET equation. Therefore, the critical pressure p_c for which the nucleus is destabilized is that given by expression (2.4). However, as soon as the nucleus is destabilized, the transformation should actually be considered as adiabatic since the volume of the nucleus increases substantially and heat exchange is no longer significant.

3.7. *STABILITY OF THE SPHERICAL INTERFACE*

The large values of the velocity $\dot{R}$ and of the acceleration $\ddot{R}$ of the interface found in the previous sections may affect the stability of the thus far assumed spherical shape of the interface.

It is well-known that a horizontal plane interface with water below and air above, with both at rest, is stable under normal conditions in a gravitational field. More generally, consider a system of two fluids of different densities, separated by a plane interface, and replace gravity by a given acceleration of the two fluids. The RAYLEIGH-TAYLOR linear stability criterion requires that the acceleration be directed from the heaviest to the lightest fluid. If this criterion, valid for a plane interface, is abruptly applied to a growing or collapsing bubble, it should be expected that the interface remain stable if the liquid acceleration is directed inwards, i.e.:

$$R\ddot{R} + \frac{3}{2}\dot{R}^2 < 0 \tag{3.60}$$

Therefore, it is expected that a negative value of $\ddot{R}$ as well as a small value of $\dot{R}$ will promote stability.

The question is to know how this stability criterion is modified in the case of a spherical interface. The problem was considered by PLESSET (1954), BIRKHOFF (1954), PLESSET and MITCHELL (1955) and later by HSIEH (1965). Here, we follow the linearized PLESSET method. Details of the calculation can be found in PLESSET's original paper. Only the main conclusions are presented here.

The liquid is assumed inviscid and the flow irrotational. The bubble interface is supposed distorted from its spherical shape $r = R(t)$ and the stability of this distortion is analyzed. As any arbitrary function of the two spherical coordinates, angles θ and ϕ, can be developed in a series of spherical harmonics, PLESSET assumed the following shape for the bubble interface:

$$\tilde{R}(r,\theta,\phi,t) = R(t) + \sum_n a_n(t)\, Y_n(\theta,\phi) \tag{3.61}$$

where the amplitudes a_n of each spherical harmonic Y_n are small in comparison with $R(t)$, consistent with a linear analysis. The kinematical condition on the interface together with the generalized BERNOULLI equation require that the amplitude coefficients $a_n(t)$ satisfy the differential equation:

$$\ddot{a}_n + \frac{3\dot{R}}{R}\dot{a}_n - \left[(n-1)\frac{\ddot{R}}{R} - \frac{(n-1)(n+1)(n+2)S}{\rho R^3}\right]a_n = 0 \qquad (3.62)$$

For simplicity, PLESSET introduced the following change of variables:

$$a_n(t) = \alpha_n(t)\left(\frac{R_0}{R}\right)^{3/2} \qquad (3.63)$$

With the new variable α_n, the differential equation (3.62) is simplified and becomes:

$$\ddot{\alpha}_n - G_n(t)\,\alpha_n = 0 \qquad (3.64)$$

with

$$G_n(t) = \frac{1}{2R^2}\left[(2n+1)R\ddot{R} + \frac{3}{2}\dot{R}^2 - (n-1)(n+1)(n+2)\frac{2S}{\rho R}\right] \qquad (3.65)$$

If we neglect the influence of the coefficient $(R_0/R)^{3/2}$ in equation (3.63), then the stability of the shape of the interface depends simply on the growth rate of α_n. Stability requires that $G_n(t)$ be negative, otherwise an exponential growth of the coefficient α_n would be expected, leading to instability. In conclusion, the spherical interface is stable if the following condition is met for all the spherical harmonics:

$$(2n+1)R\ddot{R} + \frac{3}{2}\dot{R}^2 - (n-1)(n+1)(n+2)\frac{2S}{\rho R} < 0 \qquad (3.66)$$

This condition is not very different from the approximate condition (3.60) although some new terms are present. It shows that negative values of $\ddot{R}$ associated with relatively small values of the interface velocity $\dot{R}$ are favorable to stability. On the contrary, positive values of $\ddot{R}$ and large values of the interface velocity will promote instability. However, attention must be paid to the last term, which becomes large for high order harmonics and small bubble radius. Thus, a general conclusion is not straightforward.

As an example, consider the final stage of the RAYLEIGH collapse under a constant overpressure. Equation (3.18) is written:

$$\dot{R} \cong -1.12\,\frac{R_0}{\tau}\left(\frac{R_0}{R}\right)^{3/2}$$

from which one obtains:

$$2R^2 G_n(t) \cong -3n\left[1.12\,\frac{R_0}{\tau}\right]^2 \frac{R_0^{\,3}}{R^3} - (n-1)(n+1)(n+2)\frac{2S}{\rho R}$$

This expression, which is negative, shows that the spherical shape tends to be stable in the final stages of the so-called RAYLEIGH collapse.

However, in the real case of a bubble containing a non-condensable gas, the phenomenon of regrowth will change the conclusion. Close to the end of collapse, just before any rebound, $\dot{R}$ will reach large negative values with $\ddot{R} > 0$. From the stability argument, it appears that both conditions are favorable to the development of instabilities.

On the contrary, at the end of the growth phase, before any collapse, $\ddot{R} < 0$ while $\dot{R}$ is positive and small, so that the spherical shape is likely to be stable.

In a series of experiments, JOMNI *et al.* (1999) observed successive collapses and rebounds of very small bubbles of radii of the order of a micrometer. This suggests that the bubble interface was stable throughout its lifetime. In other cases encountered in hydraulics, the explosion of nuclei moving along blades usually appears to be stable, as their interface remains smooth and regular. However, their subsequent collapse looks rather unstable, leading to the formation of several smaller bubbles and resulting in a multi-peak acoustic signal. Of course, these last observations are not easily comparable to the theoretical case analyzed here since the actual flow near a blade is turbulent and the pressure field around the bubble is obviously not uniform.

REFERENCES

BESANT W. –1859– Hydrostatics and Hydrodynamics. *Cambridge University Press.*

BIRKHOFF G. –1955– Stability of spherical bubbles. *Quart. Applied Math.* **13**, 451 *sq.*

BOROTNIKOVA M.I. & SOLOUKIN R.I. –1964– *Sov. Phys. Acoust.* **10**, 28-32.

COLE R.H. –1948– Underwater explosions. *Princeton University Press.*

CRIGHTON D.G. & FFOWCS WILLIAMS J.E. –1969– Sound generation by turbulent two-phase flow. *J. Fluid Mech.* **36**, part 3, 585-603.

FRANC J.P., AVELLAN F., BELAHADJI B., BILLARD J.Y., BRIANÇON-MARJOLLET L., FRÉCHOU D., FRUMAN D.H., KARIMI A., KUENY J.L. & MICHEL J.M. –1995– La cavitation. Mécanismes physiques et aspects industriels. *Presses Universitaires de Grenoble – Collection Grenoble Sciences*, 580 p.

HSIEH D.Y. –1965– Some analytical aspects of bubble dynamics. *J. Basic Eng.* **87**, 991-1005.

JOMNI F. –1997– Étude des phénomènes hydrodynamiques engendrés dans les liquides diélectriques par un champ électrique très intense. *Thesis,* Grenoble University (France).

JOMNI F., AITKEN F. & DENAT A. –1999– Dynamics of microscopic bubbles generated by a corona discharge in insulating liquids: influence of pressure. *J. Electrostatics* **47**, 49-59.

KELLER J.B. & MIKSIS M. –1980– Bubble oscillations of large amplitude. *J. Acoust. Soc. Am.* **68**, 628-623.

PARLITZ U., ENGLISH V., SCHEFFCZYK C. & LAUTERBORN W. –1990– Bifurcation structure of bubble oscillators. *J. Acoust. Soc. Am.* **88**, 2 *sq*.

PLESSET M.S. –1954– On the stability of fluid flows with spherical symmetry. *J. Appl. Phys.* **25**, 96-98.

PLESSET M.S. & MITCHELL T.P. –1955– On the stability of the spherical shape of a vapor cavity in a liquid. *Quart. Appl. Math.* **13**, 419-430.

PLESSET M.S. & PROSPERETTI A. –1977– Bubble dynamics and cavitation. *Ann. Rev. Fluid Mech.* **9**, 145-164.

PORITSKY H. –1952– The collapse or growth of a spherical bubble or cavity in a viscous fluid. *Proc. 1st US Nat. Congress of Appl. Mech*, 813 *sq*.

RAYLEIGH (Lord) –1917– The pressure developed in a liquid during the collapse of a spherical cavity. *Phil. Mag.* **34**, 94 *sq*.

SHIMA A. & FUJIWARA T. –1980– The collapse of bubbles in compressible hydraulic oils. *J. Acoust. Soc. Am.* **68**(5) 1509-1515.

SHU S.S. –1952– Note on the collapse of a spherical cavity in a viscous incompressible fluid. *Proc. of the 1st US Nat. Congress of Appl. Mech.*

SMEREKA P., BIRENIR B. & BANERJEE S. –1987– Regular and chaotic bubble oscillations in periodically driven pressure fields. *Phys. Fluids* **30**, 11 *sq*.

ZABABAKHIN E.I. –1960– The collapse of bubbles in a viscous liquid. *PMM* **24**(6) 1129-1131.

4. BUBBLES IN A NON-SYMMETRICAL ENVIRONMENT

4.1. INTRODUCTION

In industrial situations, the rather ideal conditions considered in chapter 3 are often far from being met. In particular, spherical symmetry can be disturbed by gravity, pressure gradients, flow turbulence, etc.

In this chapter, we emphasize inertial effects and consider two main cases in which departure from sphericity is expected:

- the case of a bubble, which moves in a static liquid and then explodes or collapses (§ 4.2),
- the case of the explosion or collapse in the vicinity of a solid wall or of a free surface (§ 4.3).

In both cases, the lack of spherical symmetry results from a non-uniform pressure distribution in the liquid domain around the interface, while the gas pressure and the vapor pressure inside the bubble are practically uniform. In general, the higher pressure region outside causes a re-entrant jet to appear inside the bubble.

The last section of this chapter is dedicated to the case of a spherical bubble that crosses a region of pressure gradient. The problem is to estimate the path of the bubble, taking into account the effects of gravity and viscous drag, and then to explain the screen effect by which bubbles can be deflected from the streamlines of the base flow as they approach a solid body.

4.2. MOTION OF A SPHERICAL BUBBLE IN A LIQUID AT REST

4.2.1. TRANSLATION OF A SOLID SPHERE IN A LIQUID AT REST

We consider first the case of a sphere whose motion is controlled by an external force $\vec{F}$ in a non-viscous liquid at rest (fig. 4.1). Gravity is ignored and the flow due to the moving sphere is assumed irrotational. If the sphere moves with velocity $\vec{W}(t)$, liquid resistance is given by (see e.g. BATCHELOR 1967):

$$\vec{R}_p = -M\dot{\vec{W}} \tag{4.1}$$

where M is the virtual mass (or added mass) of the immersed body, equal to half of the mass of the displaced fluid:

$$M = \frac{2}{3}\pi\rho R^3 \tag{4.2}$$

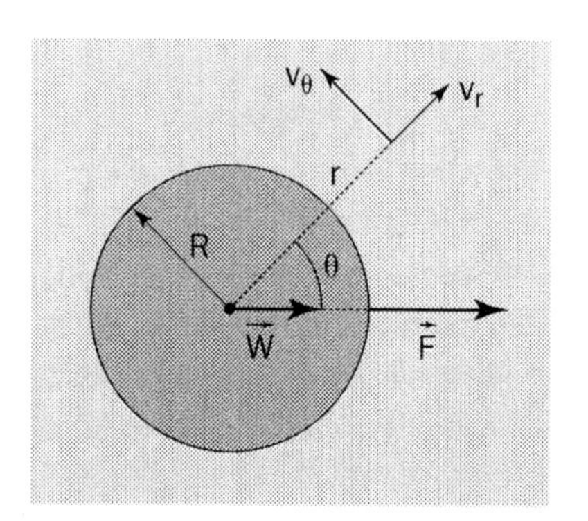

4.1
Translation of a solid sphere in a liquid at rest

The equation of motion of the sphere of mass m is:

$$(m + M)\dot{\vec{W}} = \vec{F} \tag{4.3}$$

The effect of the fluid on the movement of the sphere under the action of the applied force $\vec{F}$ is the same as if the mass of the sphere were increased by one-half the mass of the displaced fluid. Note that equation (4.3) is an illustration of d'ALEMBERT's paradox, according to which the drag of a moving body in a non-viscous fluid is zero if the translation velocity is constant.

4.2.2. *TRANSLATION WITH SIMULTANEOUS VOLUME VARIATIONS*

Suppose now that the sphere is deformable and its radius R is variable due to external effects. Deviations from symmetry in the pressure distribution will result from translation of the sphere and variation of its volume. The translation is considered linear and the translation velocity is W(t).

The velocity potential results from the superposition of the effects of volume variation and translation and takes the classic form:

$$\varphi = -\dot{R}\frac{R^2}{r} - W\frac{R^3\cos\theta}{2r^2} \tag{4.4}$$

The spherical components of the absolute velocity $\vec{V} = \vec{\text{grad}}\,\varphi$ are:

$$\begin{cases} v_r = \dot{R}\dfrac{R^2}{r^2} + W\dfrac{R^3}{r^3}\cos\theta \\ v_\theta = W\dfrac{R^3}{2r^3}\sin\theta \end{cases} \tag{4.5}$$

It is simple to check that the boundary condition on the sphere is fulfilled as the normal velocity is equal to $\dot{R} + W\cos\theta$.

The pressure is known from the generalized BERNOULLI equation:

$$\left[\frac{\partial\varphi}{\partial t} - \vec{W}.\vec{\text{grad}}\,\varphi\right] + \frac{V^2}{2} + \frac{p}{\rho} = \frac{p_\infty}{\rho}$$

in which the term $\partial\varphi/\partial t$ is the time derivative in the frame of reference moving with the bubble and $\partial\varphi/\partial t - W.\vec{\text{grad}}\,\varphi$ is the time derivative in the fixed frame of reference.

After some calculations, one obtains:

$$\begin{aligned}\frac{p(r,\theta,t)-p_\infty}{\rho} = \ddot{R}\frac{R^2}{r} + \dot{R}^2\left[\frac{2R}{r} - \frac{R^4}{2r^4}\right] + \left[\dot{W}\frac{R^3}{2r^2} + W\dot{R}\left(\frac{5R^2}{2r^2} - \frac{R^5}{r^5}\right)\right]\cos\theta \\ + W^2\frac{R^3}{r^3}\left[(1-3\sin^2\theta) - \frac{R^3}{2r^3}\left(1 - \frac{3\sin^2\theta}{4}\right)\right]\end{aligned} \tag{4.6}$$

At the surface of the sphere, the pressure distribution is given by:

$$\frac{p(R,\theta,t)-p_\infty}{\rho} = R\ddot{R} + \frac{3}{2}\dot{R}^2 + \frac{1}{2}\left(\dot{W}R + 3W\dot{R}\right)\cos\theta + \frac{1}{2}W^2\left(1 - \frac{9}{4}\sin^2\theta\right) \tag{4.7}$$

The lack of symmetry is due to the term $\cos\theta$. Integration over the surface of the sphere gives the resulting force:

$$R_p = -\frac{2}{3}\pi\rho R^2\left(\dot{W}R + 3W\dot{R}\right) = -\frac{d(MW)}{dt} \tag{4.8}$$

where M is the virtual mass already defined in equation (4.2).

The equation of motion of the sphere is then:

$$m\frac{dW}{dt} + \frac{d(MW)}{dt} = F \tag{4.9}$$

In the case of a sphere of constant radius, equation (4.9) reduces to equation (4.3).

In the case of a constant translation velocity W, if the sphere is shrinking ($\dot{R} < 0$), the resulting force R_p is positive, so that the sphere is pushed by the liquid. This is due to higher pressure values at the back of the sphere i.e. for $\theta > \pi/2$. Meanwhile, F is negative which means that the motion of a sphere at constant velocity requires an adverse force if its volume decreases.

4.2.3. *APPLICATION TO BUBBLES*

It is assumed here that the bubble remains spherical. This assumption is particularly well fulfilled for small bubbles, since surface tension, which tends to give a spherical shape to the interface, becomes more efficient for small radii of curvature.

Equation (4.9), with $F = 0$ since the bubble motion is free, and $m \cong 0$ due to the negligible mass of gas within it, becomes:

$$\frac{d}{dt}(MW) = \dot{M}W + M\dot{W} = 0 \tag{4.10}$$

Hence, the total virtual impulse MW of the bubble remains constant.

If the bubble collapses ($\dot{M} < 0$), then $\dot{W} > 0$, i.e. the bubble accelerates. This effect is called the rocket effect. Inversely, an exploding bubble will decelerate. Hence, a bubble of varying volume tends to move relative to the ambient liquid. The present effect is purely inertial, since all other effects, such as gravity, have been neglected.

One can conjecture that a bubble will depart from its spherical shape if the maximum pressure difference between two points on the interface is larger than the characteristic pressure difference $2S/R$ due to surface tension. From equation (4.7), the pressure difference between the back and the front of the bubble is $\rho\left(\dot{W}R + 3W\dot{R}\right)$, so that the condition of deviation from sphericity can be written:[1]

$$\rho\left(\dot{W}R + 3W\dot{R}\right) >> \frac{2S}{R} \tag{4.11}$$

As an example, a bubble of constant radius accelerated at $\dot{W} = 10\ \mathrm{m/s^2}$ in water will loose its spherical shape if its radius is greater than about $\sqrt{2S/\rho\dot{W}} \cong 3.8\ \mathrm{mm}$.

The problem is then to describe the subsequent evolution of the shape of the interface. This is done in the following section. However, in the present case of a collapsing, accelerated spherical bubble, it is obvious that the back of the interface, which is exposed to higher pressures, will undergo a deformation such that the liquid will tend to penetrate the bubble and give rise to a re-entrant jet.

4.3. *NON-SPHERICAL BUBBLE EVOLUTION*

4.3.1. *PRINCIPLE OF PLESSET-CHAPMAN NUMERICAL MODELING*

Consider an initially spherical bubble, close to a solid wall, which collapses under a constant pressure p_∞. The proximity of the wall will alter the sphericity of the interface. PLESSET and CHAPMAN (1971) computed the time evolution of the bubble for various distances from the wall. The principle of the simulation is presented below.

---

1. To be rigorous, the derivation of equation (4.11) should refer to a solid sphere which is removed instantaneously, generating an initially spherical cavity in the liquid. At this initial time, the velocities and acceleration in the liquid are unchanged and the initial trend for the further deformation of the cavity is given by this pressure difference.

At a given instant t, the velocity potential φ is calculated from the LAPLACE equation $\Delta\varphi = 0$ using a classical numerical method (e.g. boundary elements), together with the appropriate boundary conditions on the wall, the bubble interface and at infinity. The velocity on the interface is deduced from the velocity potential using the equation $\vec{V} = \overrightarrow{\text{grad}}\,\varphi$ and the lagrangian derivative of the potential $d\varphi/dt$ at the bubble surface is computed from the generalized BERNOULLI equation, giving:

$$\left.\frac{d\varphi}{dt}\right|_{\text{interface}} = \frac{\partial\varphi}{\partial t} + \vec{V}.\overrightarrow{\text{grad}}\,\varphi = \frac{p_\infty - p_v}{\rho} + \frac{V^2}{2} \tag{4.12}$$

Here, surface tension is ignored and the bubble is assumed to contain only vapor at vapor pressure p_v.

As the velocity of the liquid particles on the interface is known at time t, the new position of the bubble interface at $t + dt$ can be readily deduced assuming that the interface nodes are moving with fluid velocity $\vec{V}$. In addition, the new value of the velocity potential on the interface can be computed from equation (4.12). Then, the LAPLACE equation is solved at $t + dt$ using the updated boundary conditions, and so on. Therefore, computation of the time evolution of the bubble appears as a series of boundary value problems each one separated by a time step.

4.3.2. *SOME GENERAL RESULTS*

Collapse of a bubble close to a wall

Figure 4.2 presents the computational results of PLESSET and CHAPMAN (1971) for a bubble of 1 mm initial radius whose center is initially 1 mm from the wall.

During collapse, the interface close to the wall tends to flatten while a hollow develops on the opposite face with a change in curvature under very high pressure. Then, a re-entrant jet, directed towards the solid wall, develops, with a rapidly increasing velocity. Finally, the jet pierces the bubble and strikes the solid wall.

Because of the high velocities involved, the re-entrant jet is often considered as a possible hydrodynamic mechanism of cavitation erosion. In general, the lifetime of the bubble increases due to the presence of the wall. The particle paths during the growth and subsequent collapse are shown in figure 4.3 reproduced from BLAKE and GIBSON (1987).

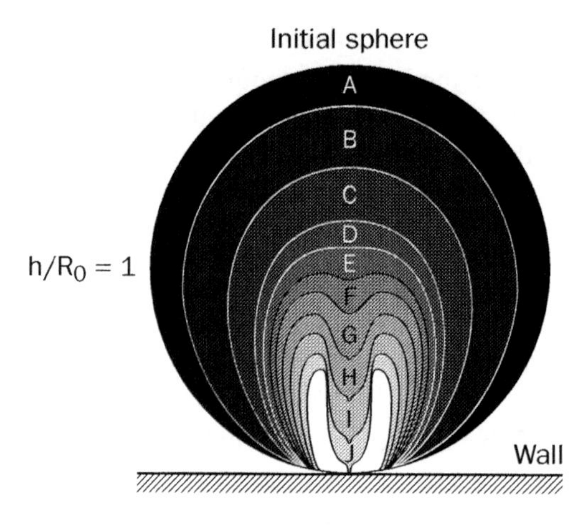

Shape	Time	Velocity (m/s)
A	0.63	7.7
B	0.885	19
C	0.986	42
D	1.013	65
E	1.033	100
F	1.048	125
G	1.066	129
H	1.082	129
I	1.098	128
J	1.119	128

4.2 - Numerical results of PLESSET and CHAPMAN (1971) for the collapse of a bubble near a wall

The time is non-dimensionalized using the reference time $R_0\sqrt{\dfrac{\rho}{p_\infty - p_v}}$*, which is close to the RAYLEIGH collapse time. The velocity given in the table is the velocity of the upper point of the interface on the axis of symmetry. The initial distance h from the bubble center to the wall is equal to the initial bubble radius* R_0*.*

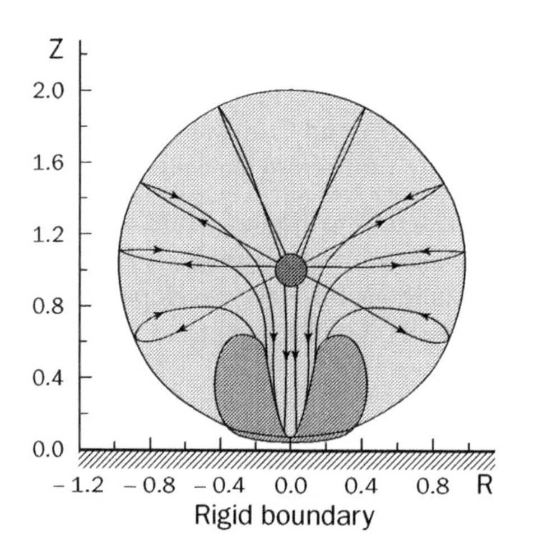

4.3
Theoretical calculations for the growth and collapse of a vapor bubble near a wall for $\gamma = h / R_m = 1$ (where h is the distance to the wall and R_m is the maximum radius)

The diagram shows particle paths during growth and collapse.
[from BLAKE & GIBSON, 1987]
Reprinted, with permission, from the Annual Review of Fluid Mechanics, Vol. 19, ©1987 by Annual Reviews (www.annualreviews.org).

Collapse of a bubble close to a free surface

If the bubble collapses close to a free surface, a re-entrant jet develops in the direction opposite to the free surface. The behavior of the free surface itself depends on the value of the ratio η, of the bubble's maximum radius to the initial distance of its center from the free surface [CHAHINE 1982]. For $\eta < 0.3$, the free surface is not greatly disturbed, whereas for larger values it is violently disturbed by a counterjet (fig. 4.4).

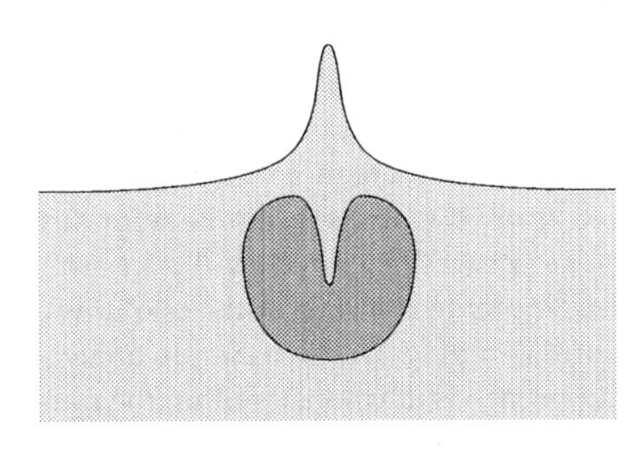

4.4

Schematic view of the collapse of a bubble near a free surface with a counterjet

At the beginning of the collapse, the liquid particles move preferentially towards regions where they encounter the weakest resistance. This is either the free surface or the face opposite the wall. This movement generates a high pressure, which triggers the re-entrant jet. Hence, it is possible to qualitatively anticipate the place where the re-entrant jet will form by identifying the less constrained region.

Collapse of a bubble confined between two walls

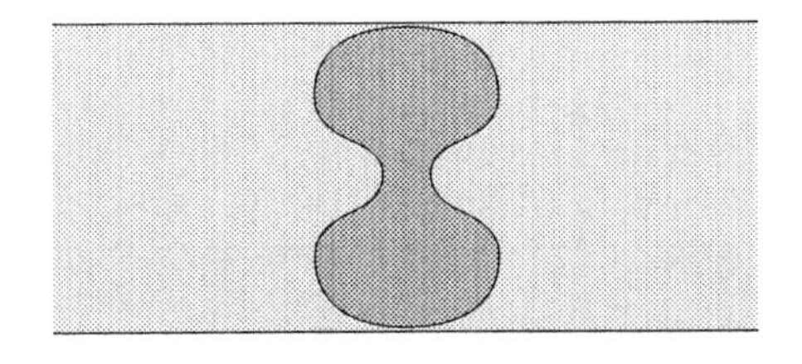

4.5

Schematic view of the collapse of a bubble between two parallel walls

In the case of a nucleus situated in the middle of a very confined domain between two parallel walls, growth is fastest in the directions parallel to the wall. During the subsequent collapse phase, the reverse occurs: there is a rapid equatorial contraction and the bubble takes the shape of an hourglass, before splitting into two symmetric bubbles. Each one then develops a re-entrant jet directed towards the nearest wall.

These results were obtained from a large number of studies, both numerical and experimental. The most important of them include: NAUDÉ & ELLIS (1961), ELLIS (1965), BENJAMIN & ELLIS (1966), PLESSET & CHAPMAN (1971), LAUTERBORN & BOLLE (1975), BLAKE & GIBSON (1981, 1987), CHAHINE (1982), TOMITA & SHIMA (1986), CHAHINE (1990a), ALLONCLE, DUFRESNE & TESTUD (1992), BLAKE (1994), ISSELIN *et al.* (1996).

Other situations were also studied, such as the collapse of a bubble in a shear layer, or on the axis of a rotating flow [CHAHINE 1990b, YAN & MICHEL 1998]. In the latter case, conservation of the moment of momentum prevents the liquid particles situated on the interface from reaching the axis of rotation, so that a complete collapse of the bubble is not possible. This point was experimentally checked by FILALI (1997).

4.3.3. *BLAKE'S ANALYTICAL APPROACH*

Principle

This approach is based on the concept of the KELVIN impulse, which was proposed to get round a theoretical difficulty encountered in hydrodynamics when the motion of a solid body in an infinite fluid domain is considered. Due to the non-uniform convergence of integrals relative to the fluid momentum, it is not possible to apply the momentum theorem directly to a fluid and body system. The impulse concept was extended by BLAKE and his co-authors (see BLAKE & GIBSON 1987, for example) to the case of the evolution of a bubble in a non-symmetric flow field, in order to give a rather simple analysis of the re-entrant jet and its direction. This approach is developed within the framework of the potential flow theory.

The liquid domain D is considered to extend between the bubble with boundary S, a large sphere S_0 at infinity and a boundary S_1 representing a wall or a free surface according to the case under examination (fig. 4.6). The total liquid momentum is:

$$\vec{M} = \iiint_D \rho \vec{V} \, dv = \iiint_D \rho \, \vec{grad}\, \varphi \, dv = \iint_{S+S_0+S_1} \rho \varphi \vec{v}_e \, dS \tag{4.13}$$

where φ is the velocity potential and $\vec{v}_e$ the outward unit vector, normal to the surface.

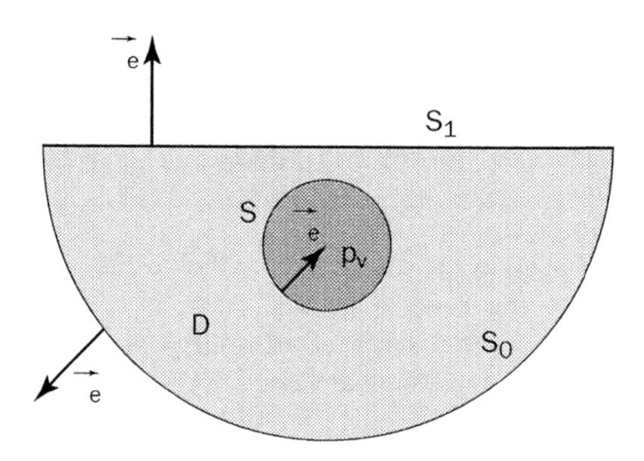

4.6
Domain of integration for BLAKE's analytical approach

The momentum equation takes the following integral form:

$$\frac{d\vec{M}}{dt} = \vec{F} \tag{4.14}$$

where $\vec{F}$ stands for the total forces resulting from the pressure on the three boundaries:

$$\vec{F} = -\iint_{S+S_0+S_1} p \, \vec{v}_e \, dS \tag{4.15}$$

The contribution from the closed surface of the bubble is zero as the pressure on the interface is assumed constant and equal to the vapor pressure.

Consider now the quantity $\vec{M}_S$ defined by:

$$\vec{M}_S = \iint_S \rho \varphi \vec{v}_e \, dS \tag{4.16}$$

With reference to equation (4.13), $\vec{M}_S$ can be interpreted as a kind of momentum attached to the bubble only. This can be termed "bubble momentum". Then, the integral momentum equation (4.14) can be written:

$$\frac{d\vec{M}_S}{dt} = \vec{F}_p \tag{4.17}$$

with

$$\vec{F}_p = -\iint_{S_0+S_1} p\,\vec{v}_e\,dS - \frac{d}{dt}\left\{\iint_{S_0+S_1} \rho\varphi\,\vec{v}_e\,dS\right\} \tag{4.18}$$

The calculation is somewhat tedious and the details are given in the appendix. Finally, $\vec{F}_p$ is given by the following equivalent expressions:

$$\vec{F}_p = \rho\iint_{S_1}\left[\varphi\frac{\partial\,\vec{\mathrm{grad}}\,\varphi}{\partial\nu} - \frac{V^2}{2}\vec{v}_e\right]dS = \rho\iint_{S_1}\left[\frac{V^2}{2}\vec{v}_e - \frac{\partial\varphi}{\partial\nu}\vec{\mathrm{grad}}\,\varphi\right]dS \tag{4.19}$$

$\vec{F}_p$ can be considered as a force due to the presence of the boundary S_1, which changes the bubble momentum $\vec{M}_S$ according to equation (4.17). Thus, the bubble momentum is related to the nature of the boundary S_1.

If the boundary S_1 is far from the bubble, the S_1 integral is zero and one obtains:

$$\frac{d\vec{M}_S}{dt} = 0 \tag{4.20}$$

which extends the validity of equation (4.10) of conservation of the bubble momentum to the case of a bubble of any shape.

4.7
Bubble near a plane boundary

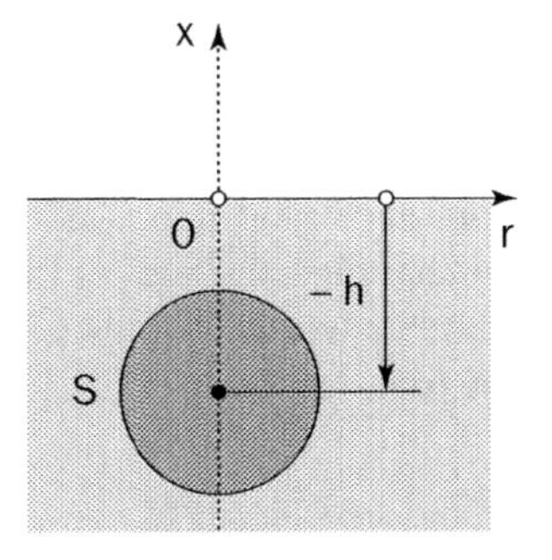

In the case of a plane boundary S_1 with unit normal vector $\vec{x}$ (fig. 4.7) and a spherical bubble, the flow is axisymmetric. From equation (4.19), the force $\vec{F}_p$ has only an axial component given by:

$$F_{px} = \pi\rho\int_{r=0}^{\infty} r\left(u_r^{\,2} - u_x^{\,2}\right)dr \tag{4.21}$$

where u_x and u_r are the axial and radial components of the velocity respectively.

The case of a solid wall

In the case of a solid wall, the flow is simulated by a source located at the bubble center and its image with respect to the wall, so that the slip condition $u_x = 0$ is fulfilled on the wall. Hence, the velocity potential is given by:

$$\varphi = -\frac{q(t)}{4\pi}\left\{\frac{1}{\left[(x+h)^2 + r^2\right]^{1/2}} + \frac{1}{\left[(x-h)^2 + r^2\right]^{1/2}}\right\} \tag{4.22}$$

where h is the distance of the bubble center from the wall and $q(t)$ is the source intensity. On the wall S_1, the axial and radial components of the velocity are given by:

$$\begin{cases} u_x = \dfrac{\partial\varphi}{\partial x} = 0 \\ u_r = \dfrac{\partial\varphi}{\partial r} = \dfrac{q(t)\,r}{2\pi(h^2 + r^2)^{3/2}} \end{cases} \tag{4.23}$$

so that

$$F_{px} = \frac{\rho q^2(t)}{4\pi}\int_0^\infty \frac{r^3\,dr}{(h^2 + r^2)^3} = \frac{\rho q^2(t)}{16\pi h^2} \tag{4.24}$$

The bubble momentum has only an axial component, which is given by:

$$M_{Sx} = \int_0^t F_{px}\,dt = \frac{\rho}{16\pi h^2}\int_0^t q^2(t)\,dt \tag{4.25}$$

Therefore, the bubble momentum is positive and directed toward the wall, which indicates that the re-entrant jet is necessarily directed toward the wall.

The case of a free surface

If the boundary S_1 is a free surface, then the boundary is free to expand in the axial direction x and the bubble collapse induces no radial velocity. Therefore, the boundary condition is assumed to be $u_r = 0$ on S_1. This condition is fulfilled simply by changing the sign of the second term of the velocity potential (expression 4.22). In other words, the image source is replaced by a sink. Then, the flow velocity on the free surface is:

$$\begin{cases} u_x = \dfrac{\partial\varphi}{\partial x} = \dfrac{q(t)\,h}{2\pi(h^2 + r^2)^{3/2}} \\ u_r = \dfrac{\partial\varphi}{\partial r} = 0 \end{cases} \tag{4.26}$$

so that the force is:

$$F_{px} = -\frac{\rho q^2(t)\,h^2}{4\pi}\int_0^\infty \frac{r\,dr}{(h^2 + r^2)^3} = -\frac{\rho q^2(t)}{16\pi h^2} \tag{4.27}$$

and the bubble momentum is:

$$M_{Sx} = \int_0^t F_{px}\, dt = -\frac{\rho}{16\pi h^2} \int_0^t q^2(t)\, dt \tag{4.28}$$

Hence, the bubble momentum M_{Sx} is negative, so that the direction of the re-entrant jet is opposite to the free surface.

Other configurations

If the plane $x=0$ is an interface between two liquids of different densities, ρ_1 for $x<0$ and ρ_2 for $x>0$, then in expression (4.25), M_{Sx} is modified by changing ρ into $(\rho_2 - \rho_1)/(\rho_2 + \rho_1)$. Thus, this factor gives the sign of M_{Sx} and is an indicator of the direction of the re-entrant jet.

Other dissymmetric effects, such as gravity, the vicinity of a stagnation point, or the effect of an elastic or visco-elastic wall, can also be analyzed by the present method (see BLAKE & GIBSON 1987).

4.4. THE PATH OF A SPHERICAL BUBBLE

It is supposed here that the sources of dissymmetry considered in section 4.3 are negligible, so that the bubble remains spherical. We look for the trajectory of the bubble center in a flowing liquid.

Let $\vec{V}_L$ be the liquid velocity in the absence of the bubble and $\vec{V}_B$ the velocity of the bubble center (fig. 4.8). The slip velocity $\vec{W}$ is defined by $\vec{W} = \vec{V}_B - \vec{V}_L$.

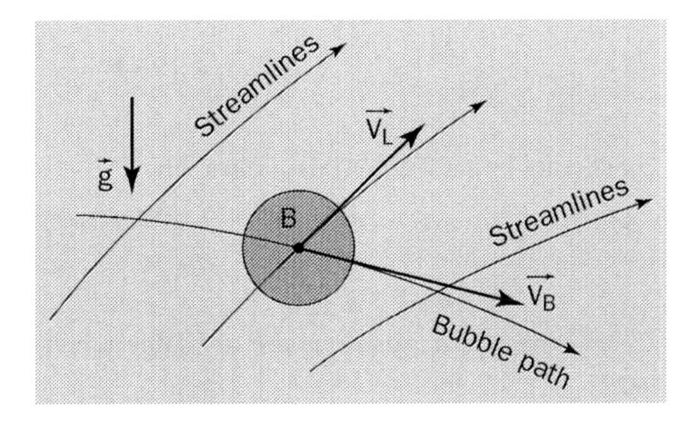

4.8
Moving bubble in a flowing liquid

The total mass of vapor and gas inside the bubble, i.e. the mass of the bubble, is assumed constant, but its volume can vary due to the pressure variations it undergoes during its movement. If the density of the gas/vapor mixture inside the bubble is ρ_g, the mass of the bubble is:

$$m_B = 2M \frac{\rho_g}{\rho} \tag{4.29}$$

The quantity M is the virtual mass defined by equation (4.2) so that the quantity $2M/\rho$ is simply the bubble volume.

The equation of motion of the bubble center is:

$$m_B \dot{\vec{V}}_B = m_B \vec{g} + \vec{P} + \vec{D} \tag{4.30}$$

The first term on the right-hand side is the weight of the bubble, the second term, $\vec{P}$, stands for the pressure forces and the third, $\vec{D}$, is the viscous drag. Such a superposition is currently used in aeronautics for the estimate of the force on a slender wing.

The viscous drag is given by the expression:

$$\vec{D} = -C_D(Re_W)\, \pi R^2 \frac{\rho W^2}{2} \frac{\vec{W}}{|W|} \tag{4.31}$$

in which $C_D(Re_W)$ stands for the drag coefficient of the bubble in its motion relative to the liquid. Re_W is the bubble REYNOLDS number based on the relative velocity W.

According to the early experiments of HABERMAN and MORTON (1953), the drag of spherical bubbles is identical to that of solid spheres with the same radius. Thus C_D is taken from data relative to solid spheres. For small bubbles whose REYNOLDS number in the relative motion is smaller than about one, the drag is given by the STOKES formula:

$$D = 6\pi\mu R|W| \tag{4.32}$$

so that the drag coefficient is:

$$C_D = \frac{24}{Re_W} \tag{4.33}$$

For slightly larger REYNOLDS numbers, the OSEEN formula can be used:

$$C_D = \frac{24}{Re_W}\left[1 + \frac{3\,Re_W}{8}\right] \tag{4.34}$$

Other empirical expressions are available when the REYNOLDS number becomes large.

The determination of the pressure force $\vec{P}$ is more complex. We adopt here a simplified and rather intuitive approach. It is assumed that the pressure force results from the superposition of two effects,
- a combined slip/rocket effect, and
- a flow effect, so that the resulting pressure force is written:

$$\vec{P} = \vec{P}_1 + \vec{P}_2 \tag{4.35}$$

To estimate the slip/rocket effect, we consider the bubble to have the slip velocity $\vec{W}$ in the liquid which is itself assumed at rest at infinity. From section 4.2.2, the pressure force acting on the bubble in this configuration is:

$$\vec{P}_1 = -\frac{d\left(M\vec{W}\right)}{dt} \tag{4.36}$$

where M is the virtual mass defined by equation (4.2). This equation is the generalization of equation (4.8) in vector form. It takes into account both the volume variation of the bubble and its slip with respect to the liquid.

To estimate the flow effect, it is assumed that the pressure force on the bubble due to the liquid movement is the same as the pressure force acting on a liquid globule of the same radius that could exactly replace the bubble and move with the liquid, without slipping. For such a configuration, the momentum balance of the liquid globule can be written:

$$\iiint \rho \frac{d\vec{V}_L}{dt} d\tau = \vec{P}_2 + 2M\vec{g} \tag{4.37}$$

The left-hand term is the variation of momentum of the liquid globule whereas the right-hand terms stand for the pressure and the gravity forces respectively. The coefficient 2 in the last term results from the fact that the virtual mass is only half the mass of the displaced fluid for a sphere.

From the previous equation, one obtains:

$$\vec{P}_2 = \iiint \rho \frac{d\vec{V}_L}{dt} d\tau - 2M\vec{g} \tag{4.38}$$

Considering the small size of the globule and assuming that the radius of curvature R' of the streamline is large enough with respect to the globule radius R, then the time derivative of the velocity can be considered constant over the entire globule. One thus obtains the approximate relation:

$$\vec{P}_2 \cong 2M\dot{\vec{V}}_L - 2M\vec{g} \tag{4.39}$$

This expression is valid to second order in R/R'.

HSIEH gives another form to the pressure force $\vec{P}_2$. Using the EULER equation:

$$\rho \frac{d\vec{V}_L}{dt} = \rho\vec{g} - \overrightarrow{\text{grad}}\, p \tag{4.40}$$

he obtained the equivalent form:

$$\vec{P}_2 = -\mathcal{V}\, \overrightarrow{\text{grad}}\, p \tag{4.41}$$

where $\mathcal{V}$ stands for the bubble volume.

Finally, the total pressure force acting on the bubble can be estimated as follows:

$$\vec{P} = -\frac{d\left(M\vec{W}\right)}{dt} + 2M\dot{\vec{V}}_L - 2M\vec{g} = -\dot{M}\left(\vec{V}_B - \vec{V}_L\right) - M\dot{\vec{V}}_B + 3M\dot{\vec{V}}_L - 2M\vec{g} \qquad (4.42)$$

Although this estimate was obtained in a rather intuitive way, it is identical to the one found by VOINOV (1973), using a more theoretical approach.

Taking into account equations (4.30) and (4.42), the following equation of motion for the bubble center is obtained:

$$M\left[1 + \frac{2\rho_g}{\rho}\right]\dot{\vec{V}}_B = -2M\left[1 - \frac{\rho_g}{\rho}\right]\vec{g} + 3M\dot{\vec{V}}_L - \dot{M}\left(\vec{V}_B - \vec{V}_L\right) + \vec{D} \qquad (4.43)$$

The first term on the right-hand side is the sum of the bubble weight and the buoyancy force. The second term is due to the pressure gradient in the liquid, the third to the effect of bubble volume variations, the so-called rocket effect, and the last term is the viscous drag on the bubble. Note that equation (4.43) has the trivial solution $\vec{V}_B = \vec{V}_L$ if it is assumed that $\rho_g = \rho$.

Usually, the terms in ρ_g/ρ are negligible, so that equation (4.43) reduces to the HSIEH equation (see the reference in chapter 3 and also JOHNSON & HSIEH 1966):

$$M\dot{\vec{V}}_B = -2M\vec{g} + 3M\dot{\vec{V}}_L - \dot{M}\left(\vec{V}_B - \vec{V}_L\right) + \vec{D} \qquad (4.44)$$

The velocity $\vec{V}_L$, which characterizes the carrier flow, is assumed to be known, while the virtual mass M, which principally depends upon the bubble radius R, is deduced from the RAYLEIGH-PLESSET equation. The solution of equation (4.43) or (4.44) coupled with the RAYLEIGH-PLESSET equation allows us to compute the bubble velocity $\vec{V}_B$ and hence, the bubble path.

Other approaches have been used to address this problem. VILLAT (1943) and FOISSEY (1972) give the following expression for the total force on a sphere of constant radius, using a no-slip condition at the boundary, at small REYNOLDS numbers:

$$\vec{T} = 6\pi\mu R\left(\vec{V}_L - \vec{V}_B\right) + M\left(3\dot{\vec{V}}_L - \dot{\vec{V}}_B\right) - 2M\vec{g} + 6R^2\sqrt{\pi\mu\rho}\int_{u=0}^{t}\frac{\dot{\vec{V}}_L(u) - \dot{\vec{V}}_B(u)}{\sqrt{t-u}}du \qquad (4.45)$$

The first term is STOKES drag. The last term is called BASSET drag. It represents the effect of the evolving viscous wake on the motion of the sphere itself. If we adopt the model of a sphere with negligible mass, as is the case for a bubble, we have $\vec{T} = 0$, which gives:

$$M\dot{\vec{V}}_B = -2M\vec{g} + 3M\dot{\vec{V}}_L - 6\pi\mu R\left(\vec{V}_B - \vec{V}_L\right) + \text{Basset drag} \qquad (4.46)$$

The term $-\dot{M}\left(\vec{V}_B - \vec{V}_L\right)$ must be added to the right-hand side of equation (4.46) in the case of a bubble of variable radius in order to take the rocket effect into account. Then, equation (4.46) reduces to the HSIEH equation (4.44) considering that the last two terms represent the total drag $\vec{D}$.

Evaluation of the BASSET and other terms in equation (4.46) can be found in an experimental study by SRIDHAR and KATZ (1995). These authors have added a term to take into account the lift caused by nuclei passing through a region of high vorticity.

REFERENCES

ALLONCLE A.P., DUFRESNE D. & TESTUD P. –1992– Etude expérimentale des bulles de vapeur générées par laser. *La Houille Blanche* **7/8**, 539-544.

BATCHELOR G.K. –1967– An introduction to fluid dynamics. *Cambridge University Press.*

BLAKE J.R. –1994– Transient inviscid bubble dynamics. *Proc. 2nd Int. Symp. on Cavitation,* Tokyo (Japan), 9-18.

BLAKE J.R. & GIBSON D.C. –1981– Growth and collapse of a vapor cavity near a free surface. *J. Fluid Mech.* **111**, 123-140.

BLAKE J.R. & GIBSON D.C. –1987– Cavitation bubbles near boundaries. *Ann. Rev. Fluid Mech.* **19**, 99-128.

BENJAMIN T.B. & ELLIS A.T. –1966– The collapse of cavitation bubbles and the pressures thereby produced against solid boundaries. *Phil. Trans. Roy. Soc. London A* **260**, 221-240.

CHAHINE G.L. –1982– Experimental and asymptotic study of non-spherical bubble collapse. *Appl. Sci. Res.* **38**, 187-197.

CHAHINE G.L. –1990a– Numerical modeling of the dynamic behavior of bubbles in non-uniform flow fields. *Proc. ASME Symp. on Numerical Methods for Multiphase Flows,* FED **91**, Toronto (Canada), 57-65.

CHAHINE G.L. –1990b– Non-spherical bubble dynamics in a line vortex. *Proc. ASME Cavitation and Multiphase Flow Forum,* FED **98**, Toronto (Canada), 121-127.

ELLIS A.T. –1965– Parameters affecting cavitation and some new methods for their study. *Cal. Inst. Techn. Hydro. Labo,* Rpt E115-1.

FILALI E.G. –1997– Étude physique de l'implosion axiale de tourbillons cavitants érosifs formés dans une chambre tournante. *Thesis,* Grenoble University (France).

FOISSEY C. –1973– Théorie asymptotique de la cavitation naissante. *Ensta,* Rpt 014, Paris (France).

HABERMAN W.L. & MORTON R.K. –1953– An experimental investigation of the drag and shape of air bubbles rising in various liquids. *DTMB*, Rpt 802.

ISSELIN J.C., ALLONCLE A.P., DUFRESNE D. & AUTRIC M. –1996– Comportement d'une bulle de cavitation à proximité d'une paroi solide: contribution à l'étude du mécanisme d'érosion. *Proc. 3e Journées Cavitation SHF*, Grenoble (France), 175-182.

JOHNSON V.E. & HSIEH T. –1966– The influence of the trajectories of gas nuclei on cavitation inception. *Trans. 6th ONR Symp. Washington DC*.

LAUTERBORN W. & BOLLE H. –1975– Experimental investigations of cavitation-bubble collapse in the neighbourhood of solid boundary. *J. Fluid Mech.* **72**, 391-399.

NAUDÉ C.F. & ELLIS A.T. –1961– On the mechanism of cavitation damage by non-hemispherical cavities collapsing in contact with a solid boundary. *Trans. ASME D*, 633-648.

PLESSET M.S. & CHAPMAN R.B. –1971– Collapse of an initially spherical cavity in the neighbourhood of a solid boundary. *J. Fluid Mech.* **47-2**, 283-290.

SHRIDAR G. & KATZ J. –1995– Drag and lift forces on microscopic bubbles entrained by a vortex. *Phys. Fluids* **7**(2), 389-399.

TOMITA Y. & SHIMA A. –1986– Mechanisms of impulsive pressure generation and damage pit formation by bubble collapse. *J. Fluid Mech.* **169**, 535-564.

VILLAT H. –1943– Leçons sur les fluides visqueux. *Gauthier-Villars Ed.*

VOINOV –1973– On the force acting on a sphere in a non-uniform stream of perfect incompressible fluid. *J. Appl. Mech. Techn. Phys.* **14**, 592-594.

YAN K. & MICHEL J.M. –1998– Numerical simulation of bubble dynamics in vortex core by the BUBMAC Method. *Proc. 3rd Int. Symp. on Cavitation*, vol. 1, Grenoble (France), 75-80.

APPENDIX TO SECTION 4.3.3

In order to evaluate the equation:

$$\frac{d\vec{M}_S}{dt} = -\iint_{S_0+S_1} p\,\vec{v}_e\,dS - \frac{d}{dt}\left\{\iint_{S_0+S_1} \rho\,\phi\,\vec{v}_e\,dS\right\} \tag{4.47}$$

and particularly the second term derivation of (4.47), let us consider the general integral:

$$\vec{J} = \iint_{\Sigma} \theta\,\vec{v}_e\,dS = \iint_{\Sigma} \theta\,v_k\,\vec{x}_k\,dS$$

where $\vec{x}_k$ stands for the unit vectors in cartesian coordinates and θ for any function. For an incompressible fluid, the derivative is (see BATCHELOR, p. 133):

$$\frac{d\vec{J}}{dt} = \iint_{\Sigma} \frac{d\theta}{dt} v_k \vec{x}_k \, dS - \iint_{\Sigma} \theta \frac{\partial u_k}{\partial x_i} v_i \, dS \, \vec{x}_k$$

which gives here, with $\theta \equiv \rho\,\phi$:

$$\frac{d}{dt}\left\{\iint_{S_0+S_1} \rho\,\phi\,\vec{v}_e \, dS\right\} = \iint_{S_0+S_1} \rho \frac{d\phi}{dt} \vec{v}_e \, dS - \iint_{S_0+S_1} \rho\,\phi \frac{\partial}{\partial v}\left(\vec{\text{grad}}\,\phi\right) dS \qquad (4.48)$$

In (4.48), the last term is obtained from:

$$\frac{\partial u_k}{\partial x_i} v_i \vec{x}_k = \frac{\partial\left(\vec{\text{grad}}\,\phi.\vec{x}_k\right)}{\partial v} \vec{x}_k = \frac{\partial\left(\vec{\text{grad}}\,\phi\right)}{\partial v}$$

Introducing the BERNOULLI equation for unsteady flow in (4.47):

$$p = p_\infty - \rho \frac{\partial \phi}{\partial t} - \rho \frac{V^2}{2}$$

and combining with (4.48) leads to the equation:

$$\frac{d\vec{M}_S}{dt} = \rho \iint_{S_0+S_1} \left(\frac{\partial\phi}{\partial t} + \frac{V^2}{2}\right) \vec{v}_e \, dS - \rho \iint_{S_0+S_1} \frac{d\phi}{dt} \vec{v}_e \, dS + \iint_{S_0+S_1} \phi \frac{\partial}{\partial v}\left(\vec{\text{grad}}\,\phi\right) dS$$

Taking the following expression for the Lagrangian derivative of ϕ:

$$\frac{d\phi}{dt} = \frac{\partial \phi}{\partial t} + \vec{V}.\vec{\text{grad}}\,\phi = \frac{\partial\phi}{\partial t} + V^2$$

one finally obtains:

$$\frac{d\vec{M}_S}{dt} = \rho \iint_{S_0+S_1} \left[\phi \frac{\partial\left(\vec{\text{grad}}\,\phi\right)}{\partial v} - \frac{V^2}{2} \vec{v}_e\right] dS \qquad (4.49)$$

At great distances from the bubble, ϕ behaves as a source term. Thus,

$$\begin{cases} \phi \approx 1/r \\ V \approx 1/r^2 \\ \dfrac{\partial\left(\vec{\text{grad}}\,\phi\right)}{\partial v} \approx 1/r^3 \end{cases}$$

and the function in the integral of equation (4.49) behaves as $1/r^2$. Two cases can be considered:

- If S_0 and S_1 are far from the bubble (case of an infinite medium), the integral in (4.49) is zero and we have:

$$\frac{d\vec{M}_S}{dt} = 0 \tag{4.50}$$

- If S_0 only goes to infinity, the integral on S_0 is zero and we have:

$$\vec{M}_S = \int_0^t \vec{F}_P \, dt \tag{4.51}$$

with

$$\vec{F}_P = \rho \iint_{S_1} \left[\phi \frac{\partial(\vec{\text{grad}}\,\phi)}{\partial v} - \frac{V^2}{2} \vec{v}_e \right] dS \tag{4.52}$$

Taking into account the fact that the Laplacian of ϕ is zero, we have (see the note below):

$$\vec{F}_P = \rho \iint_{S_1} \left[\frac{V^2}{2} \vec{v}_e - \frac{\partial \phi}{\partial v} \vec{\text{grad}}\,\phi \right] dS \tag{4.53}$$

Note on the passage from (4.52) to (4.53)

Let us write:

$$\vec{A} = \iint_{S_0+S_1} \left[\phi \frac{\partial(\vec{\text{grad}}\,\phi)}{\partial v} - \frac{V^2}{2} \vec{v}_e \right] dS = \vec{A}_1 + \vec{A}_2$$

with

$$\begin{aligned}
\vec{A}_1 &= \iint \phi \frac{\partial u_k}{\partial x_i} v_i \vec{x}_k \, dS \\
&= \iint \frac{\partial(\phi u_k)}{\partial x_i} v_i \vec{x}_k \, dS - \iint u_k \vec{x}_k \frac{\partial \phi}{\partial v} dS \\
&= \iint \vec{\text{grad}}\left(\phi \frac{\partial \phi}{\partial x_k} \right) \vec{v}_e \, dS \, \vec{x}_k - \iint \vec{\text{grad}}\,\phi \frac{\partial \phi}{\partial v} dS \\
&= \vec{A}'_1 + \vec{A}''_1
\end{aligned}$$

The GREEN-OSTROGRADSKI transformation allows us to write:

$$\vec{A}'_1 = \iiint \Delta\left(\phi \frac{\partial \phi}{\partial x_k} \right) dv \, \vec{x}_k$$

with

$$\Delta\left(\phi\frac{\partial\phi}{\partial x_k}\right)=\Delta\phi\frac{\partial\phi}{\partial x_k}+2\,\vec{\text{grad}}\,\phi\,.\,\vec{\text{grad}}\frac{\partial\phi}{\partial x_k}+\phi\Delta\left(\frac{\partial\phi}{\partial x_k}\right)=2\,\vec{\text{grad}}\,\phi\,.\,\vec{\text{grad}}\frac{\partial\phi}{\partial x_k}$$

Actually, $\Delta\phi=0$, and $\Delta\left(\frac{\partial\phi}{\partial x_k}\right)=\frac{\partial(\Delta\phi)}{\partial x_k}=0$.

Thus we have:

$$\Delta\left(\phi\frac{\partial\phi}{\partial x_k}\right)=2\vec{V}.\vec{\text{grad}}\frac{\partial\phi}{\partial x_k}=2\vec{V}.\frac{\partial\left(\vec{\text{grad}}\,\phi\right)}{\partial x_k}=2\vec{V}.\frac{\partial\vec{V}}{\partial x_k}=\frac{\partial V^2}{\partial x_k}$$

and

$$\vec{A}'_1=\iiint\frac{\partial V^2}{\partial x_k}\,\vec{x}_k\,dS=\iiint\vec{\text{grad}}\,V^2\,dS=\iint V^2\,\vec{v}_e\,dS$$

The last relation results from the GAUSS formula. Therefore:

$$\vec{A}'_1+\vec{A}_2=\iint\frac{V^2}{2}\,\vec{v}_e.dS$$

and the final result is:

$$\vec{A}=\iint_{S_0+S_1}\left[\frac{V^2}{2}\,\vec{v}_e-\frac{\partial\phi}{\partial v}\,\vec{\text{grad}}\,\phi\right]dS$$

It allows us to write relation (4.53) for $\vec{F}_p$, taking into account that the contribution from S_0 is zero.

5. FURTHER INSIGHTS INTO BUBBLE PHYSICS

The dynamics of cavitation bubbles is controlled mainly by inertia and pressure forces, as analysed in chapters 3 and 4. However, other physical phenomena may also have a non-negligible influence on their growth and collapse. These include:

- Liquid compressibility which affects the final stages of bubble collapse and causes the emission of shock waves and/or acoustic waves, essential in cavitation noise and erosion.
- Heat transfer between the entrapped gas and the surrounding liquid, which is decisive with regard to the phenomenon of sonoluminescence, i.e. light emission by collapsing bubbles, as it controls the temperature reached inside the bubble at the end of the collapse.
- Vaporization, which requires heat transfer and consequently temperature gradients between the liquid and the bubble and which is the cause of thermal delay in cavitation. This phenomenon is currently referred to as the thermodynamic effect.

The present chapter is devoted to the presentation of the fundamental ideas and classical results concerning these problems. More detailed analyses can be found in books dedicated to specialized aspects, for example LAUTERBORN (1979), TREVENA (1987), LEIGHTON (1994).

5.1. THE EFFECT OF COMPRESSIBILITY

5.1.1. TAIT'S EQUATION OF STATE

In order to account for compressibility effects, it is necessary to have an equation of state for the liquid which takes into account the variation of density with the pressure.

An appropriate law is the barotropic one proposed by TAIT, which is given by:

$$\frac{p+B}{p_0+B}=\left(\frac{\rho}{\rho_0}\right)^n \qquad (5.1)$$

where $1/nB$ is the compressibility coefficient defined by $\frac{1}{\rho}\frac{d\rho}{dp}$ and the subscript 0 refers to normal conditions.

In the case of water, $B = 304.9$ MPa and $n = 7.15$, so that the compressibility coefficient is relatively small and of the order of $4.6 \times 10^{-10}\ \text{Pa}^{-1}$.

It is useful to define the following classical quantities:
- enthalpy h:

$$h = \int_{p_0}^{p} \frac{dp}{\rho} \tag{5.2}$$

- velocity of sound c:

$$c = \sqrt{\frac{dp}{d\rho}} = \sqrt{\rho \frac{dh}{dp}} \tag{5.3}$$

Combined with TAIT's law, these take the following form:

$$h = \frac{{c_0}^2}{n-1}\left[\left(\frac{\rho}{\rho_0}\right)^{n-1} - 1\right] \tag{5.4}$$

$$c = c_0 \left[\frac{\rho}{\rho_0}\right]^{(n-1)/2} \tag{5.5}$$

where the velocity of sound c_0 in normal conditions is given by:

$$c_0 = \sqrt{\frac{n\,(p_0 + B)}{\rho_0}} \tag{5.6}$$

5.1.2. BASIC EQUATIONS

Consider a spherical bubble in an infinite medium. The radial velocity $u(r,t)$, pressure $p(r,t)$ and density $\rho(r,t)$ are governed by the mass conservation and EULER equations:

$$\frac{\partial \rho}{\partial t} + u \frac{\partial \rho}{\partial r} = -\rho \frac{1}{r^2} \frac{\partial (r^2 u)}{\partial r} \tag{5.7}$$

$$\frac{\partial u}{\partial t} + u \frac{\partial u}{\partial r} = -\frac{1}{\rho} \frac{\partial p}{\partial r} \tag{5.8}$$

The law of state of the liquid must be added in order to close the previous system of partial differential equations.

Using the enthalpy defined by equation (5.2), the mass conservation equation and the EULER equation become respectively:

$$\frac{\partial h}{\partial t} + u \frac{\partial h}{\partial r} = -c^2 \frac{1}{r^2} \frac{\partial (r^2 u)}{\partial r} \tag{5.9}$$

$$\frac{\partial u}{\partial t} + u\frac{\partial u}{\partial r} = -\frac{\partial h}{\partial r} \tag{5.10}$$

The velocity potential $\varphi(r,t)$ is defined by:

$$u = \frac{\partial \varphi}{\partial r} \tag{5.11}$$

and the EULER equation has the following first integral, which is a generalized BERNOULLI equation:

$$\frac{\partial \varphi}{\partial t} + \frac{u^2}{2} + h = C(t) \tag{5.12}$$

where $C(t)$ is a time dependant function. We consider here the collapse of a bubble under a constant pressure at infinity equal to p_0, so that C is then a constant which can be set equal to zero by a correct choice of the velocity potential φ.

The elimination of the enthalpy h between equations (5.9) and (5.12) leads to the following hyperbolic equation for the velocity potential φ:

$$(c^2 - u^2)\frac{\partial^2 \varphi}{\partial r^2} - 2u\frac{\partial^2 \varphi}{\partial r\,\partial t} - \frac{\partial^2 \varphi}{\partial t^2} = -\frac{2c^2}{r}\frac{\partial \varphi}{\partial r} \tag{5.13}$$

This equation allows us to compute the solution numerically from the initial conditions.

The characteristic lines are given by:

$$(c^2 - u^2)\,dt^2 - 2u\,dr\,dt - dr^2 = 0 \tag{5.14}$$

i.e.

$$\frac{dr}{dt} = u \pm c \tag{5.15}$$

Hence, the velocity of the characteristic lines can be very different from the velocity of sound since the radial velocity u can reach very high values.

5.1.3. THE QUASI ACOUSTIC SOLUTION [HERRING 1941 & TRILLING 1952]

In this approach, the sound velocity is assumed constant and equal to c_0 and the terms in u^2 and u are neglected on the left-hand side of equation (5.13). This equation then reduces to the classical equation for an acoustic wave:

$$\frac{\partial^2 \varphi}{\partial t^2} = {c_0}^2 \frac{1}{r^2}\frac{\partial}{\partial r}\left(r^2 \frac{\partial \varphi}{\partial r}\right) \tag{5.16}$$

whose solution is a spherical wave propagating outwards from the bubble wall:

$$\varphi = \frac{1}{r} F\left(t - \frac{r}{c_0} \right) \tag{5.17}$$

In this equation, F is an unknown function which characterizes the shape of the wave.

Moreover, the quasi acoustic solution assumes that the density is constant and equal to ρ_0, so that the relation between pressure and enthalpy is simply:

$$h = \frac{p - p_0}{\rho_0} \tag{5.18}$$

On the bubble interface, the generalized BERNOULLI equation (5.12) gives:

$$\frac{1}{R} \dot{F}\left(t - \frac{R(t)}{c_0} \right) + \frac{1}{2}\dot{R}^2 + \frac{P(t) - p_0}{\rho_0} = 0 \tag{5.19}$$

where $\dot{R}$ is the interface velocity and $P(t)$ the liquid pressure on the interface which is known from the boundary conditions. As an example, the pressure P reduces to the sum of the vapor pressure and the partial pressure of any incondensable gases, if viscosity and surface tension are neglected.

The computation of the first term in equation (5.19) needs some further development which are given in the appendix at the end of the chapter. When only the first order terms in $\dot{R}/c$ are considered, the following differential equation for the evolution of the bubble radius is obtained:

$$R\ddot{R}\left(1 - \frac{2\dot{R}}{c_0} \right) + \frac{3\dot{R}^2}{2}\left(1 - \frac{4\dot{R}}{3c_0} \right) = \frac{R}{\rho_0 c_0}\frac{dP}{dt} + \frac{P - p_0}{\rho_0} \tag{5.20}$$

This equation is a generalization of the RAYLEIGH-PLESSET equation which takes into account liquid compressibility.

5.1.4. *THE GILMORE APPROACH* (1952)

GILMORE used the KIRKWOOD-BETHE hypothesis, which was introduced for submarine explosions, according to which the quantity $r\varphi$ propagates with a velocity equal to the sum of the sound velocity and the fluid velocity. The GILMORE approach leads to the following equation of evolution for the bubble radius:

$$R\ddot{R}\left(1 - \frac{\dot{R}}{C} \right) + \frac{3\dot{R}^2}{2}\left(1 - \frac{\dot{R}}{3C} \right) = H\left(1 + \frac{\dot{R}}{C} \right) + \frac{R}{C}\left(1 - \frac{\dot{R}}{C} \right)\frac{dH}{dt} \tag{5.21}$$

In this equation, H and C stand respectively for the liquid enthalpy and the sound velocity on the bubble interface and correspond to the pressure on the bubble wall $P = p(R)$. When density is assumed constant, we simply have:

$$H = \frac{P - p_0}{\rho_0} \tag{5.22}$$

In the case of the collapse of a pure vapor bubble submitted to an external pressure greater than the vapor pressure, the pressure inside the bubble P is constant and equal to the vapor pressure. Equation (5.21) then has the following first integral:

$$\text{Log}\,\frac{R}{R_0} = -2\int_0^{\dot{R}} \frac{\dot{R}\left(\dot{R} - C\right)}{\dot{R}^3 - 3C\dot{R}^2 + 2H\left(\dot{R} + C\right)}\,d\dot{R} \tag{5.23}$$

At the end of the collapse, the interface velocity $\dot{R}$ is negative and tends to infinity. Hence, the terms in $\dot{R}$ in equation (5.23) are dominant and the integral behaves as $\text{Log}\left|\dot{R}\right|$. This leads to the following asymptotic behavior:

$$\left|\dot{R}\right| \cong R^{-1/2} \tag{5.24}$$

Therefore, in the compressible case, the GILMORE approach leads to an exponent $m = 1/2$ for the asymptotic law $\left|\dot{R}\right| \cong R^{-m}$ instead of the value $3/2$ which was obtained in the incompressible case (see § 3.2.2). The collapse velocity appears to be significantly slowed down by compressibility.

On figure 5.1, we present the results of the GILMORE model compared to the RAYLEIGH and HERRING models. Only the very last stages of the collapse are influenced by the liquid compressibility. The lifetime of the bubble is slightly increased as a result. Both the GILMORE and the HERRING and TRILLING models, are similar. The main difference lies in the exponent of the asymptotic law for the interface velocity, the GILMORE model leading to a somewhat less violent collapse. However, let us note that the last stages of the collapse which lead to very high values of the MACH number calculated with the interface velocity are beyond the assumptions of such models.

The first numerical solution which describes the bubble collapse and rebound by taking into account the liquid compressibility was given by HICKLING and PLESSET (1963). The calculation was made on the basis of the GILMORE model for MACH numbers $\dot{R}/C$ smaller than 0.1. For higher values, a Lagrangian form of the EULER equation was used.

The computational results of HICKLING and PLESSET clearly show the emission of a pressure wave from the instant of rebound. The pressure wave propagates outwards and its amplitude decreases approximately as $1/r$. Such steep-fronted waves, which can become shock waves, have been found by many authors. As an example, we present in section 5.4 the pressure distributions in the liquid obtained by FUJIKAWA and AKAMATSU (1980) on the basis of a more complicated model, but which show the same general behavior.

With velocity of the bubble wall in the final stages of collapse approximated by a power law $\left|\dot{R}\right| \cong R^{-m}$, HICKLING and PLESSET found the value 0.785 for the exponent m, whereas HUNTER (1963) found 0.801 from a direct study of the asymptotic behavior of the bubble.

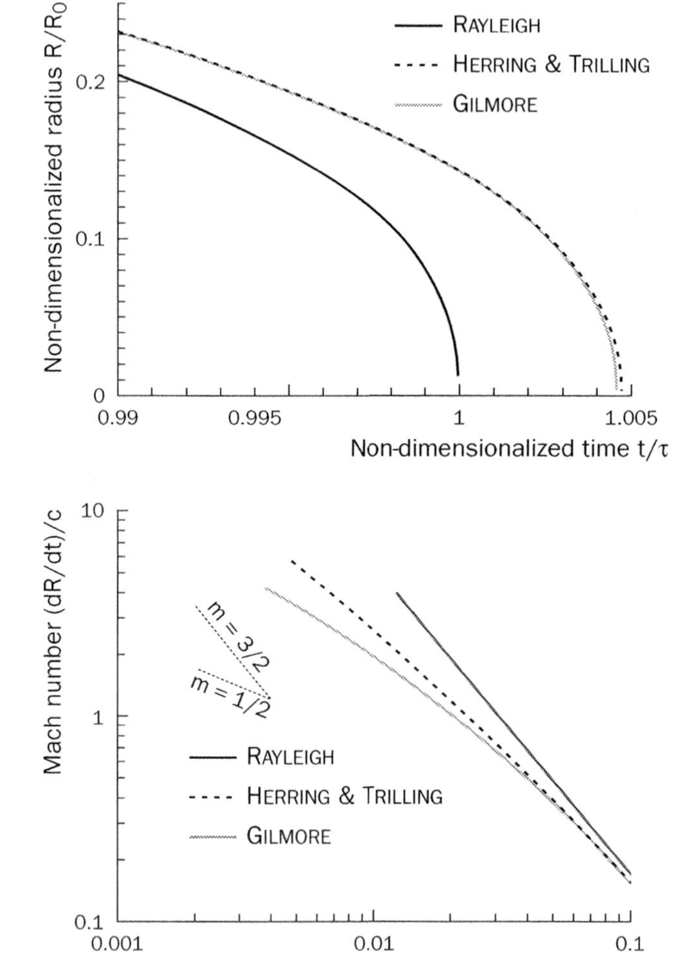

5.1 - Collapse of a 1 cm radius vapor bubble under a constant pressure difference $p_0 - p_v$ equal to 10^5 Pa

Comparison of the solutions of RAYLEIGH, HERRING & TRILLING and GILMORE.

5.2. BUBBLE NOISE

5.2.1. BASIC EQUATIONS

Consider an isolated bubble in an infinite medium of liquid which is at rest at infinity. The time-dependent law of evolution of its radius $R(t)$, or of its volume $\mathcal{V}(t)$, is taken as a solution of the RAYLEIGH equation.

The acoustic solution of the mass and momentum equations, already given in section 5.1.3, corresponds to a spherical wave propagating outwards from the bubble wall. The velocity potential is given by equation (5.17):

$$\varphi = \frac{1}{r} F\left(t - \frac{r}{c_0}\right) \tag{5.25}$$

so that the radial velocity is:

$$u = \frac{\partial \varphi}{\partial r} = -\frac{F}{r^2} - \frac{\dot{F}}{r c_0} \tag{5.26}$$

The pressure is estimated from the generalized BERNOULLI equation (5.12) in which

- the quadratic term in u is neglected according to the classical first acoustic approximation, and
- the enthalpy is estimated from equation (5.18):

$$\frac{p - p_0}{\rho_0} = -\frac{\partial \varphi}{\partial t} = -\frac{\dot{F}}{r} \tag{5.27}$$

The function F is determined by matching this compressible solution with the usual incompressible one, given by equation (3.6):

$$u = \dot{R}\frac{R^2}{r^2} = \frac{\dot{\mathcal{V}}}{4\pi r^2} \tag{5.28}$$

Combining equations (5.28) and (5.26) in which the sound velocity c_0 is made infinite (the incompressible case) allows us to define the function F:

$$F = -\frac{\dot{\mathcal{V}}}{4\pi} \tag{5.29}$$

Consequently, the solution becomes:

$$\begin{aligned} \frac{p(r,t) - p_0}{\rho_0} &= \frac{1}{4\pi r} \ddot{\mathcal{V}}\left(t - \frac{r}{c_0}\right) \\ u(r,t) &= \frac{1}{4\pi r^2} \dot{\mathcal{V}}\left(t - \frac{r}{c_0}\right) + \frac{1}{4\pi r c_0} \ddot{\mathcal{V}}\left(t - \frac{r}{c_0}\right) \end{aligned} \tag{5.30}$$

As far as noise is concerned, the interest lies in the far field. The first term in the second equation then becomes negligible with respect to the second for large enough radius r, so that the solution becomes:

$$u(r,t) \cong \frac{p(r,t) - p_0}{\rho_0 c_0} = \frac{1}{4\pi r c_0} \ddot{\mathcal{V}}\left(t - \frac{r}{c_0}\right) \tag{5.31}$$

Hence, an event produced at time t near the bubble is experienced at the distance r from the bubble, with an attenuation $1/r$ and a time delay r/c_0. Equation (5.31) also shows that the far field pressure fluctuations are proportional to the second derivative of the bubble volume.

From classical acoustics, it is known that the energy flux per unit time and through a unit surface area with outward normal vector $\vec{v}$, is given by:

$$\Phi = \rho_0 \left(h + \frac{u^2}{2} \right) u \vec{r}.\vec{v} \tag{5.32}$$

where $\vec{r}$ is the unit radial vector. As previously mentioned, the quadratic term can be neglected, so that we have:

$$\Phi \cong \rho_0 h u \vec{r}.\vec{v} = \rho_0 c_0 u^2 \vec{r}.\vec{v} \tag{5.33}$$

Taking into account equation (5.31), the flux of acoustic energy through a sphere of radius r is:

$$\frac{dE}{dt} = 4\pi r^2 \Phi \cong \frac{\rho_0 \ddot{\mathcal{V}}^2}{4\pi c_0} \tag{5.34}$$

This behaves as the second derivative of the bubble volume squared.

5.2.2. *WEAK BUBBLE OSCILLATIONS*

We consider the case of a bubble whose radius oscillates around a mean value R_0 with a small amplitude ($\varepsilon << 1$) according to the following law:

$$R(t) = R_0 (1 + \varepsilon \sin \omega t) \tag{5.35}$$

The instantaneous volume of the bubble is given, after linearization, by:

$$\mathcal{V}(t) \cong \frac{4}{3} \pi R_0^3 (1 + 3\varepsilon \sin \omega t) \tag{5.36}$$

The far field is determined by the condition:

$$\left| \frac{\dot{\mathcal{V}}}{r^2} \right| << \left| \frac{\ddot{\mathcal{V}}}{r c_0} \right|$$

which gives:

$$r >> \frac{c_0}{\omega} = \frac{\lambda}{2\pi} \tag{5.37}$$

where λ stands for the wavelength.

For bubbles oscillating at their natural frequency, the wavelength is generally of the order of the millimetre. For example, the natural frequency of a bubble of mean radius $R_0 = 10\ \mu m$ under atmospheric pressure is $f_0 = \omega / 2\pi \cong 340$ kHz (see § 3.4.1) and the corresponding wavelength is about 4 mm. Then, using expression (5.31) for the oscillating pressure in the far field, the pressure fluctuation amplitude is of the order of $\varepsilon \rho_0 \omega^2 R_0^3 / r$. For $\varepsilon = 0.1$, and $r = 1$ m, one finds an amplitude equal to 0.46 Pa or 110 dB.

5.2.3. *NOISE OF A COLLAPSING BUBBLE*

For the final phase of the collapse, the following asymptotic law is assumed (see § 5.1.4):

$$\left|\dot{R}\right| \cong R^{-m} \tag{5.38}$$

with $m = 3/2$ for the collapse in an incompressible fluid and $m \cong 0.8$ in the compressible case.

The second derivative of the volume follows the asymptotic law:

$$\ddot{\mathcal{V}} \cong R^{1-2m} \tag{5.39}$$

Hence, for both values of m, $\ddot{\mathcal{V}}$ tends to infinity. As the velocity and the pressure behave like $\ddot{\mathcal{V}}$, they both undergo very large variations which, in addition, occur during a very small lapse of time due to the rapidity of the collapse. Pressure perturbations due to the collapse and rebound of bubbles can then be regarded as shocks.

The noise emitted by a collapsing bubble is very rich in high harmonics whose frequency exceeds some hundreds of kHz. This is favorable to the detection of cavitation in a noisy environment since the ambient noise can be eliminated without difficulty by a high-pass filter.

Cavitation noise is easily measured using piezo-electric ceramics, whose natural frequencies are usually high. For example, a ceramic of 1 mm in thickness has a natural frequency of the order of 2 MHz. Such ceramics need simply to be stuck on the outer wall of the cavitating flow to pick up cavitation noise.

Generally speaking, in real flows, cavitation noise is due to the succession, at a high rate, of elementary collapses of all kinds of vapor structures such as bubbles or cavitating vortices. For a fixed value of the cavitation parameter σ_v, the noise level generally increases with the flow velocity. For a fixed velocity, when σ_v diminishes from a value corresponding to non-cavitating conditions, noise first appears at cavitation inception, then increases, reaches a maximum and finally decreases. The increase phase corresponds to the growth of isolated bubbles up to a maximum size which increases as the cavitation parameter decreases. As the cavitation number is further decreased, bubbles interact with each other, and their maximum size is smaller. At low enough values of the cavitation parameter, a large number of relatively small bubbles are present and form a kind of continuous cavity, which produces a much smaller noise level.

5.3. SOME THERMAL ASPECTS

5.3.1. THE IDEA OF THERMAL DELAY

Consider the case of a spherical nucleus in a still liquid with temperature T_∞ at infinity. Initially, the nucleus is supposed to be destabilized by a sudden pressure drop at infinity, so that its radius becomes $R(t)$, much larger than its initial radius.

The vaporization process requires the latent heat to be supplied by the liquid to the interface. Heat transfer from the liquid to the bubble is possible only if the temperature T_b inside the bubble is smaller than T_∞. Hence, the vapor pressure inside the bubble $p_v(T_b)$ is also smaller than its value $p_v(T_\infty)$ in the liquid bulk. Consequently, the pressure imbalance between the bubble and the reference point at infinity $p_v(T_b)-p_\infty$ decreases, so that the growth of the bubble is reduced. Equivalently, the reduction in bubble growth can be understood by observing that the cavitation number is increased.

In the practical case of a nucleus traveling through low pressure regions close to hydrofoils or the blades of rotating machinery, if no slip is assumed between the bubble and the liquid, heat transfer is achieved purely by conduction in the liquid. In the case of an attached cavity, it will be shown in chapter 7 that the vaporization of liquid required to feed a partial cavity which continuously sheds vapor structures at its back end involves convective heat transfer at the interface.

In the case of an isolated bubble, an estimate of the heat flux can be obtained considering that, at time t, an order of magnitude of the thermal boundary layer thickness in the liquid is $\sqrt{\alpha_\ell t}$, where $\alpha_\ell = \dfrac{\lambda_\ell}{\rho_\ell c_{p\ell}}$ stands for the thermal diffusivity of the liquid. The conductive heat flux to the interface per unit surface area can then be estimated by means of FOURIER's law:

$$q \cong \lambda_\ell \frac{\Delta T}{\sqrt{\alpha_\ell t}} \tag{5.40}$$

where $\Delta T = (T_\infty - T_b)$. The energy balance is written:

$$q \cdot 4\pi R^2 = \frac{d}{dt}\left(\frac{4}{3}\pi R^3\right)\rho_v L \tag{5.41}$$

where ρ_v is the vapor density and L the latent heat of vaporization.

The energy balance simplifies to:

$$q = \rho_v L \dot{R} \tag{5.42}$$

Using equation (5.40), this yields the following estimate of the temperature difference:

$$\Delta T \cong \frac{\dot{R}\sqrt{t}}{\sqrt{\alpha_\ell}} \frac{\rho_v L}{\rho_\ell c_{p\ell}} \tag{5.43}$$

Assuming that the initial radius of the nucleus is negligible with respect to R, the bubble growth velocity $\dot{R}$ is of the order of R/t, so that the last equation reduces approximately to:

$$\Delta T \cong \frac{R}{\sqrt{\alpha_\ell t}} \frac{\rho_v L}{\rho_\ell c_{p\ell}} \tag{5.44}$$

This rough estimate shows that the evolution law R(t) plays a role, together with other physical parameters, in fixing the temperature difference ΔT. In other words, there is a coupling between the thermal and mechanical aspects of the phenomenon.

Note that the parameter $\frac{\rho_v L}{\rho_\ell c_{p\ell}}$ has dimensions of temperature and depends only upon the fluid temperature.

To estimate the corresponding difference in vapor pressure, the CLAPEYRON-relation:

$$L = T \left[\frac{1}{\rho_v} - \frac{1}{\rho_\ell} \right] \frac{dp_v}{dT} \tag{5.45}$$

is used to get the slope of the vaporization-condensation curve. The vapor density ρ_v is generally negligible in comparison with the liquid density ρ_ℓ, so that the slope of the vapor pressure curve is given by:

$$\frac{dp_v}{dT} \cong \frac{\rho_v L}{T_\infty} \tag{5.46}$$

Hence, the thermodynamic effect in terms of vapor pressure difference is:

$$\Delta p_v = p_v(T_\infty) - p_v(T_b) \cong \frac{dp_v}{dT} \Delta T \cong \frac{\rho_v L \Delta T}{T_\infty} \tag{5.47}$$

From equation (5.44) we get:

$$\Delta p_v \cong \frac{R}{\sqrt{\alpha_\ell t}} \frac{(\rho_v L)^2}{\rho_\ell c_{p\ell} T_\infty} \tag{5.48}$$

BRENNEN introduced the parameter Σ whose units are $m/s^{3/2}$ defined by:

$$\Sigma = \frac{(\rho_v L)^2}{\rho_\ell^2 c_{p\ell} T_\infty \sqrt{\alpha_\ell}} \tag{5.49}$$

For a given fluid, this parameter depends only upon temperature. Using BRENNEN's parameter, equation (5.48) reduces to:

$$\frac{\Delta p_v}{\rho_\ell} \cong \frac{R\Sigma}{\sqrt{t}} \tag{5.50}$$

As an example, consider the growth of a bubble up to a maximum size of 1 mm in a time of 1 ms in water. Two different temperatures at infinity, 20°C and 100°C, are considered. The thermodynamic properties of water are given in the table below. The parameter $\frac{\rho_v L}{\rho_\ell c_{p\ell}}$ is multiplied by about 30 when passing from 20°C to 100°C and the parameter Σ changes by several orders of magnitude. As a consequence, the thermodynamic effect in terms of temperature difference ΔT or vapor pressure difference Δp_v is negligible at 20°C but becomes very significant at 100°C. In general, if the ambient temperature becomes close to the critical temperature, the ΔT and Δp_v-values become higher. This is mainly due to the fact that the vapor density tends to become equal to the liquid density near the critical point.

Water	20°C	100°C
$p_v(T_\infty)$ (Pa)	2,337	101,325
L (kJ/kg)	2,454	2,257
ρ_ℓ (kg/m^3)	998	987
ρ_v (kg/m^3)	0.0173	0.598
$c_{p\ell}$ (J/kg/K)	4,182	4,216
λ_ℓ (W/m/K)	0.60	0.68
$\alpha_\ell = \frac{\lambda_\ell}{\rho_\ell c_{p\ell}}$ (m^2/s)	$1.44\ 10^{-7}$	$1.63\ 10^{-7}$
$\frac{\rho_v L}{\rho_\ell c_{p\ell}}$ (K)	0.0102	0.324
Σ (m/s$^{3/2}$)	3.89	2,944
R (m)	0.001	
t (s)	0.001	
ΔT (K)	0.85	25.4
Δp_v (Pa)	123	91,900

The thermodynamic effect is often discussed in terms of the B-factor introduced originally by STEPANOFF (1961). The factor $1/B$ is defined as the ratio of the volume of liquid ϑ_ℓ to be cooled by $\Delta T = T_\infty - T_b$ to supply the heat necessary for the production of a given volume ϑ_v of vapor. The heat balance is written:

$$\rho_v \vartheta_v L = \rho_\ell \vartheta_\ell c_{p\ell} \Delta T \tag{5.51}$$

so that

$$B = \frac{\vartheta_v}{\vartheta_\ell} = \frac{\rho_\ell c_{p\ell}}{\rho_v L} \Delta T \cong \frac{\rho_\ell c_{p\ell} T_\infty}{\rho_v L^2} \Delta p_v \tag{5.52}$$

In the present case of the growth of a bubble, equation (5.44) shows that the B-factor is given by:

$$B \cong \frac{R}{\sqrt{\alpha_\ell t}} \tag{5.53}$$

5.3.2. *BRENNEN'S ANALYSIS* (1973)

A complete computation of the evolution of a spherical bubble taking into account the thermal effects requires coupling of the RAYLEIGH equation to the heat diffusion equation in the liquid, via an energy balance for the bubble. As a matter of fact, the pressure difference which appears in the RAYLEIGH equation and which is the driving parameter for the bubble evolution depends upon the temperature, whose determination requires the heat diffusion equation to be solved. This problem has been approached by PLESSET and ZWICK (1952).

A simplified and mainly dimensional approach is presented below. The objective is to point out the key parameters which allow us to ascertain if the bubble evolution is principally controlled by inertia or thermal effects.

Consider for simplicity the RAYLEIGH equation for a pure vapor bubble in an inviscid fluid without surface tension:

$$R\ddot{R} + \frac{3}{2}\dot{R}^2 = \frac{p_v(T_b) - p_\infty}{\rho_\ell} \tag{5.54}$$

This equation can be rewritten in the following form:

$$\left(R\ddot{R} + \frac{3}{2}\dot{R}^2\right) + \frac{\Delta p_v}{\rho_\ell} = \frac{p_v(T_\infty) - p_\infty}{\rho_\ell} \tag{5.55}$$

in which the last term on the left hand side is the thermal term.

Using equations (5.43), (5.47) and the definition (5.49) of the Σ parameter, the RAYLEIGH equation (including thermal effects) becomes:

$$\left(R\ddot{R} + \frac{3}{2}\dot{R}^2\right) + \Sigma \dot{R}\sqrt{t} \cong \frac{p_v(T_\infty) - p_\infty}{\rho_\ell} \tag{5.56}$$

This equation is rewritten in a non-dimensional form in a similar way to that already used in section 3.6. Let a be the characteristic scale of the bubble radius. Two characteristic time scales can be defined:

– a pressure time τ_p defined as:

$$\tau_p = a\sqrt{\frac{\rho_\ell}{p_v(T_\infty) - p_\infty}} \tag{5.57}$$

– and a "thermal" time τ_T defined as:

$$\tau_T = \left(\frac{a}{\Sigma}\right)^{2/3} \tag{5.58}$$

By introducing the following non-dimensional variables:

$$\begin{cases} \overline{R} = \dfrac{R}{a} \\ \overline{t} = \dfrac{t}{\tau_p} \end{cases} \tag{5.59}$$

equation (5.56) takes the non-dimensional form:

$$\left(\overline{R}\ddot{\overline{R}} + \frac{3}{2}\dot{\overline{R}}^2\right) + \left(\frac{\tau_p}{\tau_T}\right)^{3/2} \dot{\overline{R}}\sqrt{\overline{t}} = 1 \tag{5.60}$$

The last term on the left hand side is still the thermal term. Initially, it is zero and $\dot{\overline{R}}$ is then of the order of unity. The thermal term grows as $\sqrt{\overline{t}}$ and remains negligible as long as $\left(\dfrac{\tau_p}{\tau_T}\right)^{3/2} \sqrt{\overline{t}} \ll 1$, i.e.:

$$\overline{t} \ll \left(\frac{\tau_T}{\tau_p}\right)^3 \tag{5.61}$$

The dimensional form of the previous condition is the following:

$$t \ll \frac{\tau_T^3}{\tau_p^2} = \frac{p_v(T_\infty) - p_\infty}{\rho_\ell \Sigma^2} \tag{5.62}$$

In the practical case of a nucleus moving through the low pressure region generated by a hydrofoil or a blade, the available time for the bubble growth is the transit time D/V and this condition becomes:

$$\frac{D}{V} \ll \frac{p_v(T_\infty) - p_{min}}{\rho_\ell \Sigma^2} \tag{5.63}$$

where V is the flow velocity, D the characteristic size of the low pressure zone and p_{min} the minimum pressure. The previous condition can be rewritten as follows:

$$\Sigma << \sqrt{\frac{(-\sigma - Cp_{min})V^3}{2D}} \tag{5.64}$$

As the Σ parameter for a given fluid increases with temperature, this last condition defines a critical temperature above which thermal effects become predominant.

Figure 5.2, due to BRENNEN, shows the evolution of the thermodynamic function Σ(T) for water, liquid oxygen and liquid hydrogen. The temperature is made non-dimensional by using the difference between the temperatures at the critical point and the triple point as a reference.

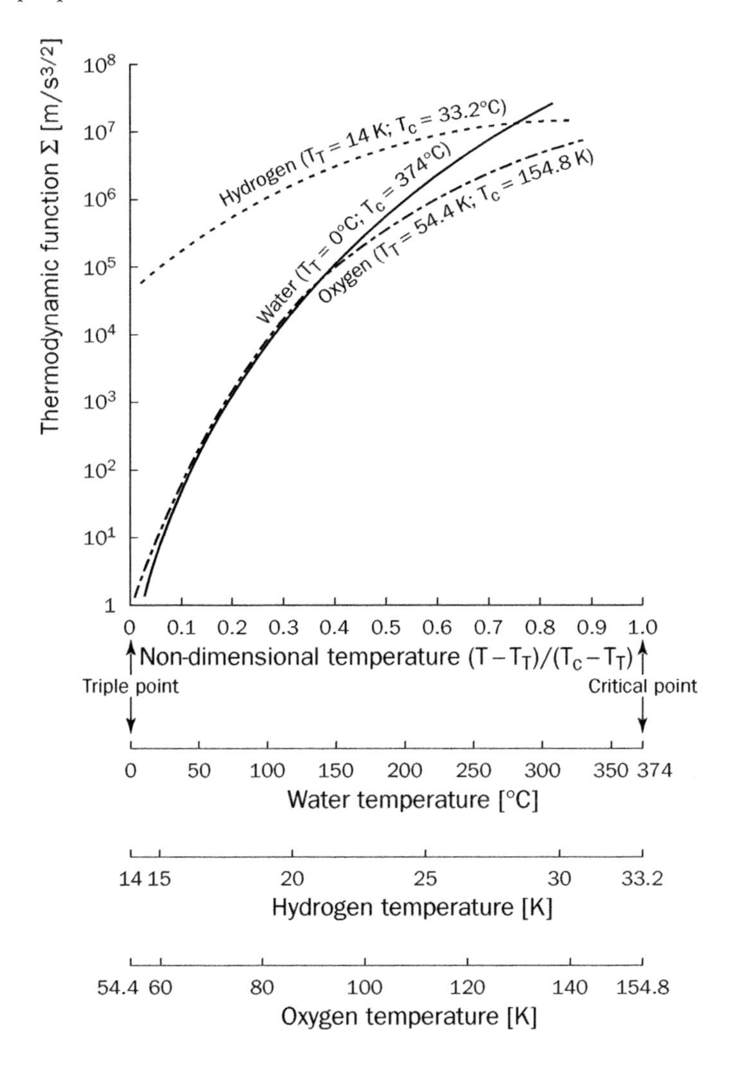

5.2 - The thermodynamic function of BRENNEN

Take, for example, a flow with a velocity of 10 m/s around a foil 10 cm in chordlength. With a cavitation number of 0.50 and a minimum pressure coefficient of –1, the right hand side of the inequality (5.64) is then $50 \text{ m}/\text{s}^{3/2}$. Furthermore, if we consider a non-dimensional temperature of 0.05 corresponding to 18.7°C for water, 59.4 K for liquid oxygen and 15 K for liquid hydrogen, then according to BRENNEN's criterion, figure 5.2 shows the thermal effects to be negligible in water and liquid oxygen, while liquid hydrogen behaves as a "hot" liquid, for which thermal delays are important.

5.4. A TYPICAL NUMERICAL SOLUTION

As the number and complexity of the physical phenomena taken into account in the description of the bubble evolution increases, it becomes necessary to develop numerical solutions. We present here the main conclusions of the numerical study of the collapse and rebound of a bubble undertaken by FUJIKAWA and AKAMATSU (1980).

In addition to liquid compressibility and conductive heat transfer outside and inside the bubble, FUJIKAWA and AKAMATSU take into account a finite rate of condensation or evaporation determined at a molecular level. This implies a thin but finite interface between liquid and vapor and a non-equilibrium in temperature across the interface, which is the driving mechanism for phase change. They compare their results with the case of thermodynamic equilibrium which corresponds to an infinite rate of phase change, assuming in addition an adiabatic transformation of the air contained in the bubble for this reference case.

Figure 5.3 presents the time evolution of the bubble radius. The first collapse appears well-approximated by the simple RAYLEIGH collapse. When compared to the adiabatic and equilibrium case, the maximum radius of the rebounding bubble appears damped by the effects of liquid compressibility, heat conduction and finite rate of evaporation or condensation. The maximum computed temperature at the bubble centre is 6,700 K. It is much lower than that for the adiabatic collapse which reaches almost 8,800 K. The temperature inside the bubble is not uniform because of heat conduction. Such high temperatures are usually considered as the origin of luminescence i.e. the emission of light pulses by collapsing bubbles.

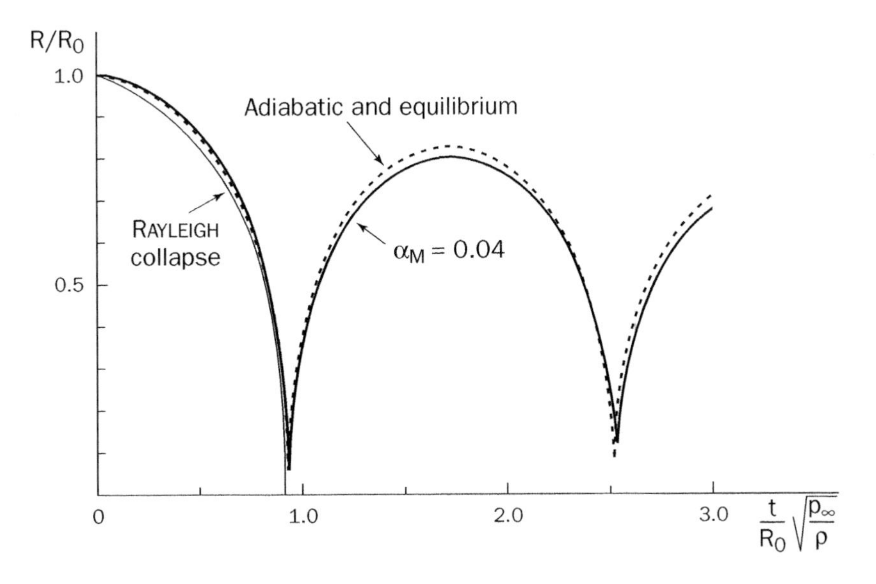

5.3 - Time evolution of the radius of a bubble (initial radius 1 mm) evolving under a pressure at infinity $p_\infty = 0.7025$ atm and containing air initially at the pressure $p_{g0} = 0.01\,p_\infty$

The accommodation coefficient α_M which controls the rate of phase change and which is defined as the ratio of the number of vapor molecules sticking to the interface to those impinging on it, is assumed constant and equal to 0.04. The temperature of the liquid at infinity is 293.15 K. The dashed line indicates the path for a bubble considered under adiabatic and equilibrium conditions (ignoring phase change kinetics).

Figure 5.4 shows the pressure distribution in the liquid during collapse and rebound. As in the incompressible case corresponding to the classical RAYLEIGH solution, a pressure maximum arises and propagates from infinity towards the bubble wall during collapse. At rebound, a pressure wave forms and propagates outwards at about the velocity of sound. Its fronts gradually steepen and the maximum pressure decreases as $1/r$.

Surprisingly, the computed evolution of pure vapor bubbles is qualitatively similar to the previous case of vapor bubbles also containing air. FUJIKAWA and AKAMATSU show that, if at the initial stages of the collapse, the vapor condenses back into the liquid, at the final stages conversely, the collapse is so rapid that most of the vapor does not have enough time to condense in comparison with the time required by the kinetics of phase change. To some extent, this remaining vapor behaves like an incondensable gas which, hence, will cause rebound and generate an outward pressure wave.

The time is non-dimensionalized using the reference time $\frac{1}{R_0}\sqrt{\frac{p_\infty}{\rho}}$. *Same conditions as in figure 5.3. The dotted lines correspond to the adiabatic and equilibrium reference case whereas the plain lines correspond to the complete calculation.*

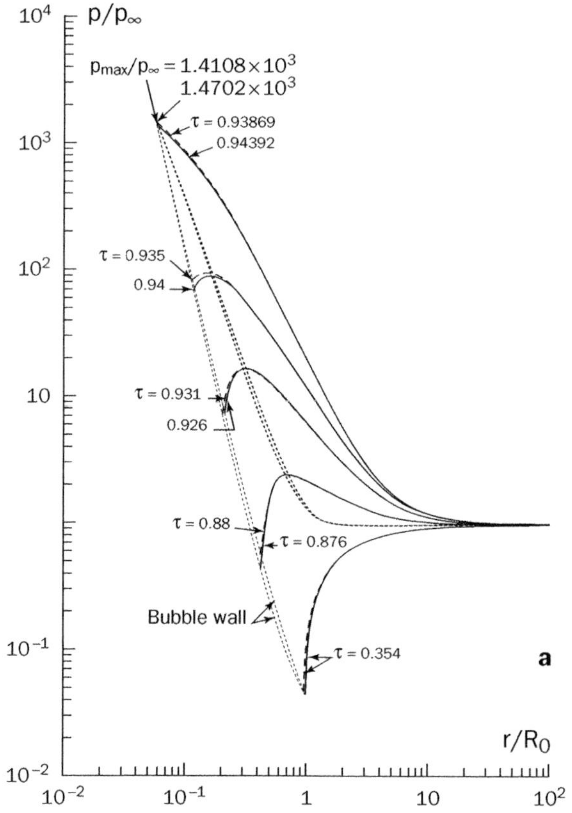

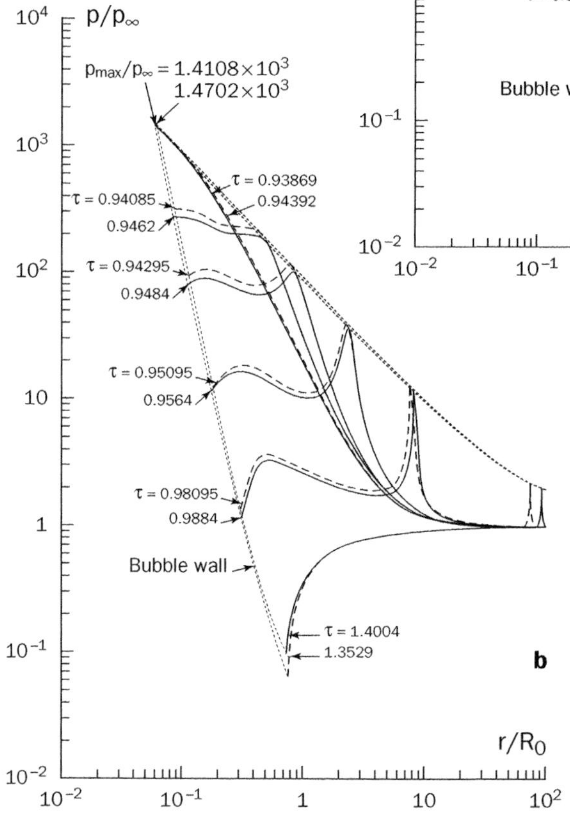

5.4 - **Pressure distributions in the liquid (a) during collapse, and (b) during rebound, at different non-dimensional times τ from the start of the collapse**

REFERENCES

BONNIN J. –1973– Thermodynamic parameters involved in boiling and cavitation. *ASME Polyphase Flow Forum,* Atlanta (USA), June 1973.

BRENNEN C. –1973– The dynamic behavior and compliance of a stream of cavitation bubbles. *J. Fluids Eng.* **95**, series I, 533-541.

FORSTER H.K. & ZUBER N. –1954– Growth of a vapor bubble in a superheated liquid. *J. Appl. Phys.* **25**, 474-478.

FUJIKAWA S. & AKAMATSU T. –1980– Effects of non-equilibrium condensation of vapor on the pressure wave produced by the collapse of a bubble in a liquid. *J. Fluid Mech.* **97**, part 3, 481-512.

GILMORE F.R. –1952– The growth or collapse of a spherical bubble in a viscous compressible liquid. *Cal. Inst. Techn. Hydro. Labo.,* Rpt 26-4.

HERRING G. –1941– Theory of the pulsations of the gas bubble produced by an underwater explosion. *OSRD,* Rpt 236.

HICKLING R. & PLESSET M.S. –1964– Collapse and rebound of a spherical bubble in water. *Phys. Fluids* **7**, 7-14.

HUNTER C. –1963– Similarity solutions for the flow into a cavity. *J. Fluid Mech.* **15**, 289.

LAUTERBORN W. (ed.) –1980– Cavitation and inhomogeneities in underwater acoustics. *Proc. 1st Int. Conf.,* Göttingen (Germany), Springer-Verlag.

LEIGHTON T.G. –1994– The acoustic bubble. *Academic Press Inc.*

PLESSET M.S. & ZWICK S.A. –1952– A non-steady heat diffusion problem with spherical symmetry. *J. Appl. Phys.* **23**, 95-98.

PLESSET M.S. & HSIEH D.H. –1960– Theory of gas bubble dynamics in oscillating pressure fields. *Physics of Fluids* **3**, 882-892.

PROSPERETTI A. –1977– Thermal effects and damping mechanisms in the forced radial oscillations of gas bubbles in liquids. *J. Acoust. Soc. Am.* **61**(1), 17-27.

SCHNEIDER A.J.R. –1949– Some compressibility effects in cavitation bubble dynamics. *PhD Thesis, Cal. Inst. Techn.*

STEPANOFF A.J. –1961– Cavitation in centrifugal pumps with liquids other than water. *J. Eng. Power* **83**, 79-90.

TREVENA D.H. –1987– Cavitation and tension in liquids. *Adam Hilger Ed.,* Bristol (England).

TRILLING L. –1952– The collapse and rebound of a gas bubble. *J. Appl. Phys.* **23**, 14.

ZWICK S.A. & PLESSET M.S. –1954– On the dynamics of small vapor bubbles in liquids. *J. Math. Phys.* **33**, 308.

APPENDIX TO SECTION 5.1.3

The interface velocity is given by:

$$\dot{R} = -\frac{F}{R^2} - \frac{\dot{F}}{cR} \tag{5.65}$$

where we note:

$$F = F\left(t - \frac{R(t)}{c}\right)$$
$$\dot{F} = \dot{F}\left(t - \frac{R(t)}{c}\right) \tag{5.66}$$

so that we have:

$$R\dot{R} = -\frac{F}{R} - \frac{\dot{F}}{c} \tag{5.67}$$

Derivation of this equation with respect to time gives:

$$\dot{R}^2 + R\ddot{R} = \frac{\dot{R}F}{R^2} - \left(1 - \frac{\dot{R}}{c}\right)\left(\frac{\dot{F}}{R} + \frac{\ddot{F}}{c}\right) \tag{5.68}$$

The second derivative $\ddot{F}$ is estimated by deriving equation (5.19).

$$-\frac{\dot{R}F}{R^2} + \left(1 - \frac{\dot{R}}{c}\right)\frac{\dot{F}}{R} + \dot{R}\ddot{R} + \frac{\dot{P}}{\rho} = 0 \tag{5.69}$$

Here $\dot{P}$ is the time derivative of the liquid pressure on the bubble interface.

If the expression of $\ddot{F}$ deduced from equation (5.69) is substituted into equation (5.68), we obtain:

$$\dot{R}^2 + R\ddot{R} = \frac{\dot{R}F}{R^2} - \frac{\dot{F}}{R} + \frac{R\dot{R}\ddot{R}}{c} - \frac{R\dot{P}}{\rho c} \tag{5.70}$$

In the previous equation, F is eliminated using equation (5.67). The final result is:

$$\left(1 + \frac{\dot{R}}{c}\right)\frac{\dot{F}}{R} = -2\dot{R}^2 - R\ddot{R} + \frac{R\dot{R}\ddot{R}}{c} - \frac{R\dot{P}}{\rho c} \tag{5.71}$$

This equation allows us to compute $\dot{F}$. When only the first order terms in $\dot{R}/c$ are kept, we obtain:

$$\frac{\dot{F}}{R} \cong -2\dot{R}^2\left(1 - \frac{\dot{R}}{c}\right) - R\ddot{R}\left(1 - \frac{2\dot{R}}{c}\right) + \frac{R\dot{P}}{\rho c} \tag{5.72}$$

This expression for $\dot{F}$ is introduced in equation (5.19) to give equation (5.20).

6. SUPERCAVITATION

As the cavitation parameter is decreased, a small cavity attached to a hydrofoil will extend and grow longer and longer. It becomes a supercavity as soon as it ceases to close on the cavitator wall but inside the liquid, downstream of the cavitator. Simultaneously, the lift of the hydrofoil decreases while its drag increases.

6.1 - Supercavity behind a two-dimensional NACA 16012 hydrofoil
(REYNOLDS number 10^6, cavitation parameter 0.07, angle of attack 17 deg.)

For very high relative velocities between the liquid and the body, it is practically impossible to use non-cavitating foils, such as the conventional ones used in aerodynamics. In such cases, different types of supercavitating foils have been designed for better efficiency, such as truncated foils with a base cavity or supercavitating foils with non-wetted uppersides.

This chapter begins with a presentation of the main physical aspects of supercavities (§ 6.1). Although the background of applications was chosen rather on the side of two-dimensional, lifting bodies, most of the features are applicable to axisymmetric supercavities. After a section devoted to the basis of flow modeling (§ 6.2), some typical results are given in section 6.3. The case of axisymmetric supercavities is considered at the end of the chapter (§ 6.4).

6.1. PHYSICAL ASPECTS OF SUPERCAVITIES

6.1.1. CAVITY PRESSURE

As an example, consider a supercavity attached to a two-dimensional foil, as schematically shown in figure 6.2. The cavity is made of a mixture of vapor and non-condensable gas, so that the pressure inside is:

$$p_c = p_v + p_g \tag{6.1}$$

In this equation, as usual, p_v stands for the vapor pressure and p_g, for the partial gas pressure. The cavity pressure p_c is generally considered as constant in time and uniform throughout the cavity.

The presence of gas in the cavity is due to diffusion through the interface of gases dissolved in the liquid. If the concentration of gas at saturation in the liquid is large, the partial pressure of gas in the cavity will be large. On the contrary, with deaerated water at room temperature, which is usually the case in hydrodynamic tunnels, the value of p_g is small with respect to p_v. Another case which deserves attention is ventilated cavities for which the gas pressure can be high, depending mainly on the ambient pressure and the injection flowrate. Ventilated cavities will be considered more especially in chapter 9.

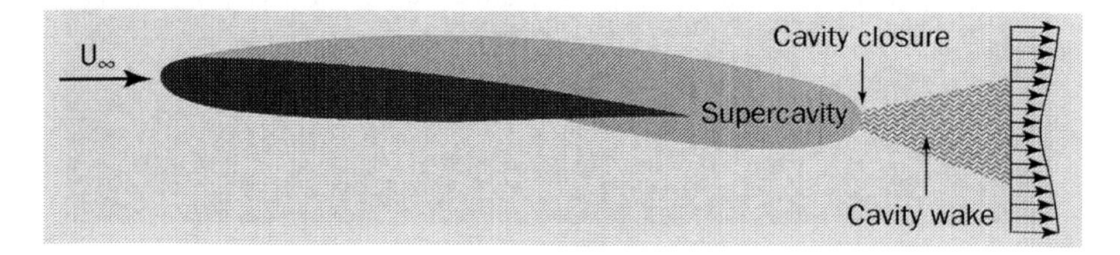

6.2 - Scheme of a supercavitating flow

The tangential shear stress on the cavity interface is due to the friction between the vapor-gas layers of the cavity and the external flow. Its value is generally small since the density of the fluid inside the cavity is much smaller than that of the liquid. Hence, the shear stress on the cavity is usually negligible.

6.1.2. CAVITY DETACHMENT

If no special element, such as a step or a sharp edge, is designed to fix the detachment point of the cavity (fig. 6.3), the position of detachment is unknown *a priori* and a detachment criterion is necessary to predict the location of the cavity detachment. Two main criteria are available.

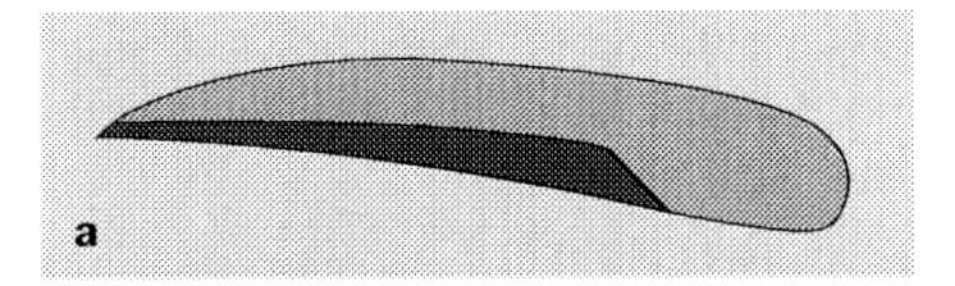

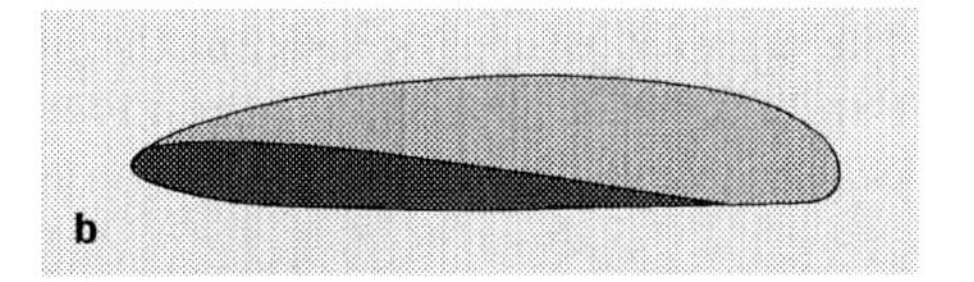

6.3 - Detachment of a supercavity
(a) detachment at a geometrical singularity - (b) detachment from a smooth wall

Villat-Armstrong criterion

The VILLAT (1911) and ARMSTRONG (1953) criterion was established on the basis of inviscid two-dimensional flow theory. Within the framework of this theory, the cavity must detach tangentially to the solid wall (fig. 6.4). If not, the velocity on the free streamline would tend to zero at detachment, which is incompatible with the condition of constant pressure or constant velocity along the whole free streamline.

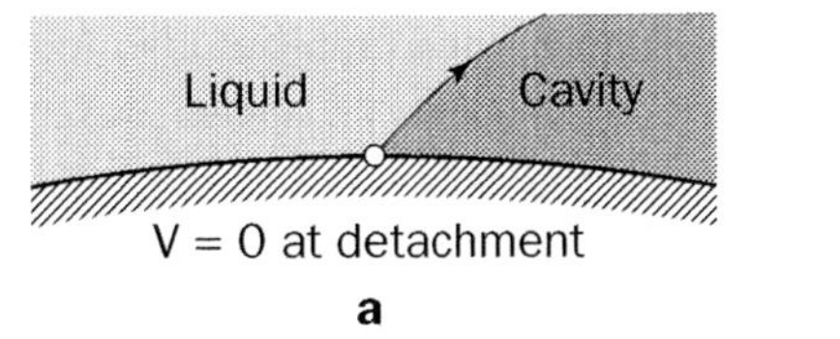

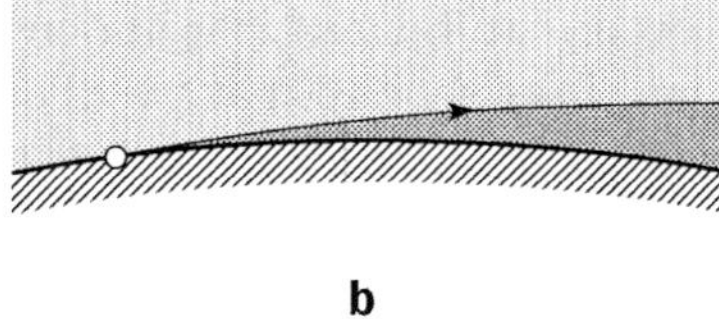

6.4 - Cavity detachment
(a) non-tangential detachment - (b) tangential detachment

The VILLAT-ARMSTRONG criterion assumes that the cavity is a zone of minimum pressure for the whole flowfield. According to this criterion, the detachment point is then the point of minimum pressure on the wall. Its determination is often calculated via an iterative procedure. Initially, it can be chosen at the minimum pressure point taken from non-cavitating conditions. However, this initial guess should be corrected to take into account the change in the pressure distribution due to the development of cavitation and to make sure that the detachment point is actually a point of minimum pressure of the supercavity flow.

The VILLAT-ARMSTRONG criterion has a local interpretation. It is equivalent to assuming that the position of detachment is determined by the non-singular behavior of the mathematical solution. Indeed, for any position of the detachment point, the change in boundary condition from a slip condition on the wetted wall to a constant pressure condition on the cavity, induces a velocity singularity at detachment. The smooth detachment criterion of VILLAT and ARMSTRONG consists in selecting the particular detachment point which removes this mathematical singularity and allows firstly the cavity line curvature to be regular and equal to the solid body curvature at detachment (VILLAT) and secondly the pressure gradient to be zero at this point, as it is on the cavity line (ARMSTRONG) (see appendix).

The criterion is usually used in the numerical modeling of cavity flows. Being a purely inviscid approach, it does not account for viscous effects and ignores the behavior of the upstream boundary layer, particularly separation.

Laminar separation criterion

From the examination of cavitation inception on spheres, circular cylinders and ogives, ARAKERI and ACOSTA proved that viscous effects are predominant in cavitation inception. They showed that laminar separation on a wall provides a site for inception and that the elimination of an existing laminar separation by stimulating the boundary layer with a trip has a strong effect on inception. From those experiments, ARAKERI (1975) proposed a criterion to predict the position of detachment of a cavity. This criterion also takes into account the local effects of surface tension which obliges the free streamline to be non-tangential to the body at detachment (see fig. 6.5).

FRANC and MICHEL (1985) confirmed this viscous effect in the case of hydrofoils and proved that it is relevant not only for cavitation inception but also for supercavitation. They showed that a well developed cavity always detaches downstream of laminar separation of the boundary layer. The existence of separation, which generates a relatively dead zone, is the only opportunity for a cavity to remain attached to the wall and to be sheltered from the incoming flow. If the boundary layer does not separate, the cavity is swept away by the flow and cannot attach to the smooth wall.

As a corollary, transition to turbulence, by allowing the boundary layer to overcome the adverse pressure gradient without separating, prevents steady cavities attaching to a wall. FRANC and MICHEL also showed that, if a cavity is attached while the fully wetted flow does not separate, it is because the changes in the pressure distribution generated by the development of the cavity forces the boundary layer to separate just upstream the cavity.

In conclusion, the laminar separation criterion assumes that a supercavity detaches at the laminar separation point of the boundary layer. The position of laminar separation, and even its existence in some cases, may strongly differ from the case of the fully wetted flow because of the changes in pressure distribution due to developed cavitation.

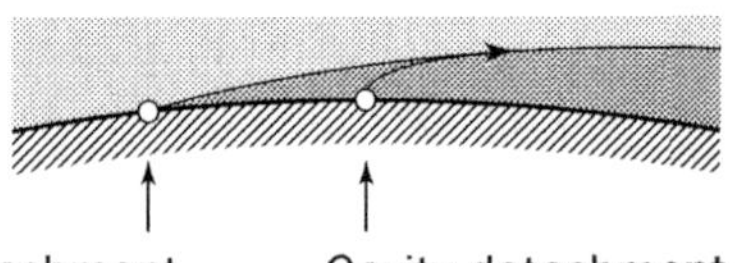

6.5
Local effect of surface tension on cavity detachment

As surface tension forces the cavity to bend locally (fig. 6.5), cavity detachment actually takes place slightly downstream of laminar separation. ARAKERI (1975) proposed a correlation to estimate the actual distance between boundary layer separation and cavity detachment (see § 8.1.4).

In most cases and in particular for hydrofoils at large angles of attack, the minimum pressure point and the laminar separation point are close to the leading edge and so close each other that the laminar separation criterion leads almost to the same prediction as the VILLAT-ARMSTRONG criterion.

It is only in a few special cases that both criteria do not agree. For example, in the case of hydrofoils at very low angle of attack, the minimum pressure point is located close to the point of maximum thickness, whereas the laminar separation point is located in the rear. Only the laminar separation criterion is able to account for the cavities which detach in the aft part of a foil at low angles of attack (see chap. 8).

As a boundary layer requires an adverse pressure gradient to separate, it is clear that the detachment point calculated via the laminar separation criterion is different from the point of minimum pressure, which is located somewhat upstream of laminar separation. The liquid then experiences pressure levels lower than the vapor pressure without phase change upstream of detachment. From a physical viewpoint, the liquid is in a metastable state as described in section 1.1.2. This is possible only if no nucleus is activated by the minimum pressure, i.e. if their critical pressure is below the minimum pressure. If not, traveling bubble cavitation will develop upstream of the attached cavity and interact with it (see chap. 8).

6.1.3. CAVITY CLOSURE

As the cavity pressure p_c is lower than the surrounding pressure, the balance of inertia and pressure forces gives a curvature oriented towards the cavity, as can be seen from the localized EULER equation:

$$\frac{V^2}{R}\vec{n} = -\frac{1}{\rho}\,\vec{\text{grad}}\,p \tag{6.2}$$

where $\vec{n}$ is the unit vector normal to the cavity and directed inwards.

This obliges the liquid to penetrate into the cavity and to form a re-entrant jet (fig. 6.6).

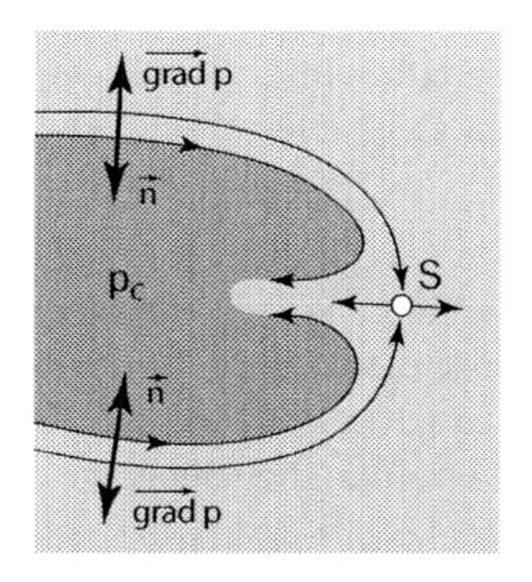

6.6
Cavity closure and re-entrant jet

In an ideal steady flow, the starting point S of the jet should be a stagnation point, where the local pressure coefficient should be unity. In fact, even for a globally steady cavity flow, the stagnation point is highly unstable [GILBARG *et al.* 1950], so that the maximum value of the mean pressure coefficient at the rear of the cavity is actually far from unity and does not exceed in fact about 0.1 [MICHEL 1973, 1977].

The instability affects the re-entrant jet and the whole closure region. Two main regimes occur and alternate continuously with each other:

- the re-entrant jet actually develops and tends to confine the gas and vapor mixture inside the cavity;
- there is an emission of limited coherent trains of alternate vortices which take off gas and vapor from the cavity and entrain any excess liquid.

Thus the rear part of the cavity alternately plays the role of a valve and a pump. On the whole, the suction or pumping effect is dominant and constitutes the driving phenomenon for vaporization at the cavity interface. Vaporization, which takes place mainly at the front part of the cavity, continuously feeds the cavity with vapor and counterbalances the amount of vapor entrained at the rear.

In the wake of the cavity, i.e. just downstream of the attached cavity, the flow contains many bubbles which are released from the cavity and appear more or less entrapped in the core of alternate vortices. This region is always highly turbulent. It may happen that large and smooth supercavities are formed behind small sized bodies, on which the boundary layer remains laminar. Turbulence in the wake is then due to the instability of the cavity closure.

6.1.4. *CAVITY LENGTH*

The length of a supercavity is one of the most important parameters of the cavity flow. It is measured from detachment to closure and may be affected, experimentally, by large uncertainties due to the instability of the closure region.

The length of a supercavity increases when the relative cavity underpressure σ_c decreases. This is easily understandable since, in that case, the pressure difference between the reference point and the cavity decreases, resulting in smaller pressure gradients in the whole flowfield except in the vicinity of cavity closure. Then the streamlines tend to have a smaller curvature and to become closer to straight lines parallel to the upstream velocity.

In many cases, it is possible to model the experimental dependence of the cavity length with the cavity underpressure, for low values of that parameter, by a power law:

$$\frac{\ell}{c} \cong A\sigma^{-n} \tag{6.3}$$

where c stands for a characteristic size of the body. Throughout this chapter, we use σ to define the cavity underpressure rather than σ_c. The exponent n is found equal to 2 if the body is located in an infinite medium (see eq. 6.27). As for the A-values, they depend on the body shape and position.

As an example, figure 6.7 presents the evolution of the non-dimensionalized cavity length with the relative underpressure of the cavity in logarithmic coordinates for a symmetrical wedge in a free surface channel [MICHEL 1973] for three values of the submersion depth h.

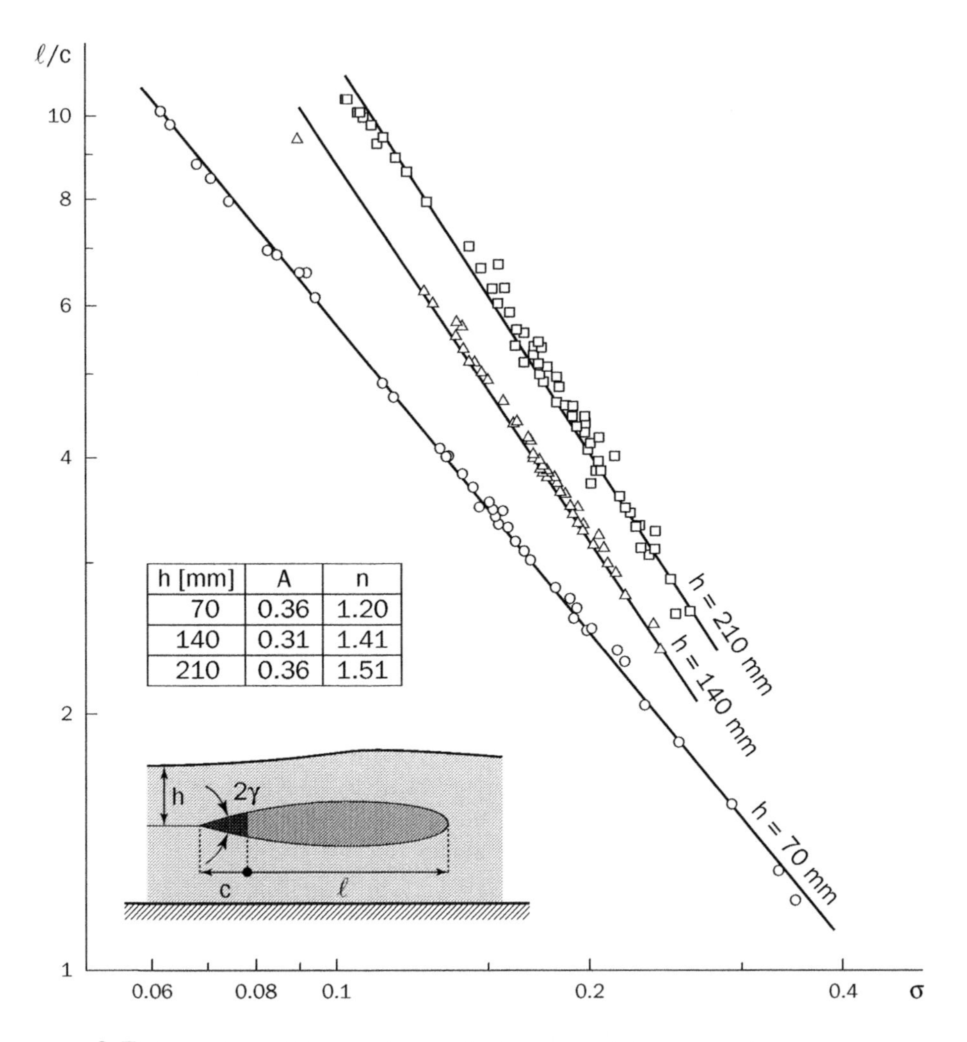

6.7 - Evolution of the relative cavity length with the cavitation number for three values of the submersion depth (vapor cavities)

Case of a symmetrical wedge of chord $c = 60$ mm, base height 17 mm and vertex angle $2\gamma = 16$ deg., in a free surface channel of height 280 mm. The three values of the submersion depth h are shown [from MICHEL, 1973].

The smaller the submersion depth h, the smaller is the exponent n. For small values of the submersion depth h, the exponent n approaches unity. In that case, the liquid flow above the cavity can be considered as a circular jet submitted to pressures p_0 and p_c on its two faces, where p_0 is the pressure on the free surface and p_c the cavity pressure. The normal EULER equation gives:

$$\frac{p_0 - p_c}{h} \approx \frac{\rho V^2}{R} \tag{6.4}$$

where R is the mean radius of curvature of the circular jet. Assuming that the cavity free streamline originates from the body at an angle γ equal to half the wedge vertex angle (fig. 6.7), we obtain, from purely geometrical considerations:

$$\frac{\ell}{2R} \approx \gamma \tag{6.5}$$

By combining equations (6.4) and (6.5), we get:

$$\frac{\ell}{h} \approx \frac{4\gamma}{\sigma} \tag{6.6}$$

Equation (6.6) shows that the cavity length varies as σ^{-1} for small submersion depths.

By using a linearized theory with the TULIN wake model, ROWE and MICHEL (1975) obtained variations of the cavity length which follow a law of the type $\ell / c = A\sigma^{-n}$, with n-values close to experimental results but with A-values about 2.4 times too small. Meanwhile, the calculated cavity was too thick, for the same value of σ, with respect to the experimental shape. By modifying the wake velocity distribution, a better agreement was found between theory and experiment for both the length and the shape of the cavity [MICHEL 1977].

Blockage

Contrary to the previous cases, in a flow confined by solid walls, the cavity becomes infinite for a value of the cavity pressure p_c smaller than the pressure at infinity p_∞ i.e. for a non-zero value of the cavitation parameter (fig. 6.8-b). This phenomenon corresponds to blockage of the flow.

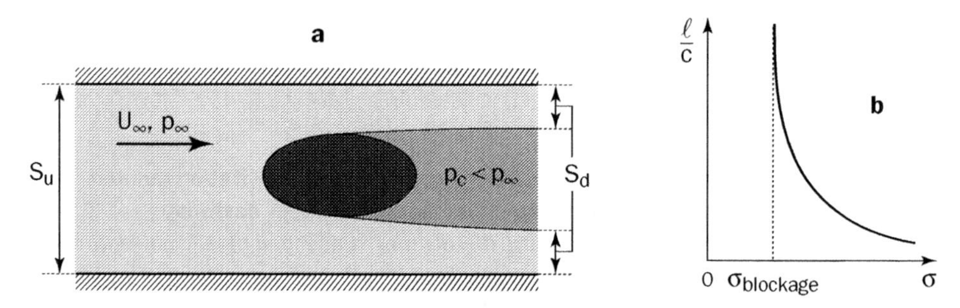

6.8 - **Illustration of blockage in supercavitating flows**

From the mass conservation and the BERNOULLI equations, the critical value of the cavitation parameter corresponding to blockage can be estimated by:

$$\sigma_{blockage} = \frac{S_u^{\,2}}{S_d^{\,2}} - 1 \tag{6.7}$$

where S_u and S_d stand respectively for the cross-sectional areas of the upstream and downstream regions of the liquid flow (fig. 6.8-a).

For a symmetrical wedge of chord c and small vertex angle 2γ, in the center of a channel of height H, the linearized theory [COHEN *et al.* 1958] gives:

$$\frac{\sigma_{\text{blockage}}}{1+\dfrac{\sigma_{\text{blockage}}}{2}} \cong \frac{4\gamma}{\pi} \operatorname{ch}^{-1}\left[e^{\pi \frac{c}{H}}\right] \tag{6.8}$$

This equation agrees fairly well with experiments. For example, for $c/H = 0.2$ and $2\gamma = 16$ deg., the theoretical value of the blockage cavitation number σ_{blockage} is 0.250 whereas the experimental value is 0.254 [MICHEL 1973].

From an experimental viewpoint, the closeness of blockage conditions results in a critical behavior of the flow, since small pressure variations induce large variations in cavity length.

6.2. SUPERCAVITY FLOW MODELING USING STEADY POTENTIAL FLOW THEORY

In this section, attention is particularly focused on the hydrodynamics of cavity flows disregarding viscous and surface tension effects. Viscous effects can easily be taken into account in a second step by conducting a classical boundary layer calculation on the wetted walls, on the basis of the pressure distribution obtained from potential flow modeling. Such a procedure is necessary if the cavity detachment point is unknown and if the laminar separation criterion is used for its determination. This requires an iterative process which will be discussed in chapter 8. In the present section, the position of the detachment point is supposed to be known, a priori, for simplicity.

6.2.1. THE MAIN PARAMETERS

As an example, consider the case of a supercavitating steady horizontal flow around a two-dimensional hydrofoil (fig. 6.9).

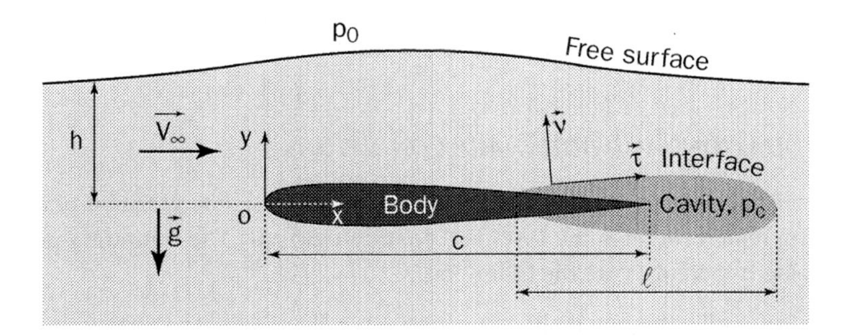

6.9 - Case of a horizontal supercavitating flow with an external free surface

In addition to the geometrical parameters defining the shape of the foil, its chordlength c, its angle of attack and the position of the external boundaries, the main parameters which influence the cavity flow are:

- the relative cavity underpressure:

$$\sigma = \frac{p_\infty - p_c}{\frac{1}{2}\rho V_\infty^2} \tag{6.9}$$

- and, if gravity effects are included, the FROUDE number based either on the chord length c:

$$Fr_c = \frac{V_\infty}{\sqrt{gc}} \tag{6.10}$$

  or on the cavity length ℓ:

$$Fr_\ell = \frac{V_\infty}{\sqrt{g\ell}} \tag{6.11}$$

In the previous relations, V_∞ is the free-stream velocity, g the acceleration due to gravity, ρ the liquid density, p_c the absolute pressure inside the cavity and p_∞ a reference pressure at upstream infinity conventionally chosen at the ordinate $y = 0$ (fig. 6.9).

The modeling of the potential flow allows us to calculate the shape y_c of the cavity and particularly its length ℓ as well as the velocity and pressure fields and more especially the drag and lift coefficients:

$$C_D = \frac{D}{\frac{1}{2}\rho V_\infty^2 c} = \frac{\oint_{body} p.dy}{\frac{1}{2}\rho V_\infty^2 c} \tag{6.12}$$

$$C_L = \frac{L}{\frac{1}{2}\rho V_\infty^2 c} = \frac{-\oint_{body} p.dx}{\frac{1}{2}\rho V_\infty^2 c} \tag{6.13}$$

where D and L stand respectively for the drag and the lift forces per unit span length. The integrals are taken clockwise around the body.

6.2.2. *EQUATIONS AND BOUNDARY CONDITIONS*

For an inviscid fluid in irrotational flow, the velocity $\vec{V}$ is the gradient of a velocity potential φ which satisfies the LAPLACE equation $\Delta\varphi = 0$. The boundary conditions on the velocity potential are the following.

On the solid walls, the usual slip condition $\vec{V}.\vec{\nu} = 0$ or $\partial\varphi/\partial\nu = 0$ has to be satisfied where $\vec{\nu}$ is a unit vector normal to the wall.

On the cavity interface, the shear stress and the mass transfer due to phase change are neglected. The pressure is assumed constant, uniform and equal to the cavity pressure p_c and the velocity is tangential (see § 1.3.2). Using the BERNOULLI equation between upstream infinity and a point of ordinate y_c on the cavity interface where the velocity is assumed to be V_c, we get:

$$p_\infty + \rho \frac{V_\infty^2}{2} = p_c + \rho g y_c + \rho \frac{V_c^2}{2} \tag{6.14}$$

so that the tangential velocity on the cavity $V_c = \partial\varphi/\partial\tau$ is given by (τ is the curvilinear distance along the free streamline):

$$\frac{V_c}{V_\infty} = \sqrt{1 + \sigma - \frac{2 g y_c}{V_\infty^2}} \tag{6.15}$$

In the case of large supercavities, i.e. for small σ-values, the maximum value of y_c is of the order of $\ell\sigma/2$ (see eq. 6.29). Thus, we have approximately:

$$\frac{V_c}{V_\infty} \cong \sqrt{1 + \sigma \left[1 - \frac{1}{Fr_\ell^2}\right]} \tag{6.16}$$

According to this equation, gravity terms can be neglected if the FROUDE number based on cavity length Fr_ℓ is much larger than one. If so, the condition of constant pressure on the cavity interface reduces to a condition of constant tangential velocity:

$$\frac{V_c}{V_\infty} = \sqrt{1 + \sigma} \tag{6.17}$$

6.2.3. *CAVITY CLOSURE MODELS*

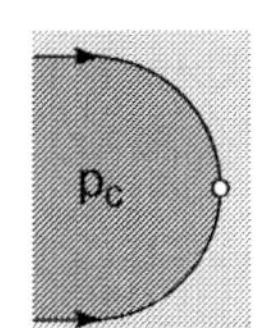

6.10

Firstly, it should be noted that the stagnation point cannot be the junction of two free streamlines as schematically shown in figure 6.10 since the overpressure at this point would not be consistent with the constant pressure condition in the cavity. Therefore, several cavity closure models have been proposed in the past to close the cavity more or less artificially.

6.11
RIABOUCHINSKY model

In the RIABOUCHINSKY model (1920), the cavity is closed by a solid body, which is supposed to be held in the flow from outside. According to d'ALEMBERT's paradox, the cavitator, the cavity and the closure body form a closed body on which the drag is zero.

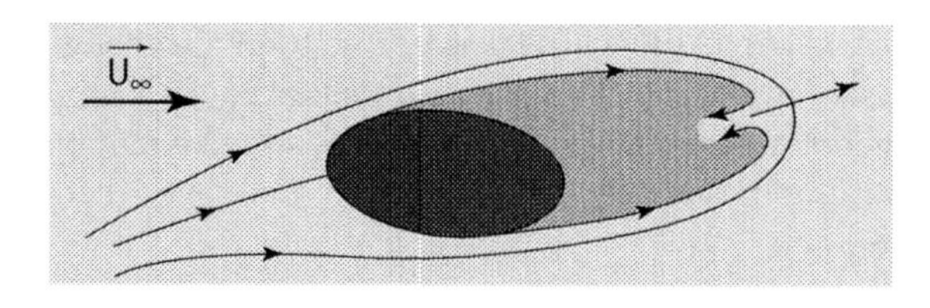

6.12

Re-entrant jet model

The re-entrant jet model [GILBARG & SERRIN 1950] assumes that a re-entrant jet develops at cavity closure. It is supposed steady and to evolve on a RIEMANN sheet, which allows the flowrate coming from the re-entrant jet to be continuously withdrawn from the cavity. Indeed, the stagnation point formed in the fluid is highly unstable as already mentioned in section 6.1.3.

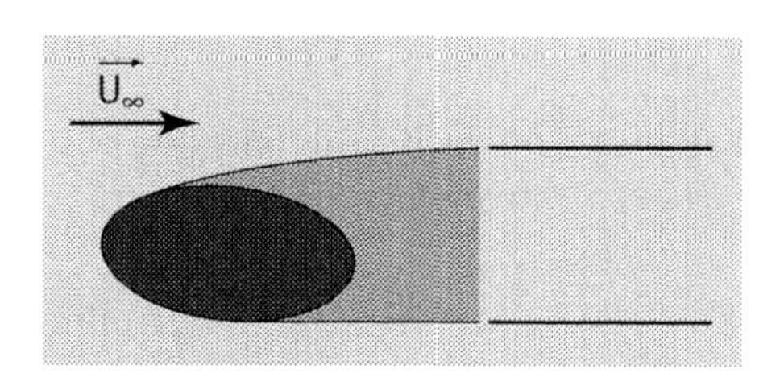

6.13

WU model

In the WU model (1956), the upper and lower free streamlines of the cavity are supposed continued by two solid plates, parallel to the velocity at infinity and starting at the rear part of the cavity.

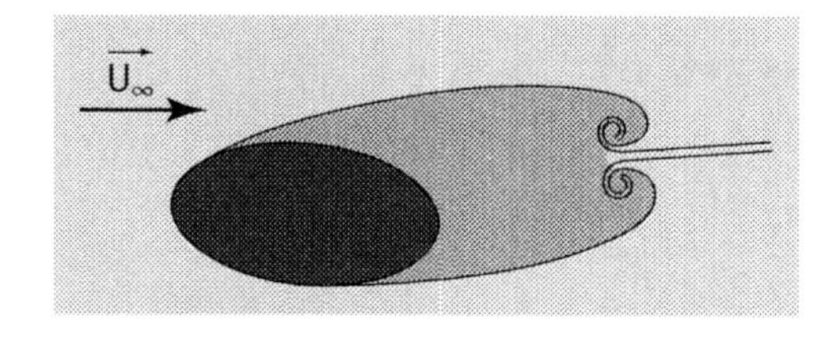

6.14

TULIN spiral vortices model

Finally, the TULIN model (1964) assumes that the cavity is closed by two spiral vortices which result from conditions superimposed on the cavity $\left(\|\vec{V}\| = V_c\right)$ and its wake $\left(\|\vec{V}\| = V_\infty\right)$.

Such models are necessary to close the cavity within the framework of a steady potential approach. If the steady restriction is released and if an unsteady computation is conducted, it is no longer necessary to use such closure models and the cavity naturally exhibits an unsteady behavior with a more or less cyclic development of the re-entrant jet.

6.2.4. *OVERVIEW OF CALCULATION TECHNIQUES*

Different techniques can be used to calculate steady supercavity flows. In the past, non-linear, analytic techniques have been used, principally for bodies limited by a few straight lines, such as the flat plate considered in section 6.3.1. Theoretical results concerning the conditions of existence and uniqueness of solutions were also obtained that way. Further information on these techniques can be found in the book of JACOB (1959) referenced in chapter 1 and also in BIRKHOFF & ZARANTONELLO (1957), MILNE-THOMSON (1960) and LOGVINOVICH (1969).

For bodies with continuous curvature, the technique leads to integrals which have to be evaluated numerically and requires an iterative procedure, whose convergence is not guaranteed. The solution is made even more complicated if a criterion for the cavity detachment has to be used. Thus, at the present time, the non-linear, analytic technique remains a reference rather than an operative method to solve practical problems.

Linearized, analytic methods were developed to model supercavitating flows around slender lifting bodies. Such methods assume that the perturbed velocity components remain small compared to the free-stream velocity. This condition cannot be met in the vicinity of the leading edge and the cavity closure, so that singularities of the solution appear at those points.

Two classical problems can be solved easily on the basis of the linearized theory:

- The inverse problem which consists in computing the shape of the foil and the cavity, given the pressure coefficient on the cavity interface and on the external boundaries (DIRICHLET problem).
- The direct problem which consists in computing the pressure coefficient anywhere in the flow field and the cavity shape and length (alternatively the σ-value) given the foil shape and the relative underpressure σ (or alternatively the cavity length ℓ) (mixed NEUMANN problem).

A detailed example of the use of the linearized method can be found in ROWE and MICHEL (1975). In this work, the singularity at the leading edge is removed, using the method of matched asymptotic expansions, which allows us to obtain a uniformly valid solution near the rounded nose of a truncated foil. It turns out that the numerical results agree pretty well with the experimental ones. This is particularly the case for the range of attack angles for which cavitation does not occur near the leading edge.

Finally there are the purely numerical techniques such as the boundary element method which is particularly suitable to the modeling of supercavity flows (see e.g. LEMONNIER & ROWE 1988). More recently, other techniques of direct interface tracking have been developed on the basis of the resolution of the EULER or NAVIER-STOKES equations, such as marker techniques or volume of fluid (VOF) techniques (see e.g. SCARDOVELLI & ZALESKI 1999). These methods are inherently unsteady and allow the time evolution of the cavity interface to be followed.

6.3. TYPICAL RESULTS

In this section, some results taken from analytic models are given. They allow us to point out the main trends of supercavity flows. The following assumptions are common to all models:

- flows are irrotational, steady and two-dimensional while the liquid is incompressible and inviscid;
- gravity and surface tension are neglected;
- the stress at the liquid-vapor interface has a unique normal component equal to the cavity pressure p_c, supposed uniform throughout the cavity.

In most cases, the relative cavity underpressure σ is assumed much smaller than 1, so that the cavity is considered long.

6.3.1. INFINITE CAVITY BEHIND A FLAT PLATE IN AN INFINITE FLOW FIELD ($\sigma = 0$)

Let c the chord AB of the plate. The cavity underpressure σ is assumed to be zero, so that the pressure p_c in the cavity is equal to the reference pressure p_∞ far upstream the foil (fig. 6.15).

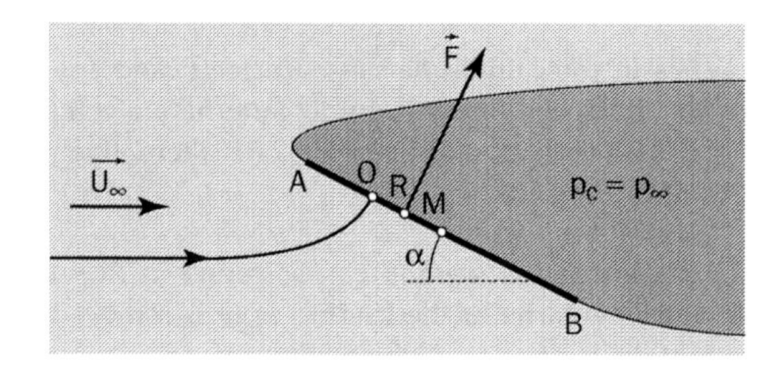

6.15
Flat plate in an infinite medium ($\sigma = 0$)

The force F applied by the flow on the plate, per unit span length, is given by:

$$\frac{F}{\rho U_\infty^2 c} = \frac{\pi \sin\alpha}{4 + \pi \sin\alpha} \tag{6.18}$$

The distance between its application point R and the middle M of the plate is:

$$\frac{MR}{c} = \frac{3}{4} \frac{\cos\alpha}{4 + \pi \sin\alpha} \tag{6.19}$$

The lift and drag coefficients are respectively:

$$\begin{cases} C_L = \dfrac{F\cos\alpha}{\frac{1}{2}\rho U_\infty^2 c} = \dfrac{2\pi \sin\alpha \cos\alpha}{4 + \pi \sin\alpha} \\ C_D = \dfrac{F\sin\alpha}{\frac{1}{2}\rho U_\infty^2 c} = \dfrac{2\pi \sin^2\alpha}{4 + \pi \sin\alpha} \end{cases} \tag{6.20}$$

The asymptotic shape of the free streamlines is parabolic:

$$\left[\frac{y}{c}\right]^2 \approx \frac{16\sin^2\alpha}{4+\pi\sin\alpha}\frac{x}{c} \tag{6.21}$$

If the plate is held perpendicular to the incident velocity ($\alpha = \pi/2$), the drag coefficient is equal to $\dfrac{2\pi}{4+\pi} \cong 0.88$, whereas the experimental value in the very different case of a fully wetted flow is about 1.95.

At small angles of attack α, we have:

$$\begin{cases} C_L \cong \dfrac{\pi}{2}\alpha \\ C_D \cong \dfrac{\pi}{2}\alpha^2 \end{cases} \tag{6.22}$$

The slope of the curve $C_L(\alpha)$ is $\pi/2$.

The distance between the stagnation point O and the leading edge A is about $\frac{2}{3}c\,\alpha^4$.

6.3.2. *FINITE CAVITY BEHIND A SYMMETRICAL BODY IN AN INFINITE FLOW FIELD*

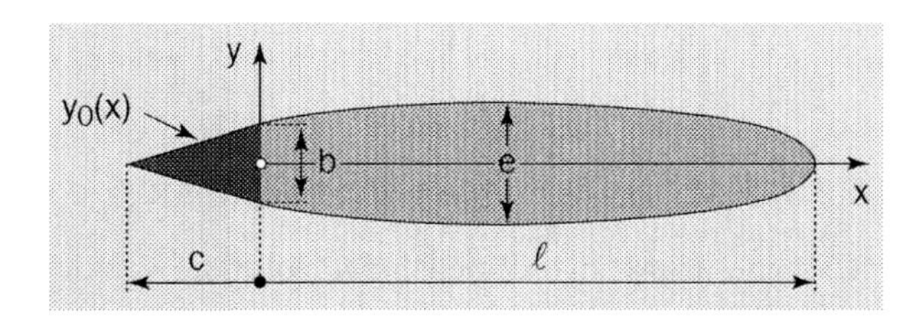

6.16
Symmetrical body with finite cavity in infinite medium

This model was developed by TULIN (1953) on the basis of a linearized theory in which the cavity was considered closed downstream at first order. Let b the wedge base and c its chord (fig. 6.16). The linearization procedure requires the condition $b \ll c$ to be fulfilled.

For small σ-values, the drag coefficient defined by:

$$C_D = \frac{D}{\frac{1}{2}\rho U_\infty^2\, b} \tag{6.23}$$

can be approximated by:

$$C_D(\sigma) \cong C_D(0)\,[1+\sigma] \tag{6.24}$$

with

$$C_D(0) = \frac{2}{\pi b}\left[\int_{-c}^{0} \frac{dy_0}{dx}\frac{dx}{\sqrt{-x}}\right]^2 \tag{6.25}$$

The cavity length is given by:

$$\ell = \frac{4}{\pi}\frac{1+\frac{\sigma}{2}}{\sigma}\left[\int_{-c}^{0}\frac{dy_0}{dx}\frac{\sqrt{\ell - x}}{\sqrt{-x}}\,dx\right] \tag{6.26}$$

and is approximated, when σ is small, by:

$$\frac{\ell}{b} \cong \frac{8C_D(0)}{\pi\sigma^2} \tag{6.27}$$

Hence, the cavity length varies as σ^{-2} for small σ-values. As for the cavity thickness e, it is given by:

$$\frac{e}{b} \cong \frac{4C_D(0)}{\pi\sigma} \tag{6.28}$$

The cavity is approximately elliptic, except in the vicinity of the body, and its relative thickness is:

$$\frac{e}{\ell} \cong \frac{\sigma/2}{1+\sigma/2} \tag{6.29}$$

Finally, for $\sigma = 0$, the asymptotic shape of the cavity is given by:

$$\frac{y_c(x)}{b} \cong \sqrt{\frac{2C_D}{\pi}\frac{x}{b}} \tag{6.30}$$

6.3.3. *FINITE CAVITY BEHIND A CIRCULAR ARC IN AN INFINITE FLOW FIELD*

This non-linear, two-dimensional, solution was developed by WU (1956). The far wake is represented by two rigid plates parallel to the incident velocity. The near wake is made up of the cavity limited by two free streamlines which link the leading and the trailing edges of the circular arc to the two plates (fig. 6.17).

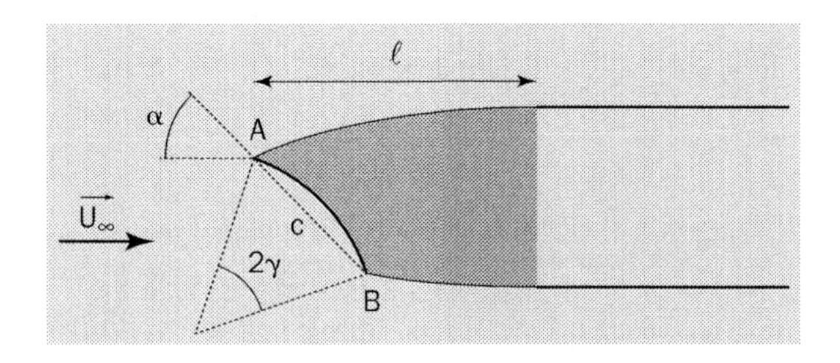

6.17 - Circular arc with finite cavity in infinite medium

For small values of σ, α and γ, the main results are the following:

$$\begin{cases} C_L \cong \dfrac{\pi}{2}[1+\sigma]\left[\left(\alpha+\dfrac{7\gamma}{8}\right)+\dfrac{3\sigma^2}{16}\left(\alpha+\dfrac{\gamma}{4}\right)^{-1}+\ldots\right] \\ C_D \cong \dfrac{\pi}{2}[1+\sigma]\left[\left(\alpha+\dfrac{\gamma}{4}\right)^2+\dfrac{3\sigma^2}{16}\left(\alpha+\dfrac{7\gamma}{4}\right)+\ldots\right] \end{cases} \tag{6.31}$$

$$\begin{cases} \dfrac{\ell}{c} \cong 1+\dfrac{8C_D(0)}{\pi\sigma^2}=1+\dfrac{4}{\sigma^2}\left(\alpha+\dfrac{\gamma}{4}\right)^2 \\ \dfrac{e}{c} \cong \dfrac{4C_D(0)}{\pi\sigma} \end{cases} \tag{6.32}$$

6.3.4. *VARIATION OF LIFT AND DRAG COEFFICIENTS WITH CAVITY UNDERPRESSURE*

In the various cases considered above, corresponding to supercavity flows in an infinite medium, the drag and lift coefficients appear to obey the following asymptotic law, for small σ-values:

$$\begin{cases} C_D(\sigma) \cong C_D(0)\,[1+\sigma] \\ C_L(\sigma) \cong C_L(0)\,[1+\sigma] \end{cases} \tag{6.33}$$

This means that, starting from an infinite cavity, if the cavity underpressure σ is increased, resulting in a finite although long cavity, the relative velocity distribution around the body is unchanged. To first order of approximation in σ, the velocity on the body and the cavity is simply multiplied by $\sqrt{1+\sigma}$, so that the drag and lift coefficients are multiplied by $1+\sigma$. In particular, the position of the stagnation point is unchanged.

If the flow is limited by solid or free external boundaries, this result no longer holds since the position of the stagnation point changes together with the velocity distribution. Then, a change in σ-value strongly modifies the variations of the global force coefficients with the cavity underpressure.

The effect of solid walls on the force coefficients is an essential topic in the literature due to its importance in testing facilities. It was studied by BIRKHOFF *et al.* (1950, 1952), FABULA (1964), WU *et al.* (1971), and numerically calculated by BRENNEN (1969) in the case of a disk and a sphere. An empirical method of correction was proposed by MEIJER (1967).

In the case of a supercavitating foil with wetted upper side and truncated base held under a free surface, the evolution of the lift-curve with the cavity underpressure is schematically given in figure 6.18. It starts with a negative slope for small σ-values, which is very different from the classical law discussed above [ROWE *et al.* 1975]. A close examination of the leading edge region showed that this is actually due to the displacement of the stagnation point which is correlated to a change in circulation around the body. Let us note that, in the present case of a truncated hydrofoil, the two detachment conditions are equivalent to the classical JOUKOWSKI condition in non-cavitating flow, which determines the circulation around the foil and its lift.

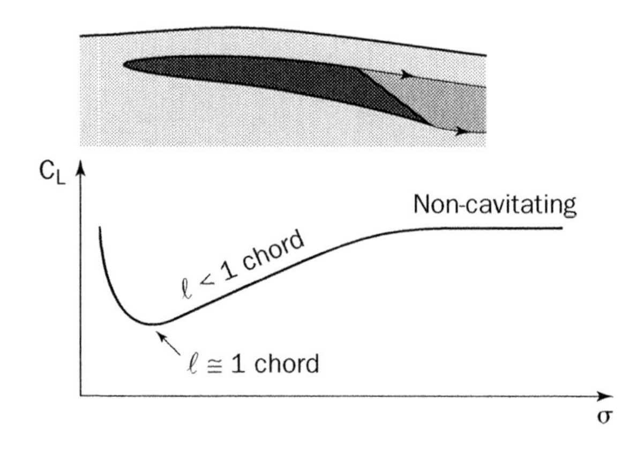

6.18 - Typical evolution, with the cavitation number, of the lift coefficient of a supercavitating foil in a free surface channel

6.3.5. *EFFECT OF SUBMERSION DEPTH ON THE SLOPE OF THE CURVE* $C_L(\alpha)$

Within the framework of the linearized theory and for infinite cavities ($\sigma = 0$), MICHEL and ROWE (1974) computed the slope $\frac{\partial C_L}{\partial \alpha}$ of the lift coefficient C_L versus the angle of attack α. The calculation was made for hydrofoils with a truncated base (as shown on figure 6.18) under a free surface and for different submersion depths. The chord length of the lower side is taken as unity whereas the chord length e of the upper side is variable from 0 to 1.

Figure 6.19 shows the evolution of the slope $\partial C_L / \partial \alpha$ versus the submersion depth h, for different values of the upper side chord length e. The theoretical results agree fairly well with the experimental ones. For example, with $h = 1$ and $e = 1$, the experimental value of $\frac{\partial C_L}{\partial \alpha}$ is 5.23, in good agreement with the theoretical prediction 5.32.

For small values of the submersion depth, the slope tends to the limit value π. On the other hand, for an infinite submersion depth, the asymptotic trend is given by:

$$\frac{\partial C_L}{\partial \alpha} = \frac{\pi}{2}\left(1 + \sqrt{e}\right)^2 \tag{6.34}$$

from which the classical value 2π is found for $e = 1$.

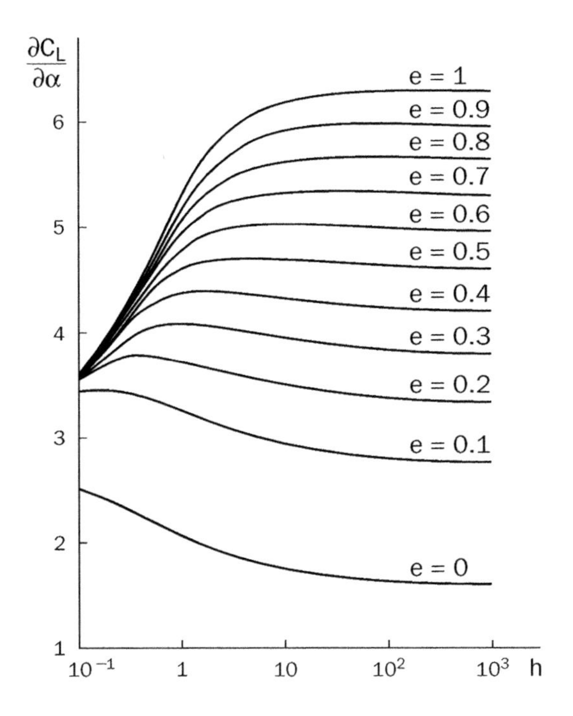

6.19 - Evolution of the slope of the lift coefficient curve with the submersion depth h for various lengths e of the upper side
(the chord length of the lower side is taken as unity)

6.4. AXISYMMETRIC CAVITIES

6.4.1. THE GARABEDIAN ASYMPTOTIC SOLUTION FOR STEADY SUPERCAVITIES

In most cases of axisymmetric supercavitation, the shape of the forebody (or cavitator) can be approximated by a disc or a cone. In that case, if gravity is neglected and for small enough values of the relative underpressure σ of the cavity (typically for $\sigma < 0.1$), the shape of the cavity is close to an ellipsoid whose length ℓ and maximum diameter d_c are expressed versus σ by the following asymptotic formulae [GARABEDIAN 1956]:

$$\begin{cases} \dfrac{d_c}{d} = \sqrt{\dfrac{C_D}{\sigma}} \\ \dfrac{\ell}{d} = \dfrac{1}{\sigma}\sqrt{C_D \ln \dfrac{1}{\sigma}} \end{cases} \tag{6.35}$$

in which d stands for the cavitator diameter. The drag coefficient C_D defined by:

$$C_D = \frac{D}{\frac{1}{2}\rho V_\infty^2 \frac{\pi d^2}{4}} \tag{6.36}$$

is connected to its value $C_D(0)$ for $\sigma = 0$ by the classic relation (6.33):

$$C_D(\sigma) \approx (1+\sigma)\, C_D(0) \tag{6.37}$$

For a disc, we have $C_D(0) \cong 0.82$.

6.4.2. *MOMENTUM BALANCE AND DRAG*

The first equation of (6.35) can be obtained from the momentum balance of the liquid contained in the domain shown in figure 6.20 limited by a surface S_u at upstream infinity, the surface Σ_∞ made up of a tube of streamlines at infinity and the annular surface S corresponding to an arbitrary cross-section along the cavity. This liquid domain is closed by the cavitator and part of the cavity interface.

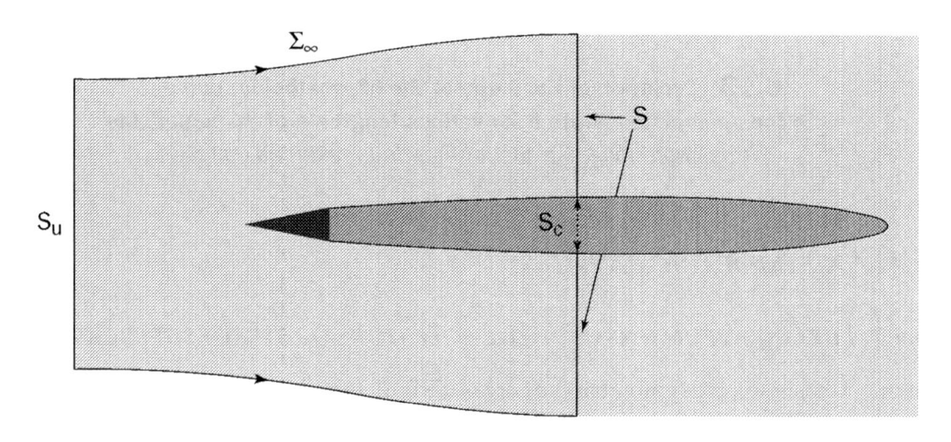

6.20

The momentum balance in this domain leads to the calculation of the drag D:

$$D = (p_\infty - p_c)\, S_c - \iint_S (p - p_\infty)\, dS + \rho V_\infty^2\, S_u - \iint_S \rho v_x^2\, dS \tag{6.38}$$

In this equation, S_c is the cross-sectional area of the cavity and p and v_x the pressure and axial velocity respectively. By introducing the mass conservation equation between cross-sections S_u and S:

$$\rho V_\infty\, S_u = \iint_S \rho v_x\, dS \tag{6.39}$$

the momentum balance becomes:

$$D = (p_\infty - p_c)\, S_c - \iint_S (p - p_\infty)\, dS + \iint_S \rho v_x (V_\infty - v_x)\, dS \tag{6.40}$$

From the BERNOULLI equation, we can calculate the pressure:

$$p - p_\infty = \frac{1}{2}\rho \left(V_\infty^{\ 2} - v_x^{\ 2}\right) - \frac{1}{2}\rho v_r^{\ 2} \tag{6.41}$$

where v_r is the radial velocity. The momentum balance finally gives:

$$D = (p_\infty - p_c)\, S_c - \iint_S \frac{1}{2}\rho\, (V_\infty - v_x)^2\, dS + \iint_S \frac{1}{2}\rho v_r^{\ 2}\, dS \tag{6.42}$$

Now, if the cross-section S is chosen to be at the location of the maximum cavity thickness, so that S_c is the maximum cavity area, then, it follows that the radial velocity v_r is zero on the cavity and zero everywhere on S as will be seen later from equation (6.47). Hence, the last integral vanishes. As for the first integral, it is usually negligible with respect to the first term since the axial velocity v_x is close to the velocity at infinity V_∞. Hence, the drag can be estimated by the following formula:

$$D \cong (p_\infty - p_c)\, S_c \tag{6.43}$$

By introducing the drag coefficient and the usual cavity underpressure σ, this equation appears to be strictly equivalent to the first equation of the GARABEDIAN solution (6.35). The second one which gives the cavity length is not so easy to get. It will be obtained in next section, on the basis of an approximate method.

Finally, let us note that, near the body, matching formulae are required to connect the cavitator to the ellipsoid (see e.g. LOGVINOVICH 1969, SEMENENKO 2001).

6.4.3. *APPROXIMATE, ANALYTIC SOLUTION FOR STEADY SUPERCAVITIES*

The GARABEDIAN solution was obtained on the basis of important mathematical developments. An effort was made, especially in Russia and Ukraine, but also in Japan and in US, to look for approximate but efficient methods of calculation of axisymmetric cavities. When necessary, empirical coefficients are introduced to meet the experimental results. Purely numerical methods of solution were also used in the past (for example, BRENNEN 1969). We present here the method derived by SEREBRYAKOV (1972), LOGVINOVICH and SEREBRYAKOV (1975), VASIN (2001) and SEMENENKO (2001).

Derivation of the solution

Assuming that the velocity field has only radial and axial components v_r and v_x respectively but no tangential component (fig. 6.21), the mass conservation equation, in steady conditions, is:

$$\frac{1}{r}\frac{\partial(rv_r)}{\partial r}+\frac{\partial v_x}{\partial x}=0 \tag{6.44}$$

The calculation of the flow around a slender body consists in approximating the axial component v_x by the velocity at infinity V_∞, so that the second term in the previous equation is neglected. Hence, the radial velocity behaves as $1/r$:

$$v_r \cong \frac{C(x)}{r} \tag{6.45}$$

The function $C(x)$ is determined from the boundary condition on the cavity interface $r=R(x)$ where the radial velocity is:

$$\frac{v_r(R,x)}{V_\infty}=\frac{dR}{dx} \tag{6.46}$$

This condition simply means that the velocity on the cavity is tangential to the interface. Therefore, the radial velocity is given by:

$$v_r=\frac{V_\infty}{2}\frac{dR^2}{dx}\frac{1}{r} \tag{6.47}$$

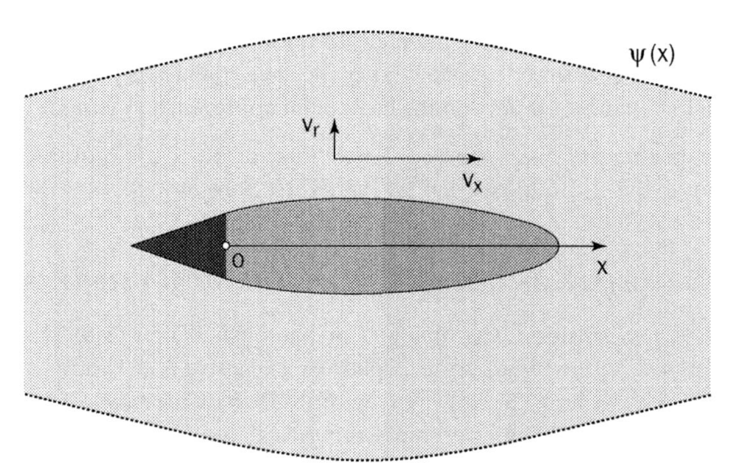

6.21

The steady state EULER equation in the radial direction is written:

$$-\frac{1}{\rho}\frac{\partial p}{\partial r}=v_r\frac{\partial v_r}{\partial r}+v_x\frac{\partial v_r}{\partial x} \tag{6.48}$$

Using equation (6.47) for the radial velocity v_r and again assuming that $v_x \cong V_\infty$, equation (6.48) becomes:

$$\frac{1}{\frac{1}{2}\rho V_\infty^2}\frac{\partial p}{\partial r} = R^2\left(\frac{dR}{dx}\right)^2\frac{2}{r^3} - \frac{d^2R^2}{dx^2}\frac{1}{r} \tag{6.49}$$

This equation is integrated between the radius R of the cavity where the pressure is equal to the cavity pressure p_c and a radius $\Psi(x)$ far from the cavity which corresponds to an external hypothetical surface on which the pressure at infinity $p_\infty(x)$ is applied. The introduction of this function $\Psi(x)$ is necessary in an axisymmetric flow because of the emergence of a logarithmic singularity (see eq. 6.50) which prevents us letting Ψ tend to infinity. The integrated equation is:

$$\left(\frac{dR}{dx}\right)^2 - \frac{d^2R^2}{dx^2}\ln\frac{\Psi}{R} = \frac{p_\infty - p_c}{\frac{1}{2}\rho V_\infty^2} \tag{6.50}$$

In the first term on the left-hand side, Ψ was set equal to infinity, which does not pose any mathematical difficulty.

The flow appears to be formed by the piling-up of liquid annular layers which evolve with an internal radius $R(x)$ and an external radius $\Psi(x)$ under the pressure difference $p_\infty - p_c$. The liquid particles which are outside the surface $\Psi(x)$ have no influence on the flow. The parameter:

$$\mu = \ln\frac{\Psi}{R} \tag{6.51}$$

characterizes the inertial properties of the liquid annular layer and plays the role of an added mass.

For small values of the slenderness parameter measured by the ratio of the maximum diameter of the cavity d_c to its length ℓ:

$$\delta = \frac{d_c}{\ell} \tag{6.52}$$

the parameter μ can be considered as constant and dependent only upon the cavity slenderness δ.

Returning to equation (6.50), the first term on the left-hand side is of the order δ^2. It will be shown later (eq. 6.57) that Ψ is of the order of the cavity length, so that the second term of equation (6.50) is of the order $\delta^2 \ln(1/\delta)$. Then, the first term can be neglected and the equation simplifies to [SEREBRYAKOV 1972]:

$$\mu\frac{d^2R^2}{dx^2} + \frac{p_\infty - p_c}{\frac{1}{2}\rho V_\infty^2} = 0 \tag{6.53}$$

The second term of this equation is the cavity underpressure σ. Assuming that the cavity radius is zero at both ends of the cavity $x = 0$ and $x = \ell$, the solution of this differential equation, with μ constant, is the ellipsoid given by:

$$\frac{R^2}{\frac{\sigma}{2\mu}\frac{\ell^2}{4}} + \frac{\left(x - \frac{\ell}{2}\right)^2}{\frac{\ell^2}{4}} = 1 \quad (6.54)$$

The origin of the x coordinate is taken at the middle of the cavitator base (fig. 6.21). The agreement with experimental results improves as σ is made smaller.

The cavity slenderness is given by:

$$\delta = \sqrt{\frac{\sigma}{2\mu}} \quad (6.55)$$

This parameter, which can easily be measured, is used for the determination of the μ-value and therefore of the external radius Ψ.

Asymptotic behavior

The μ-parameter, which is assumed constant here for a given cavity, can be estimated on the basis of the maximum cavity diameter d_c using equation (6.51) combined with equations (6.52) and (6.55) to give:

$$\mu = \ln\frac{2\Psi}{d_c} = \ln\frac{2\Psi}{\delta\ell} = \ln\left(\frac{2\Psi}{\ell}\sqrt{\frac{2\mu}{\sigma}}\right) \quad (6.56)$$

By comparison with experiments, it is found that the χ-parameter defined by:

$$\chi = \frac{2\Psi}{\ell} \quad (6.57)$$

lies in the range 0.54-0.64 [SEREBRYAKOV 1972]. Introducing this χ-parameter in equation (6.56), one obtains:

$$\mu = \ln\left(\chi\sqrt{\frac{2\mu}{\sigma}}\right) = \frac{1}{2}\ln\frac{1}{\sigma} + \frac{1}{2}\ln\mu + \ln\left(\chi\sqrt{2}\right) \quad (6.58)$$

Hence, when σ approaches zero, μ tends to infinity and the term in $\ln\mu$ can be neglected with respect to μ. It is similar for the last term $\ln\left(\chi\sqrt{2}\right)$ as χ is almost constant. Hence, we have the following asymptotic behavior:

$$2\mu \approx \ln\frac{1}{\sigma} \quad (6.59)$$

From equation (6.55), for small values of the cavity underpressure, the cavity slenderness δ behaves as:

$$\delta \approx \sqrt{\frac{\sigma}{\ln(1/\sigma)}} \tag{6.60}$$

It can easily be checked that the present approximate solution is strictly equivalent to the GARABEDIAN solution expressed by equation (6.35).

6.4.4. *UNSTEADY AXISYMMETRIC SUPERCAVITIES*

Unsteady, axisymmetric supercavity flows are encountered especially when the relative cavity underpressure is time-dependent, either because of external pressure changes (as in the case of vertical motion of the cavitator in a gravity field), or because the pressure inside the cavity is time-dependent, which is generally the case for ventilated cavities. The approach already used in the previous section 6.4.3 and based on the slender body approximation can be generalized to the unsteady case.

Derivation of the solution

The mass conservation equation for an incompressible liquid (6.44) is unchanged. In this equation, the axial velocity is still approximated by the velocity at infinity so that the radial velocity still behaves as $1/r$. It is determined from the boundary condition on the cavity interface, where the kinematic condition is:

$$v_r(r=R,x,t) = \frac{\partial R}{\partial t} + V_\infty \frac{\partial R}{\partial x} \tag{6.61}$$

Here $R(x,t)$ is the radius of the cavity at station x and time t. Hence, the radial velocity $v_r(r,x,t)$ is given by:

$$v_r \cong \frac{1}{2\pi r}\left[\frac{\partial S}{\partial t} + V_\infty \frac{\partial S}{\partial x}\right] \tag{6.62}$$

where $S(x,t) = \pi R^2$ is the cross-sectional area of the cavity.

Introducing the above expression for the velocity in the unsteady radial EULER equation:

$$-\frac{1}{\rho}\frac{\partial p}{\partial r} = \frac{\partial v_r}{\partial t} + \left(v_r \frac{\partial v_r}{\partial r} + v_x \frac{\partial v_r}{\partial x}\right) \tag{6.63}$$

and again approximating the axial velocity v_x by V_∞, one obtains:

$$-\frac{1}{\rho}\frac{\partial p}{\partial r} = \frac{1}{2\pi r}\left[\frac{\partial^2 S}{\partial t^2} + 2V_\infty \frac{\partial^2 S}{\partial x\,\partial t} + V_\infty{}^2 \frac{\partial^2 S}{\partial x^2}\right] - \frac{1}{4\pi^2 r^3}\left[\frac{\partial S}{\partial t} + V_\infty \frac{\partial S}{\partial x}\right]^2 \tag{6.64}$$

By integrating this equation between the radius R of the cavity where the pressure is assumed to be $p_c(t)$ and a radius Ψ far from the cavity (see § 6.4.3 for details), one has:

$$\frac{1}{8\pi^2 R^2}\left[\frac{\partial S}{\partial t}+V_\infty\frac{\partial S}{\partial x}\right]^2-\left[\frac{1}{2\pi}\ln\frac{\Psi}{R}\right]\left[\frac{\partial^2 S}{\partial t^2}+2V_\infty\frac{\partial^2 S}{\partial x\,\partial t}+V_\infty{}^2\frac{\partial^2 S}{\partial x^2}\right]=\frac{p_\infty-p_c(t)}{\rho} \tag{6.65}$$

This equation, which governs the time-dependent evolution of the cavity, reduces to equation (6.50) for the steady state flow ($\partial/\partial t=0$). As in the steady state case, the first term on the left-hand side is neglected in comparison with the second one, so that this equation finally reduces to:

$$\left[\frac{\partial^2 S}{\partial t^2}+2V_\infty\frac{\partial^2 S}{\partial x\,\partial t}+V_\infty{}^2\frac{\partial^2 S}{\partial x^2}\right]=-\kappa\frac{p_\infty-p_c(t)}{\rho} \tag{6.66}$$

where the parameter κ is defined by:

$$\kappa=\frac{2\pi}{\ln\frac{\Psi}{R}} \tag{6.67}$$

Equation (6.66) can be rewritten in the following form:

$$\left[\frac{\partial}{\partial t}+V_\infty\frac{\partial}{\partial x}\right]^2 S=-\kappa\frac{p_\infty-p_c(t)}{\rho} \tag{6.68}$$

As in the steady state case for which the μ-parameter was considered constant, it is usually assumed that the κ-parameter is constant for elongated cavities. The κ- and μ-parameters are connected by:

$$\kappa=\frac{2\pi}{\mu} \tag{6.69}$$

Using equations (6.59) and the second equation of (6.35), the asymptotic value of κ is given by:

$$\kappa\approx\frac{4\pi}{\ln\frac{1}{\sigma}}\approx\frac{4\pi C_D d^2}{\sigma^2\ell^2} \tag{6.70}$$

Lagrangian approach

Equation (6.68) is more easily interpreted using a Lagrangian system of coordinates.

Consider a given cross-section of the cavity and suppose that it is convected by the axial flow at the velocity V_∞. For this cross-section, the origin of time t_0 is chosen as the instant when it coincides with the cavitator base. At time t, this section is located at a distance $x = V_\infty (t - t_0)$. Another cross-section would be characterized by a different initial time t_0. The law of variation of its area, $S_{t_0}(t)$, is given by:

$$S_{t_0}(t) = S\,[V_\infty (t - t_0), t] \tag{6.71}$$

where $S(x,t)$ is the cross-sectional area given in Eulerian coordinates (see the previous section).

In Lagrangian terms, equation (6.68) reduces then to:

$$\frac{d^2 S_{t_0}}{dt^2} = -\kappa \frac{p_\infty - p_c(t)}{\rho} \tag{6.72}$$

This shows that the law of expansion of any cross-section of the cavity is almost independent –say at the order $1/\ln\delta$– from the neighbouring ones. It depends only on the difference between the pressure at infinity and the pressure inside the cavity at the same position. This result is known as the LOGVINOVICH independence principle of cavity expansion [LOGVINOVICH 1969].

The integration of equation (6.72) leads to the following solution for the time-dependent evolution of the cross-section:

$$S_{t_0}(t) = S_0 + (t - t_0)\,\dot{S}_{t_0}(t_0) - \kappa \int_{\tau = t_0}^{\tau = t} (t - \tau)\,\frac{p_\infty - p_c(\tau)}{\rho}\,d\tau \tag{6.73}$$

where S_0 is the area of the cavitator basis and $\dot{S}_{t_0}(t_0)$ the initial rate of expansion of the cross-section in question.

Case of oscillating cavity pressure

As an example, consider the case of ventilated cavities for which the cavity pressure undergoes harmonic oscillations:

$$p_c = \overline{p}_c + \tilde{p} \sin \omega t \tag{6.74}$$

while the pressure at infinity p_∞ remains constant. Here, $\overline{p}_c$ is the steady part of the cavity pressure and $\tilde{p}$ represents the amplitude of the cavity pressure oscillations. As a result, the cross-sectional area of the cavity can be considered as the superposition of two components:

$$S(x,t) = \overline{S}(x) + \tilde{S}(x,t) \tag{6.75}$$

$\overline{S}$ is the steady, elliptic term already calculated in section 6.4.3. It corresponds to the pressure difference $p_\infty - \overline{p}_c$ and is the solution of the steady state equation:

$$\frac{\partial^2 \overline{S}}{\partial x^2} = -\kappa \frac{p_\infty - \overline{p}_c}{\rho V_\infty^{\;2}} \tag{6.76}$$

The fluctuating term $\tilde{S}$ is the solution of the unsteady part of equation (6.68) i.e.:

$$\left[\frac{\partial}{\partial t}+V_\infty\frac{\partial}{\partial x}\right]^2\tilde{S}=\frac{\kappa\tilde{p}}{\rho}\sin\omega t \tag{6.77}$$

Using equation (6.73) which is given in Lagrangian coordinates, the solution of (6.77) in Eulerian terms is:

$$\tilde{S}=\frac{\kappa\tilde{p}}{\rho\omega^2}\left[-\sin\omega t+\sin\omega\left(t-\frac{x}{V_\infty}\right)+\frac{\omega x}{V_\infty}\cos\omega\left(t-\frac{x}{V_\infty}\right)\right] \tag{6.78}$$

Thus the fluctuations $\tilde{S}$ of any cavity cross-section have three components. The first affects all cross-sections uniformly and its phase is in opposition to the excitation. The second corresponds to a progressive wave. At the point of detachment from the body ($x=0$), it is in phase with the excitation. The third term corresponds to a progressive wave whose amplitude increases linearly with the distance x. At the point of detachment from the body, the fluctuation is zero at all times.

6.5. SPECIFIC PROBLEMS

6.5.1. UNSTEADY 2D SUPERCAVITIES

Apart from their theoretical interest (in stability analysis for instance), unsteady supercavitating flows were mainly considered in the past for their application to hydrofoils under transient or periodic conditions.

On the whole, the basis of the modeling of unsteady 2D supercavities is the same as for steady flows. The main analysis methods are:

- analytical, non-linear methods [VON KARMAN 1949, WOODS 1955, 1964, 1966, WU 1958, WANG *et al.* 1963, 1965, BENJAMIN 1964],
- analytical, linearized methods [TIMMAM 1958, GEURST 1960, CUMBERBATCH 1961, ACOSTA *et al.* 1979, TULIN *et al.* 1980], and
- numerical, non-linear methods, which are currently employed and usually use the scheme initially proposed by PLESSET *et al.* (1971) (see § 4.3.1), to track the time-dependent evolution of the free surface which is considered as a material one. A survey of numerical techniques for unsteady cavity flow modeling was given by KINNAS (1998).

Due to the difficulty of conducting tests under unsteady conditions, experimental data in this field are rather rare. TSEN and GUILBAUD (1971) determined the coefficients of added (or virtual) mass for supercavitating foils with non-wetted uppersides. The global behavior of cavities attached to an oscillating NACA 16012 foil was described by FRANC and MICHEL (1988). They showed that the collapse of largely developed cavities can be forced by the oscillation and generate severe perturbations of the lift together with significant erosion, although the mean velocity was by far too small to cause any erosion under steady conditions.

Furthermore, from the generalized BERNOULLI equation, the pressure far from the cavitating body behaves as the time derivative of the velocity potential $\partial\varphi/\partial t$. If the volume of the cavity changes, the potential at infinity behaves as a source or a sink of variable intensity $Q(t)$.

In a three-dimensional flowfield where $\varphi \approx Q(t)/r$, we have $p \approx \rho\dot{Q}/r$ and the pressure is non-singular at infinity. However, for a two-dimensional flowfield, the velocity potential behaves as $(Q(t)/2\pi).\ln r$ and the pressure exhibits a logarithmic singularity $\left(\rho\dot{Q}/2\pi\right).\ln r$ at infinity. An example of this 2D logarithmic behavior is found in equations (6.50) and (6.65): according to the Logvinovich independence principle, the radial motion of each cross-section is two-dimensional although the axisymmetric flow is three-dimensional.

This fundamental difficulty, originally pointed out by WU (1958), can be overcome in two ways:
- by considering the 2D flow as an approximation of real flows which actually are three-dimensional [WU 1958, BENJAMIN 1964],
- or by considering a wake for the cavity and assuming that the total volume of the cavity and its wake remain constant. This condition was used by WOODS (1966) in order to model pulsating ventilated cavity flows.

Another method, largely used as a short cut in numerical modeling of unsteady 2D cavity flows, is to consider only the vicinity of the body and its attached cavity, which allows us to ignore the pressure singularity in the far field.

6.5.2. *COMPRESSIBLE EFFECTS IN SUPERCAVITATING FLOWS*

For classical applications of supercavitating flows, compressibility effects in the liquid are negligible since the order of magnitude of the velocities, say some tens of meters per second, results in very small values of the MACH number, for example 0.02 for $V_\infty = 30\ m/s$. Thus, the assumption of an incompressible liquid is correct.

It is not the case when the velocity approaches sound velocity, i.e. 1,500 m/s in water. By launching small projectiles in still water, very high velocities together with very long cavities can be obtained. Flow velocities greater than the velocity of sound can be reached, such that the body has a shock wave in front of it, and a classical cavity behind.

Theoretical work on this topic can be found in VASIN (1998) and SEREBRYAKOV (1998). The main differences with classical compressible aerodynamics lie in the presence of the cavity and in the form of the compressibility relation (TAIT's law, see § 5.1.1), which gives the water a greater "stiffness" compared to gases.

REFERENCES

ACOSTA A.J. & FURUYA O. –1979– A brief note on linearized, unsteady supercavitating flows. *J. Ship Res.* **2**, 85-88.

ARAKERI V.H. –1975– Viscous effects on the position of cavitation separation from smooth bodies. *J. Fluid Mech.* **68**, 779-799.

ARMSTRONG A.H. –1953– Abrupt and smooth separation in plane and axisymmetric flow. *Memo. Arm. Res. Est., G.B.*, n° 22-63.

BENJAMIN T.B. –1964– Note on the interpretation of two-dimensional theories of growing cavities. *J. Fluid Mech.* **19**(1), 137-144.

BIRKHOFF G., PLESSET M.S. & SIMMONS N. –1950– Wall effects in cavity flows I. *Quart. Appl. Math.* **8**, 161 *sq.*

BIRKHOFF G., PLESSET M.S. & SIMMONS N. –1952– Wall effects in cavity flows II. *Quart. Appl. Math.* **9**, 413 *sq.*

BIRKHOFF G. & ZARANTONELLO E.H. –1957– Jets, wakes, and cavities. *Academic Press Inc.*

BRENNEN C.E. –1969– A numerical solution of axisymmetric cavity flows. *J. Fluid Mech.* **37**(4), 671-688.

COHEN H. & DI PRIMA R.C . –1958– Wall effects in cavitating flows. *Proc. 2nd ONR Symp. on Naval Hydrodynamics.*

CUMBERBATCH E. –1961– Accelerating, supercavitating flow past a thin two-dimensional wedge. *J. Ship Res.* **1**(8), 1-8.

EFROS D. –1946– Hydrodynamical theory of two-dimensional flow with cavitation. *CR Acad. Sci.*, vol. LI, n° 4, URSS, 267-270.

FABULA A.G. –1964– Choked flow about vented or cavitating hydrofoil. *Trans. ASME D* **86**, 561 *sq.*

FRANC J.P. & MICHEL J.M. –1985– Attached cavitation and the boundary layer: experimental investigation and numerical treatment. *J. Fluid Mech.* **154**, 63-90.

FRANC J.P. & MICHEL J.M. –1988– Unsteady attached cavitation on an oscillating hydrofoil. *J. Fluid Mech.* **193**, 171-189.

GARABEDIAN P.R. –1956– Calculation of axially symmetric cavities and jets. *Pac. J. Math.* **6**, 611-684.

GEURST J.A. –1960– Some investigations of a linearized theory for unsteady cavity flows. *Arch. Rat. Mech. Anal.* **5**(4), 316-346.

GILBARG D. & SERRIN J. –1950– Free boundaries and jets in the theory of cavitation. *J. Math. Phys.* **29**, 1-12.

VON KARMAN T. –1949– Accelerated flow of an incompressible fluid with wake formation. *Ann. Mathem. Pura Applic.* **IV**(29), 247-249.

KINNAS S.A. –1998– The prediction of unsteady sheet cavitation.
Proc. 3rd Int. Symp. on Cavitation, vol. 1, Grenoble (France), 19-36.

KNAPP R.T., DAILLY J.W. & HAMMITT F.G. –1970– Cavitation.
McGraw-Hill Book Company Ed.

LEMONNIER H. & ROWE A. –1988– Another approach in modeling cavitating flows.
J. Fluid Mech., vol. 195, 557-580.

LOGVINOVICH G.V. –1969– Hydrodynamics of free surface flows (in Russian).
Nauvoka Dunka Ed., Kiev (Ukraine).

LOGVINOVICH G.V. & SEREBRYAKOV V.V. –1975– The methods for the calculation of the shape of slender axisymmetric cavities (in Russian).
Hydromechanics, Nauvoka Dunka Ed., Kiev (Ukraine).

MEIJER M.C. –1967– Pressure measurements on flapped hydrofoils in cavity flows and wake flows. *J. Ship Res.* **11**, 170 *sq.*

MICHEL J.M. –1973– Sillages plans supercavitants : étude physique.
Thesis, Grenoble University (France).

MICHEL J.M. & ROWE A. –1974– Profils minces supercavitants à arrière tronqué. Définition et étude théorique de profils portants à nombre de ventilation nul en présence d'une surface libre. *La Houille Blanche* **217**(3), 205-214.

MICHEL J.M. –1977– Wakes of developed cavitities. *J. Ship Res.* **21**(4), 225-238.

MILNE-THOMSON L.M. –1960– Theoretical hydrodynamics. *Macmillan Ed.*, London (England).

RIABOUCHINSKY D. –1920– On steady fluid motion with free surface.
Proc. Math. Soc. London **19**, 206-215.

ROWE A. –1974– Profils minces supercavitants à arrière tronqué. Influence de la ventilation sur les caractéristiques de profils portants évoluant au voisinage d'une surface libre. *La Houille Blanche* **217**(3), 215-230.

ROWE A. & MICHEL J.M. –1975– Two-dimensional base-vented hydrofoils near a free surface: influence of the ventilation number. *J. Fluids Eng.* **97**, 465-474.

SCARDOVELLI R. & ZALESKI S. –1999– Direct numerical simulation of free-surface and interfacial flow. *Ann. Rev. Fluid Mech.* **31**, 567-603.

SEMENENKO V.N. –2001– Artificial supercavitation. Physics and calculation.
VKI/RTO Special Course on Supercavitation. Von Karman Institute for Fluid Dynamics, Brussels (Belgium).

SEREBRYAKOV V.V. –1972– The annular model for calculation of axisymmetric cavity flows. *Hydromechanics, Nauvoka Dunka Ed.*, Kiev (in Russian), **27**, 25-29.

SEREBRYAKOV V.V. –1998– Asymptotic approach for problems of axisymmetric supercavitation based on the slender body approximation.
Proc. 3rd Int. Symp. on Cavitation, vol. 2, Grenoble (France), 61-70.

TIMMAM R. –1958– A general linearized theory for cavitating hydrofoils in non-steady flow. *Proc. 2nd Int. Symp. on Naval Hydrodynamics*, 559-579.

TSEN L.F. & GUILBAUD M. –1971– Calcul et essais des ailes supercavitantes finies en mouvement harmonique. *Proc. IUTAM Symp. on Non-Steady Flow of Water at High Speeds*, Leningrad (Russia), 427-447.

TULIN M.P. –1964– Supercavitating flows. Small perturbation theory. *J. Ship Res.* **8**, 16-37.

TULIN M.P. –1953– Steady two-dimensional cavity flows about slender bodies. *DTMB* Rpt 834.

TULIN M.P. & HSU C.C. –1980– New applications of cavity flow theory. *Proc. 13th Int. Symp. on Naval Hydrodynamics*, Tokyo (Japan), 107-132.

VASIN A. –1998– Supercavities in compressible fluid. *Proc. 3rd Int. Symp. on Cavitation*, vol. 2, Grenoble (France) 3-8.

VASIN A. –2001– The principle of independence of the cavity sections expansion as the basis for investigation on cavitation flows. *VKI/RTO Special Course on Supercavitation*. Von Karman Institute for Fluid Dynamics, Brussels (Belgium).

VILLAT H. –1911– Sur la résistance des fluides. *Annales de l'Ecole Normale Supérieure*, 203 *sq*.

WANG D.P. & WU T.Y.T. –1963– Small-time behavior of unsteady cavity flows. *Arch. Rat. Mech. Anal.* **14**(2), 127-152.

WANG D.P. & WU T.Y.T. –1965– General formulation of a perturbation theory for unsteady cavity flows. *Appl. Mech. Fluid Eng. Conf.*, June 1965.

WOODS L.C. –1955– Unsteady plane flow past curved obstacles with infinite wakes. *Proc. Roy. Soc. A* **229**, 152-180.

WOODS L.C. –1964– On the theory of growing cavities behind hydrofoils. *J. Fluid Mech.* **19**(1), 124-136.

WOODS L.C. –1966– On the instability of ventilated cavities. *J. Fluid Mech.* **26**(3), 437-457.

WU T.Y.T. –1956– A free streamline theory for two-dimensional fully cavitated hydrofoils. *J. Math. Phys.* **35**, 236-265.

WU T.Y.T. –1958– Unsteady supercavitating flows. *Proc. 2nd Int. Symp. on Naval Hydrodynamics*, 293-313.

WU T.Y.T., WHITNEY A.K. & BRENNEN C. –1971– Wall effects in cavity flows and their correction rules. *Proc. IUTAM Symp. on Non-Steady Flow of Water at High Speeds*, Leningrad (Russia), 461-470.

APPENDIX: SINGULAR BEHAVIOR AT DETACHMENT

In the complex system of coordinates $z = x + iy$ (fig. 6.22), consider the complex velocity $w = Ve^{-i\theta}$ where V is the modulus and θ the slope of the velocity. Taking the reference conditions on the cavity where the velocity is V_c, we introduce the classical logarithmic function:

$$\omega = \ln \frac{w}{V_c} = \ln \frac{V}{V_c} - i\theta \tag{6.79}$$

In the vicinity of cavity detachment, the change in boundary conditions assigns the following form to ω:

$$\omega \approx -ik\sqrt{z} \tag{6.80}$$

where k is a real constant. The following table gives the real and imaginary parts of ω on both sides of the detachment point (x is a positive real number):

	z	**ω**	**V/V$_c$**	**θ**
Cavity	x	$-ik\sqrt{x}$	1	$k\sqrt{x}$
Wall	$-x$	$k\sqrt{x}$	$e^{k\sqrt{x}}$	0

This form for ω satisfies the boundary conditions i.e. the condition of constant velocity ($V = V_c$) on the cavity and the slip condition ($\theta = 0$) on the wall.

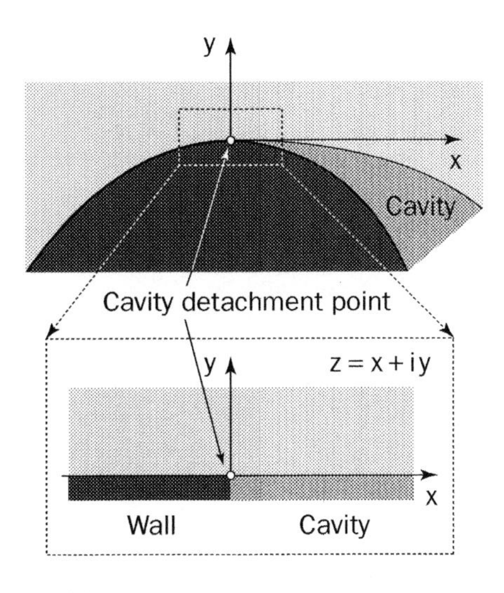

6.22 - Zoom of the detachment area

Moreover, the previous table shows that:

- on the cavity, the curvature of the free line behaves like:

$$\frac{d\theta}{dx} \approx \frac{k}{2\sqrt{x}} \tag{6.81}$$

and hence presents a singularity at detachment $x = 0$;

- on the wall, the pressure coefficient behaves like:

$$C_p = 1 - \left(\frac{V}{V_\infty}\right)^2 = 1 - (1+\sigma)\left(\frac{V}{V_c}\right)^2 \approx 1 - (1+\sigma)\, e^{2k\sqrt{x}} \cong -\sigma - 2(1+\sigma)\, k\sqrt{x} \tag{6.82}$$

and the streamwise pressure gradient presents a singularity:

$$\frac{\partial C_p}{\partial x} \approx -\frac{(1+\sigma)\, k}{\sqrt{x}} \tag{6.83}$$

which gives a vertical tangent to the pressure curve at detachment (fig. 6.23).

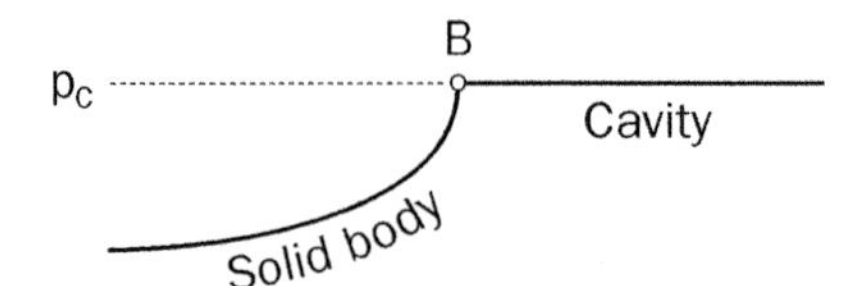

6.23

Sketch illustrating the singular behavior of the pressure gradient at cavity detachment

VILLAT and ARMSTRONG proposed choosing the point of detachment such that the solution is regular or in other words the constant k vanishes. This condition ensures the continuity of the curvature and of the pressure gradient at detachment. As the pressure gradient is zero on the cavity, it is also zero on the wall at detachment.

7. Partial Cavities

Partial cavities usually develop in regions of separated flow. Initially, cavitation appears in the shear layers limiting such a region and, as the cavitation number is decreased, it spreads over the full separated region and often extends its original size.

Partial cavities are encountered in two main practical situations, either on the upperside of foils and blades or in internal flows, such as Venturi nozzles.

In the case of hydrofoils, as the cavitation number is further decreased, a partial cavity turns into a supercavity as soon as it no longer closes on the wall but downstream of the trailing edge. Hence, partial cavitation can be regarded as an intermediate stage of development of cavitation. On the contrary, partial cavities in internal flows are the ultimate stage since they obviously always close on the solid wall, whatever their size.

Compared to supercavities, it is the aft region of partial cavities where the liquid flow reattaches to the wall which is of specific interest. While the conditions of detachment are the same for partial and supercavities, the presence of the wall makes the closure region different. The relatively small size of partial cavities can even allow the influence of the closure region to extend up to the front section. It is especially the case for cloud cavitation which occurs when a partial cavity oscillates in length and periodically sheds clouds of small vapor structures.

The main features of partial cavities are described in sections 7.1 and 7.2 whereas the cloud cavitation instability is presented in more detail in section 7.3. Section 7.4 is devoted to a description of the wake of partial cavities and section 7.5 to the prediction of thermal effects in partial cavity flows. General considerations are given on system instabilities and on the modeling of partial cavity flows at the end of this chapter.

7.1. Partial cavities on two-dimensional foils

7.1.1. Main patterns

Various cavitation patterns can occur on a two-dimensional hydrofoil according to the operating conditions. As a specific example, we consider the case of a plano-circular foil here [Le *et al.* 1993a]. Nevertheless, the observations reported are qualitatively the same for almost any type of hydrofoil.

Such observations are usually obtained from visualization in a cavitation tunnel at constant velocity or REYNOLDS number. The results are presented in the form of a map indicating the various cavitation patterns as a function of the angle of attack and the cavitation number (fig. 7.1).

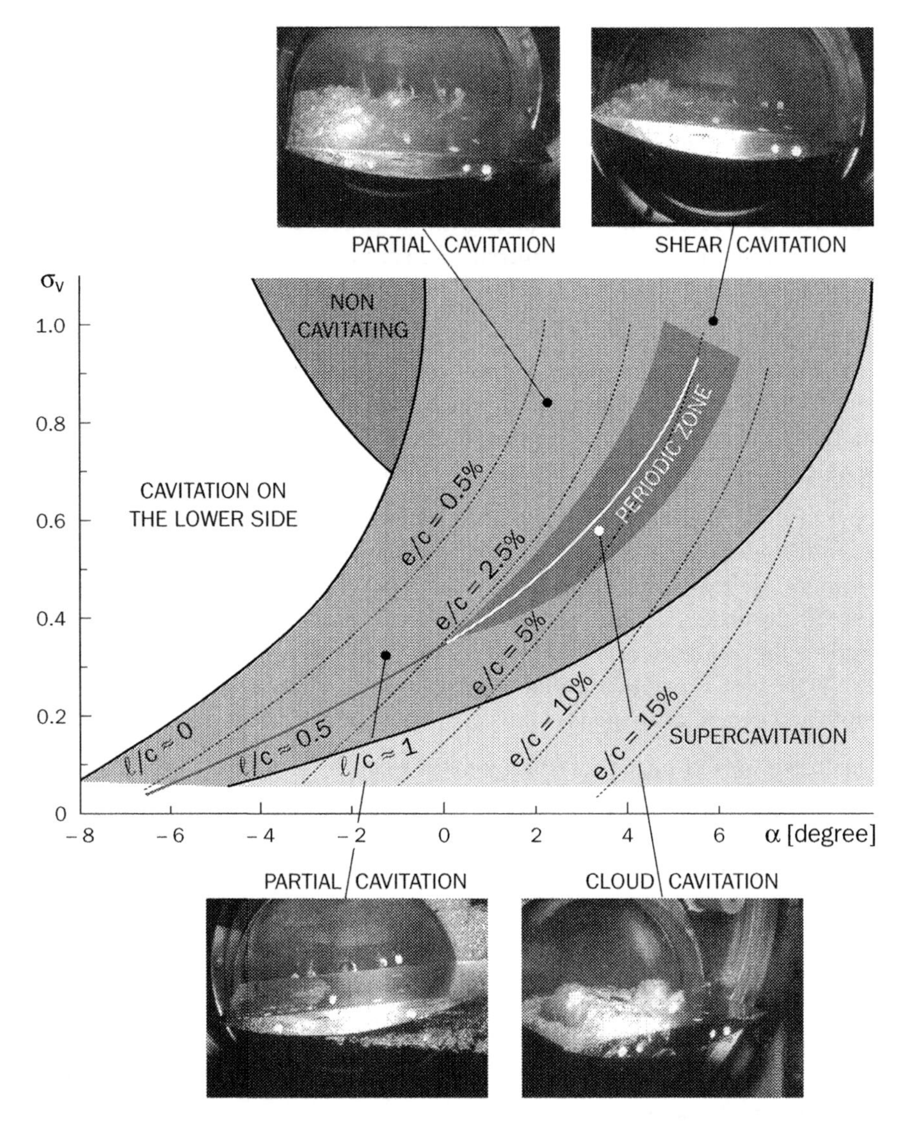

7.1 - Main cavity patterns at Re = 2×10^6 (V_∞ = 10 m/s) on the upperside of a plano-circular hydrofoil

(chord length c = 200 mm, maximum thickness 20 mm, radius of curvature at the leading edge 1 mm)

ℓ is the cavity length (the maximum in the case of an unsteady cavity), and e the maximum thickness of the cavity.

On this map, the domain of partial cavitation on the flat upperside of the foil extends between cavitation inception, limited by a cavity length ℓ equal to zero, and supercavitation, which corresponds to a cavity length ℓ equal to the chord length c. In the case of an unsteady cavity, ℓ stands for the maximum length of the cavity.

Generally speaking, there are two types of behavior observed in this partial cavitation domain:

- for small values of both the angle of attack and the cavitation number, the cavities are rather short and thin; their lengths are fairly constant and the flow is on the whole stable;
- in the upper part of the domain i.e. for large enough values of both the angle of attack and the cavitation number, the cavities are thicker and become unstable. Their lengths are variable because of the shedding of part of the cavities, entrained by the main flow. The shedding can be either random or periodic.

The region of periodic oscillations, commonly called cloud cavitation, is indicated in figure 7.1 by the shaded area. Here, it is roughly centered on the line $\ell/c = 0.5$, which indicates that this peculiar instability develops for partial cavities of medium length. On the left hand side, it is limited by a line corresponding approximately to a minimum cavity thickness (around 2.5% of the chord length here). Such a limit suggests that a minimum value of the cavity thickness is required for the periodic regime to develop. It will be seen in section 7.2 that the cavity thickness must actually be significantly larger than the re-entrant jet thickness for this instability to occur. On the right hand side, the periodic regime is bounded by a maximum value of the cavity length, which indicates that this instability should not occur for very long cavities. This condition is connected to a minimum threshold value of the adverse pressure gradient which is necessary for the re-entrant jet to gain enough impulse (see § 7.3.1). The pulsation frequency decreases when the cavity length is increased and their product tends to remain constant (see § 7.3.3).

7.1.2. *CAVITY CLOSURE*

As the minimum pressure occurs inside the cavity itself, the curvature of the surrounding streamlines tends to be directed towards the cavity (fig. 7.2). Hence, in most cases, a partial cavity reattaches to the solid wall by splitting the surrounding liquid flow into two parts:

- the re-entrant jet which travels upstream, carrying a small quantity of the liquid to the inside the cavity, and
- the outer flow which reattaches to the wall.

Both parts are separated by a streamline which, if the flow were steady, would impact upon the wall perpendicularly and ideally give rise to a stagnation point.

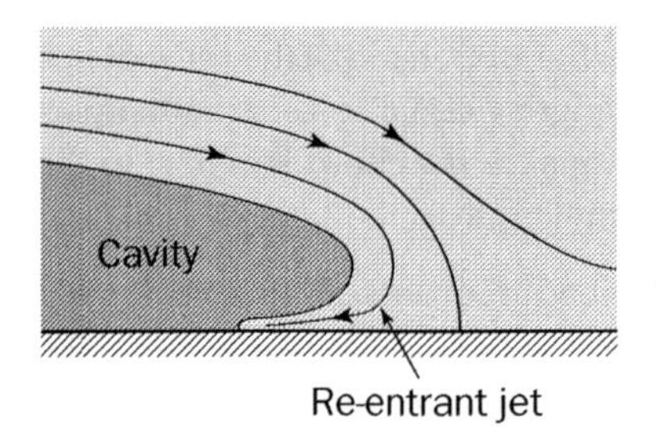

7.2
Closure region of a partial cavity

This is particularly the case with thin cavities, for which the liquid counterflow is very small and can easily be re-entrained locally by the main flow. The flow tends to be stable or at least any unsteadiness remains confined to a limited region, whose characteristic length scale is much smaller than the cavity length.

However, for thick cavities, the liquid flowrate associated with the re-entrant jet is more important and the jet may even have enough impulse to reach the front section of the cavity. Such a configuration cannot be steady, or else the cavity would be filled with liquid. Therefore, the jet from time to time strikes the front section of the cavity interface. This leads to the separation of part of the cavity which is entrained downstream by the main flow. At the instant of shedding, a circulation arises around this vapor structure, which takes the form of a spanwise vortex. It is broken up into numerous smaller vapor structures such as bubbles or cavitating vortex filaments. A new cavity then develops and grows, and a new re-entrant jet forms. As previously mentioned, this process, which is mainly controlled by inertia, can be either random or periodic (see § 7.3 for the latter).

7.1.3. *CAVITY LENGTH*

Figure 7.3 shows the variations of the upper side cavity length versus the cavitation number for the plano-circular hydrofoil previously considered. There is no difference in trends between the supercavity and the partial cavity regimes and the curves $\ell(\sigma)$ can be considered as continuous through the transition between both regimes.

These curves are re-plotted using the parameter:

$$\frac{\sigma_v}{\alpha - \alpha_i(\sigma_v)} \tag{7.1}$$

A similar parameter, σ_v / α, is classically introduced in linearized theories to take into account the almost identical effect of either a decrease in the cavitation number or an increase in the angle of attack on cavity length. Here the angle α is replaced by its difference with the angle $\alpha_i(\sigma_v)$ which characterizes cavitation inception at a given value of the cavitation number and which is known from the curve $\ell / c = 0$ of figure 7.1. The result of the scaling shows a good correlation of the cavity length with this new parameter for all operating conditions.

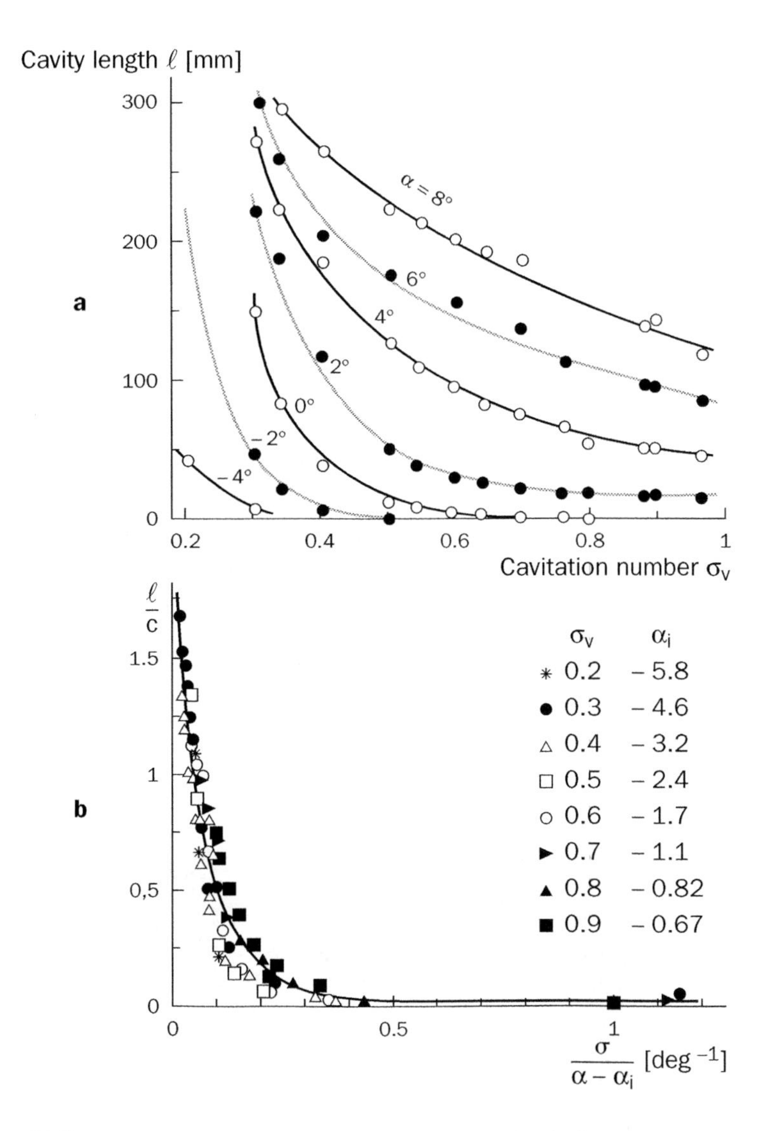

7.3 - Length of the cavity developing on a plano-circular hydrofoil (see figure 7.1)

(a) cavity length ℓ versus cavitation number σ_v
(b) non-dimensional cavity length versus the parameter $\sigma_v/(\alpha - \alpha_i)$

7.1.4. THREE-DIMENSIONAL EFFECTS DUE TO AN INCLINATION OF THE CLOSURE LINE

Even in two-dimensional configurations, it often happens that the cavity closure line is not a straight line perpendicular to the channel walls, but is curved due to the effect of the walls (see e.g. fig. 7.6-a). Three-dimensionality is then expected.

Consider the case of a cavity whose closure line is straight, but not perpendicular to the incident velocity. Assuming that the pressure gradient is zero along the closure line, DE LANGE and DE BRUIN (1998) predicted that the tangential component of the velocity along the closure line should remain constant (see also DE LANGE 1996). Hence, the re-entrant jet should simply be reflected at the closure line and so gain a spanwise velocity component (fig. 7.4).

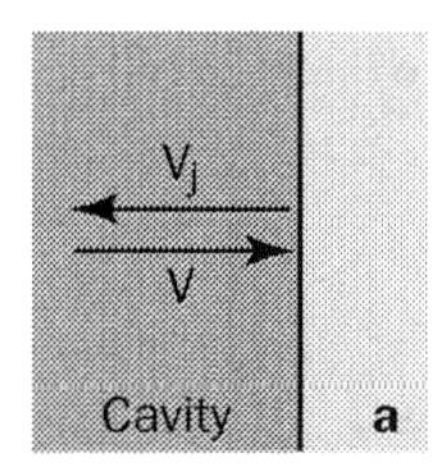

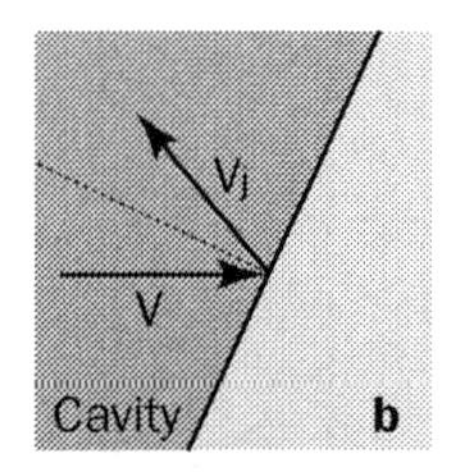

7.4 - Local 3D behavior of the re-entrant jet in the case of (a) a cavity closure line perpendicular to the free stream velocity and (b) an inclined cavity closure line
(V: oncoming flow velocity; V_j: re-entrant jet velocity) [from DE LANGE et al., 1998]

Using the previous argument, DUTTWEILER *et al.* (1998) explained how, in the case of a swept hydrofoil, the re-entrant jet can be strongly deviated and the breaking-off mechanism deeply altered by three-dimensional effects. Due to the inclination of the closure line, the re-entrant jet does not remain counter current to the flow everywhere, as shown in figure 7.5. For the upstream part of the cavity where the

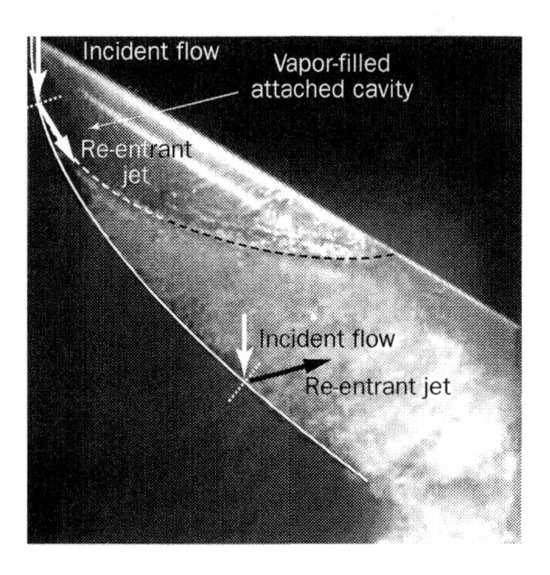

7.5 - Cavitation on a three-dimensional hydrofoil with 30° sweep, and angle of attack 2° for a flow velocity of 10.1 m/s and a cavitation number of 0.7
The flow is downward.
[from LABERTEAUX & CECCIO, 2001 – interpretation from DUTTWEILER & BRENNEN, 1998]

closure line is very inclined, the re-entrant jet is directed downstream. It never reaches the leading edge and the corresponding part of the cavity is quite stable, glossy and vapor filled. Conversely, in the downstream part where the closure line is less inclined, the re-entrant jet is directed upstream and can trigger a breaking-off mechanism comparable to the one observed in two-dimensional cases by impinging on the cavity interface. As a result, the corresponding part of the cavity appears rather frothy and unsteady.

7.1.5. *MULTIPLE SHEDDING ON 2D HYDROFOILS*

KAWANAMI *et al.* (1998) observed several regimes of vortex shedding on two-dimensional cavitating hydrofoils. The photographs in figure 7.6 present various cases of partial cavities which shed either a unique vapor structure or several along the span.

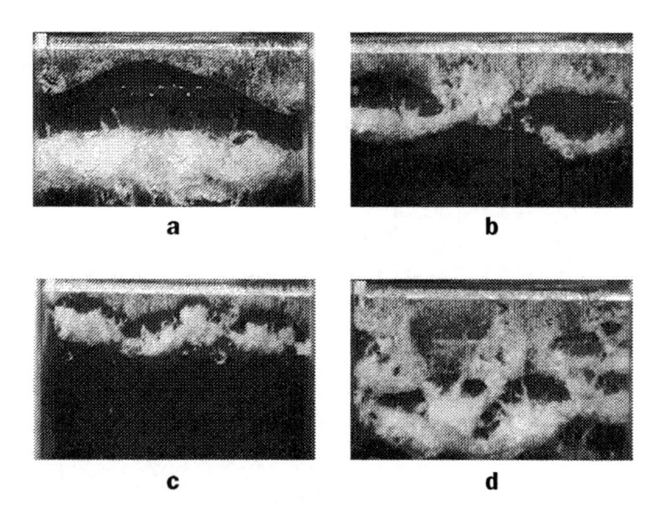

7.6 - Multiple shedding on a NACA 0015 two dimensional hydrofoil
The flow is downward.
(a) 1 cloud - (b) 2 clouds - (c) 3 clouds - (d) irregular break-off
[from KAWANAMI et al., 1998]

The regime essentially depends upon the cavity length. The spanwise length of these structures is roughly proportional to the streamwise length of the partial cavity. Multiple shedding occurs when the cavity length is smaller than the channel width. The shorter the cavity, the larger is the number of sub-vortices. An irregular break-off mechanism is also possible (fig. 7.6-d) when there is no matching of the shedding to the channel width.

7.2. PARTIAL CAVITIES IN INTERNAL FLOWS

Partial cavitation in a Venturi-type nozzle was firstly studied by FURNESS and HUTTON (1975) from both the experimental and numerical points of view. The dynamic behavior of the cavity was modeled using a two-dimensional, unsteady, potential flow theory, from which the authors succeeded in predicting the early formation of the re-entrant jet and the roller-type motion of the rear part of the cavity during its growth phase.

We present here the results of an experimental study of partial cavities [CALLENAERE *et al.* 2001] formed behind a diverging step whose geometry is shown on figure 7.7.

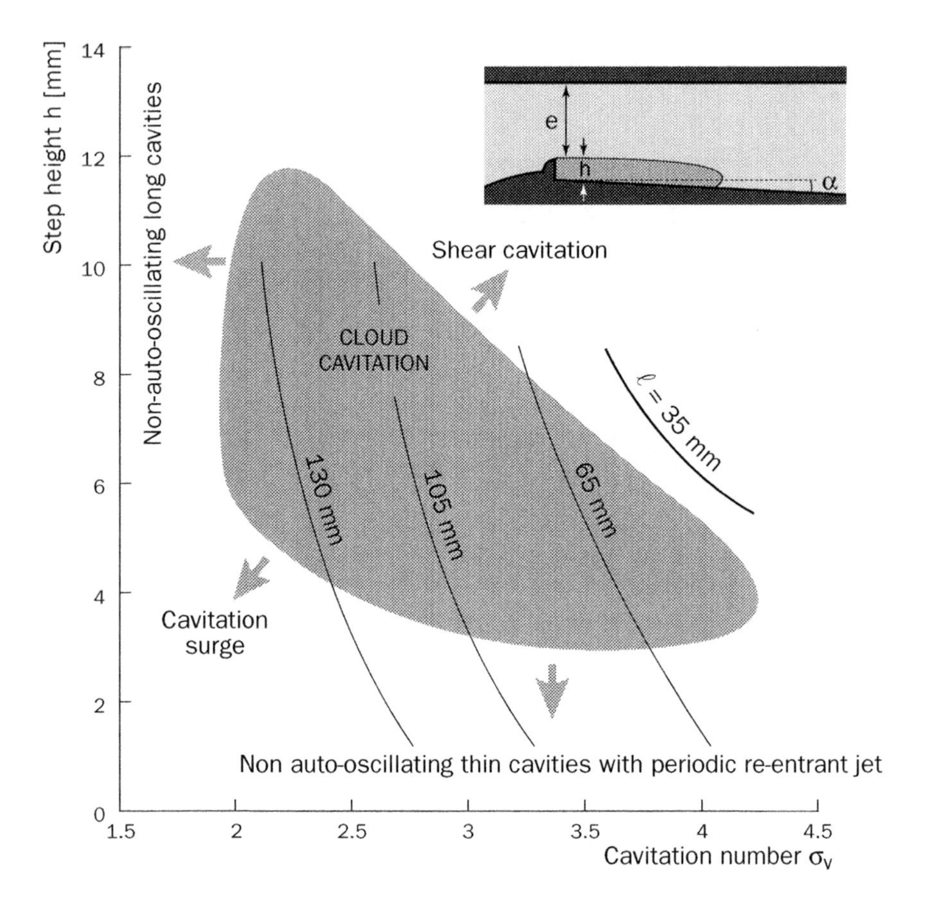

7.7 - The domain of cloud cavitation on a diverging step as a function of the cavitation number and the height h of the step

(flow velocity 11.6 m/s, confinement height e = 20 mm, divergence angle α = 4.2 deg.)
[from CALLENAERE et al., 2001]

Three geometric parameters could be adjusted, along with the usual cavitation and REYNOLDS numbers. These are the step height h, the slope α of the lower wall and the minimum distance e to the horizontal upper wall. This allowed the authors to change almost independently the cavity thickness, which depends directly on the step height h, and the adverse pressure gradient downstream of the throat which primarily depends upon the two other parameters.

The main partial cavity patterns observed in such an internal flow are presented on figure 7.7 for different values of the cavitation parameter and the step height, all other parameters (flow velocity, confinement height and angle of divergence) being constant.

The central region corresponds to cloud cavitation, with a periodic re-entrant jet and a periodic shedding of vapor clouds. The maximum cavity length increases when the cavitation number decreases, as expected, and it diminishes slightly when the step height increases. This self-oscillating regime, already observed on two-dimensional foils, will be studied in more detail in section 7.3. It is surrounded by three main regions:

- For small values of the step height h (for h smaller than about 3 mm), a periodic re-entrant jet still exists, as in the cloud cavitation regime, but the length of the cavity is almost constant. In this region, where the cavity thickness is small and close to the re-entrant jet thickness, a strong interaction exists between the re-entrant jet and the main cavity interface through their surface irregularities, leading to an almost continuous production of small scale vapor structures.
- For small σ_v-values, the cavities are longest. Basically they do not exhibit any regular oscillating behavior. From time to time, and irregularly, these long cavities are broken and a variable volume of vapor detaches and is entrained by the main flow. It may happen that such long cavities oscillate periodically, not naturally, but because of their interaction with the experimental facility. Such an oscillation is called cavitation surge. It usually occurs at low frequencies, smaller than the characteristic frequency of cloud cavitation (see § 7.6).
- The cloud cavitation domain is limited on its upper right side by shear cavitation. This pattern corresponds to the development of cavitation in the shear layer which limits the recirculating zone behind the diverging step. Its frontier with the cloud cavitation domain corresponds approximately to the filling up of the recirculating zone by cavitation. Along this frontier, the length of the cavity is of the order of 6 to 10 times the step height, as can be verified by the lines of equal cavity length given in figure 7.7. This value compares well to the mean length of the reattachment bubble behind a rearward facing step in non-cavitating flows (see for example DRIVER & SEEGMILLER 1985).

7.3. THE CLOUD CAVITATION INSTABILITY

7.3.1. CONDITIONS FOR THE ONSET OF THE CLOUD CAVITATION INSTABILITY

Strong similarities exist between the frontiers of the cloud cavitation domains in the internal configuration (see fig. 7.7) and in the case of a hydrofoil (see fig. 7.1), regarding their respective physical roles.

For both configurations, three conditions control the onset of the cloud cavitation instability and give the limits of the corresponding domain:

- firstly, the cavity thickness must be significantly larger than the re-entrant jet thickness so that the re-entrant jet is not broken during its upstream movement by interaction with the downward mean flow on the cavity interface;
- secondly, the cloud cavitation instability does not affect very long cavities. This condition is discussed in more detail below;
- finally, the cloud cavitation domain is naturally limited by shear cavitation as the non-cavitating regime is approached.

Consequently the physics of the cloud cavitation instability appears to be comparable for external and internal flow configurations. This conclusion is corroborated by the comparison of STROUHAL numbers whose range appears comparable in both cases.

For both configurations, long cavities are not affected by the cloud cavitation instability. In fact, CALLENAERE *et al.* (2001) have shown that the key parameter is not the cavity length but actually the adverse pressure gradient imposed by the external flow around cavity closure.

Indeed, the cavity length and the external pressure gradient are closely linked. Whatever the configuration (2D hydrofoil or internal flow), the longer the cavity, the smaller the adverse pressure gradient becomes at closure.

This point is particularly clear in the case of a diverging channel. On the basis of a simplified one-dimensional approach and using the mass conservation and BERNOULLI equations, the pressure distribution along a non-cavitating divergent is given by:

$$C_p = \frac{p(x) - p(0)}{\frac{1}{2}\rho V(0)^2} = 1 - \left[\frac{S(0)}{S(x)}\right]^2 \tag{7.2}$$

where $p(x)$, $V(x)$ and $S(x)$ are the pressure, velocity and cross-sectional area of the divergent respectively at the distance x. Hence, the pressure gradient is given by:

$$\frac{dC_p}{dx} = \frac{2S(0)^2}{S^3}\frac{dS}{dx} \tag{7.3}$$

Because of the S^{-3} term, the pressure gradient clearly decreases downstream. Although the pressure distribution is, to some extent, changed by the development of cavitation, this argument remains qualitatively applicable to cavitating conditions and we can expect that the longer the cavity, the smaller the adverse pressure gradient at closure.

This point remains valid for partial cavities on a hydrofoil. The typical L-shape of the curves $\ell(\sigma)$ (see fig. 7.3) shows that, for large values of the cavitation number (i.e. for short cavities), small variations of σ (caused for example by variations of the ambient pressure) result in very small variations in cavity length ℓ. However, long cavities are much more sensitive to external pressure fluctuations and exhibit large variations in length for even small variations in pressure.

This difference in behavior is an indicator of the mean pressure distribution in the closure region. For small cavities, the adverse pressure gradient at closure is high enough to prevent the cavity from extending significantly after a small pressure drop. Conversely, for long cavities, the pressure hardly varies at closure, so that a small decrease in upstream pressure causes a substantial part of the wetted wall to fall below the vapor pressure. This leads to a dramatic extension of the cavity, with a possible transition to supercavitation. The cavity length and the adverse pressure gradient are then strongly linked.

It is clear then that a strong adverse pressure gradient at closure is very favorable to the development of the re-entrant jet. A simple approach shows that its thickness will increase proportionally to the adverse pressure gradient (see § 7.3.4). Consequently, this will promote the cloud cavitation instability.

Long cavities, which generally close in a region of small adverse pressure gradient, do not exhibit the cloud cavitation instability. On the contrary, they are more sensitive to system instabilities. Because of the relatively flat pressure distribution around the closure point of long cavities, even small external pressure fluctuations can make them oscillate very significantly in length. Such a situation is typical of a system instability, in so far as the cavitation behavior depends upon the upstream pressure and therefore upon the whole system comprising the partial cavity coupled with its surroundings and in particular with the circuit (see § 7.6).

In the experiments of CALLENAERE *et al.* (2001), the domain of the cloud cavitation instability is considerably reduced if the angle of the divergent is decreased or if the thickness of the channel is increased. It may even completely disappear for a large enough value of the channel thickness. These observations confirm the major role of the adverse pressure gradient on the cloud cavitation instability.

7.3.2. GLOBAL BEHAVIOR

A schematic description of cavity evolution during one period of oscillation is presented in figure 7.8, as reconstructed from high speed movies. It represents partial cavitation on a hydrofoil, although the physics of the phenomenon is qualitatively similar in a Venturi.

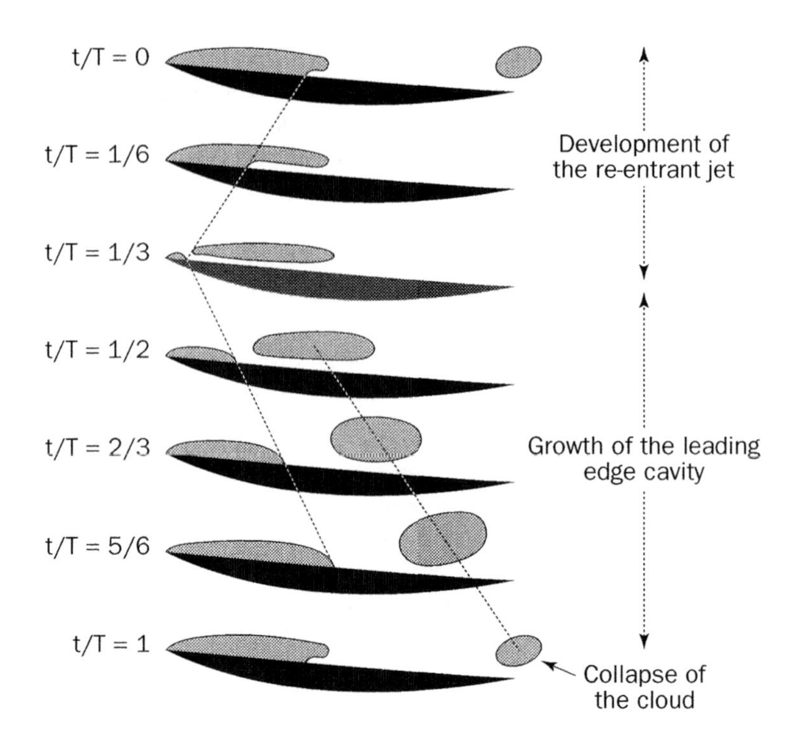

7.8 - Typical unsteady behavior of a partial cavity with the development of a re-entrant jet and the periodic shedding of cavitation clouds
[from LE et al., 1993a]

The re-entrant jet flows upwards and it requires about one third of the period for the re-entrant jet to reach the front part of the cavity. From that moment, a new leading edge cavity grows over the rest of the period, while the vapor cavity shed near the leading edge is advected downstream. It loses its slenderness very quickly [KUBOTA *et al.* 1987] and becomes a two-phase vortex, before collapsing quasi simultaneously with the onset of the re-entrant jet.

According to the configuration and the operating conditions, some differences have been observed, for example in the travel time of the jet. Differences have also been reported in the velocity of the re-entrant jet although it is always of the order of the flow velocity V_∞, and generally a little smaller. In the framework of steady potential flow theory, the velocity of the re-entrant jet should be equal to the velocity on the cavity interface, i.e. $V_\infty\sqrt{1+\sigma_v}$.

The previous description of the phenomenon shows that the re-entrant jet plays a prominent role in the periodic behavior of partial cavities. This point was demonstrated by KAWANAMI *et al.* (1997). They showed that a small obstacle placed on the wall prevents the generation of cloud cavitation by stopping the progression of the re-entrant jet.

The physical reason for the onset of the re-entrant jet and consequently for the onset of cloud cavitation is not yet quite clear. Two explanations have been put forward. On one hand, the moment of birth of the jet often coincides with the collapse of the previously shed cloud. Hence, it is conjectured that the overpressure caused by the collapse could trigger the jet. On the other hand, after its period of growth, the leading edge cavity keeps an almost constant length during part of the period (approximately one third according to figure 7.8). This should be long enough for steady flow conditions to be restored and the re-entrant jet to develop naturally, with negligible influence from the cloud collapse.

During the growth phase of the leading edge cavity, the interface usually appears glossy and the cavity clear, so that the cavity is probably filled with vapor only. This is not the case for the shed cloud which is more of a bubbly mixture. In some cases, the growing cavity itself may be filled with a mixture of liquid and vapor. This was the case for the cavities investigated by STUTZ *et al.* (1997a, b) in a Venturi. Using optical probes, they measured void fractions of only 10 to 20% depending on the location in the cavity and the time in the period. The cavity content is important when it comes to estimating thermal effects, as mentioned in section 7.5.

7.3.3. *PULSATION FREQUENCY*

From the previous description, the period $T = 1/f$ of the phenomenon appears to be correlated to the time required for the re-entrant jet to cover the cavity length ℓ. As the re-entrant jet velocity is of the order of the flow velocity V_∞, this time is of the order of ℓ / V_∞. The ratio of this characteristic time to the period of oscillation is the STROUHAL number:

$$S = \frac{f\ell}{V_\infty} \tag{7.4}$$

where f denotes the frequency of cloud shedding.

Usually, the STROUHAL number lies in the range 0.25-0.35. A major reason for the non-negligible dispersion in STROUHAL number is probably the large uncertainty on the determination of the cavity length. A typical value is 0.3, which means that the rise time of the re-entrant jet is about 30% of the shedding period.

Compared to other self-oscillating flows, (e.g. the BÉNARD-KARMAN wake vortices for which the STROUHAL number is based on the crosswise characteristic length of the body), here, the reference length is a streamwise length scale, the cavity length.

The re-entrant jet contributes to the shedding of circulation into the cavity wake, together with other classical mechanisms related to liquid viscosity. The order of magnitude of this contribution can be easily estimated. The circulation around each vapor cloud, at the instant of shedding, is $2\ell V_\infty$ as the cloud has a typical perimeter length 2ℓ and the fluid velocity is of the order of V_∞ on the upper and lower parts of the cloud. Thus the production rate of circulation is given by:

$$\frac{2\ell V_\infty}{T} = 2S V_\infty^{\;2} \tag{7.5}$$

The coefficient 2S on the right hand side is approximately 0.6 for cavities with a well formed re-entrant jet.

7.3.4. *JET THICKNESS*

Measurements of the thickness of the re-entrant jet were carried out by CALLENAERE *et al.* (2001) using the reflection of ultrasonic waves on a divergent step (see § 7.2). The thickness of the jet lays between 15 and 30% of the maximum cavity thickness. The same order of magnitude was found by STUTZ and REBOUD (1997a, b) for the reverse flow detected in their Venturi by means of a double optical probe.

Such values are much larger than the typical values which are obtained from non-linear potential flow modeling of a steady cavity behind a backward facing step in a semi-infinite flow field (see e.g. MICHEL 1978), which is of the order of a few percent at most.

Indeed, the adverse pressure gradient imposed by the external flow increases the impulse of the re-entrant jet and enhances very significantly its thickness. A simple phenomenological approach can be derived to demonstrate this.

If the mean pressure gradient imposed by the external flow is $\frac{dp}{dx}$, the order of magnitude of the corresponding pressure difference along a characteristic length chosen as the cavity length ℓ is $\ell\frac{dp}{dx}$. This pressure difference is typically applied on a characteristic length of the order of the cavity thickness t, so that the corresponding pressure force, per unit span length, is of the order of $t\ell\frac{dp}{dx}$. This force contributes to an increase in the re-entrant jet momentum. As its velocity V_j is almost constant and remains of the order of $V_\infty\sqrt{1+\sigma}$, the increase in momentum results mainly in an increase in thickness $\Delta\delta$. The momentum balance of the re-entrant jet is approximately:

$$\rho V_j^2 \Delta\delta \approx t\ell\frac{dp}{dx} \tag{7.6}$$

This equation can be re-written in non-dimensional form as:

$$\frac{\Delta\delta}{t} \approx \frac{1}{2}\frac{dC_p}{d(x/\ell)} \tag{7.7}$$

A more precise derivation of the previous equation can be found in CALLENAERE *et al.* (2001).

Hence, the re-entrant jet thickness increases proportionally to the adverse pressure gradient. The measurements of CALLENAERE *et al.* on their configuration show that the non-dimensional gradient of the pressure coefficient in equation (7.7) is of the order of 0.6 so that the increase in jet thickness due to the adverse pressure gradient is typically of the order of some tens of percent of the cavity thickness. Although such an estimation is very crude, it shows that the effect of the adverse pressure gradient must be considered in order to obtain orders of magnitude of the jet thickness consistent with experimental data.

7.4. WAKES OF PARTIAL CAVITIES

In industrial situations, the wakes of partial cavities are often the location of intense erosion and noise. In recent years, a substantial effort has been devoted to the measurement of the production rate of vapor structures downstream of partial cavities in order to estimate their erosion and noise potential. Mean pressure distributions in the wakes of cavities were also investigated as backup to the modeling of partial cavitation on the basis of steady state approaches.

7.4.1. MEAN PRESSURE DISTRIBUTION

As mentioned in section 7.1.2, the flow in the closure region of a partial cavity tends to develop a stagnation point. In the case of stable cavities, an overpressure actually exists near the cavity end, as shown by the measured pressure distribution on figure 7.9 and corresponding to the smallest σ_v-value. However, the mean value of the maximum pressure coefficient does not exceed about 0.2, whereas it should reach unity in the ideal case of a stagnation point.

The maximum value in the mean pressure distribution progressively decreases as the cavitation parameter increases and the cavity becomes more unsteady (see fig. 7.9). It even vanishes for high enough values of the cavitation parameter.

A pressure distribution with a maximum is typical of closed cavities, i.e. cavities which present few vapor structures in their wake. On the contrary, open cavities, followed by a wake with a significant amount of vapor, exhibit a gradually increasing pressure in their wake.

This distinction between closed and open cavities is commonly used in the numerical modeling of partial cavitation. Various models have been developed to represent the closure region of partial cavities within the conditions of steady flow. If such conditions are relaxed, a special model is no longer needed to account for closure of the cavity which does so naturally through unsteady processes.

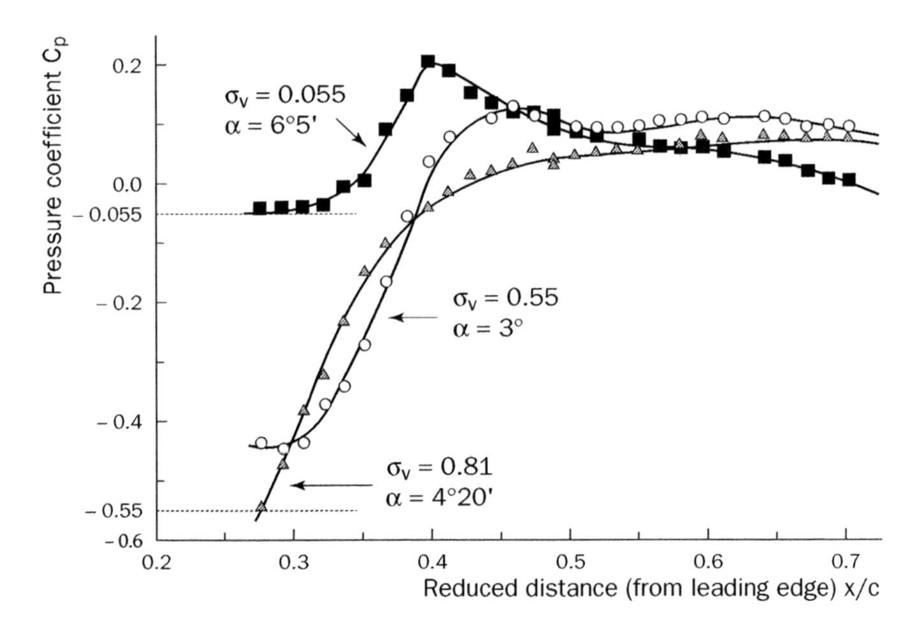

7.9 - Influence of partial cavitation patterns on the mean pressure distribution at constant cavity length for a plano-circular hydrofoil (see figure 7.1) for different combinations of the angle of attack and the cavitation number, all leading to the same reduced cavity length $\ell/c \cong 0.37$ (REYNOLDS number 2.10^6)
[from LE et al., 1993a]

For small enough values of the cavitation number, i.e. for closed cavities, the pressure coefficient is almost equal to $-\sigma_v$ in the cavity (see fig. 7.9). Therefore, the cavity pressure is close to the vapor pressure and the cavity is filled with effectively pure vapor. As the cavitation number is increased, the cavity pressure progressively departs from the vapor pressure. The cavity turns into an open one and the void fraction inside the cavity decreases, leading to a two-phase mixture rather than a vapor cavity.

7.4.2. *PRODUCTION OF VAPOR BUBBLES*

Partial cavities produce vapor structures which are entrained in the wake by the main liquid flow. Thin and stable cavities generate a large number of very small bubbles whereas thick cavities shed large scale clouds which subsequently transform into smaller structures such as bubbles and vapor vortices.

Using a holographic method, MAEDA *et al.* (1991) measured the bubble population in cavitation clouds. A typical result is given in figure 7.10. The total concentration of bubbles measured in the range of radius 10-100 μm is about 4,000 bubbles/cm^3. Under non-cavitating conditions, the bubble concentration is about two orders of magnitude smaller. The concentration in the range 30-50 μm, which is a typical range for cavitation nuclei, is about 330 nuclei/cm^3. Such high values show that partial cavities can be the source of an abundant production of cavitation nuclei.

Other results of bubble measurements downstream of partial cavities, aimed at estimating the influence of dissolved air on the production of nuclei, can be found in PO and CECCIO (1997).

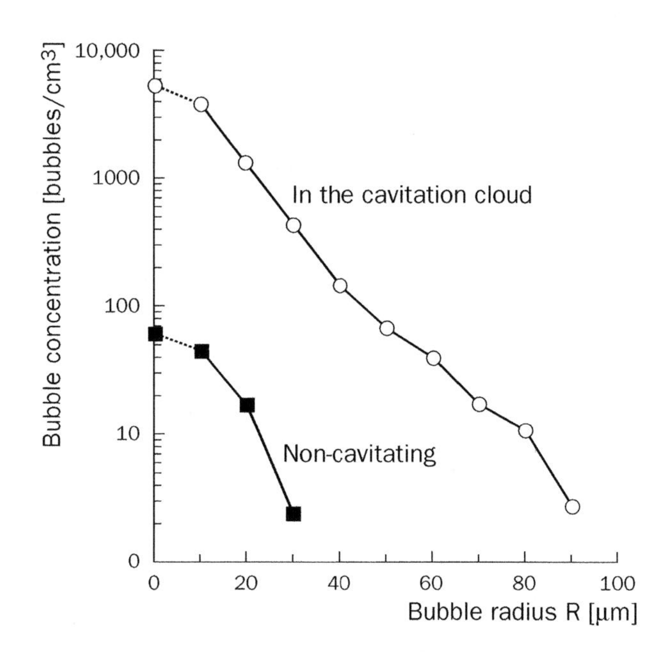

7.10 - Cumulative histogram of bubble concentration as a function of bubble radius in a cavitation cloud

Cavitation was generated by a NACA 0015 foil (chord length 80 mm, angle of attack 8.36°, flow velocity 7.8 m/s, $\sigma_v = 1.96$). The ordinate gives the concentration of bubbles whose radius is greater than that indicated on the abscissa. As a reference, measurements in non-cavitating conditions are also shown [from MAEDA et al., 1991].

7.4.3. PRESSURE FLUCTUATIONS

Pressure fluctuations in the cavity wake have been measured by several investigators using flush-mounted pressure transducers. According to the frequency domain considered, different phenomena are revealed.

In the case of cloud cavitation, the low-frequency content of the pressure signal exhibits oscillations at the same frequency as the cavity. On the pressure signal, it is possible to distinguish low and high pressure levels, corresponding approximately to the presence, above the transducer, of either the leading edge cavity or the cavity wake.

If the rise time of the pressure transducer is short enough, typically smaller than the microsecond, the high-frequency content of the pressure fluctuations can be investigated. REISMAN *et al.* (1998) have identified two kinds of pressure pulses which are the result of either global or local events.

Global events are typically connected to the coherent collapse of large scale clouds. When such a cloud is convected to a region of high pressure, it collapses as a whole and generates an instantaneous overpressure which is simultaneously detected by all the transducers located in the collapsing zone.

As well as these global events, local events have also been observed. They are randomly distributed and generally correspond to the passage of a front of strongly varying void fraction on the sensitive surface of the pressure transducer. REISMAN *et al.* interpreted these as shock waves in the bubbly mixture.

It should be noted that the sonic speed inside such a medium is very small (typically of the order of 0.1 m/s) in the absence of non-condensable gas (see appendix). Then, any perturbation which propagates at the speed of sound through the cloud can be considered as almost frozen with respect to the cloud. Hence, it is nearly impossible to distinguish its motion from the motion of the cloud itself. Nevertheless, the existence of shock waves is an important feature of cavitation clouds as their possible focusing can generate very high local pressures and contribute strongly to noise and erosion [REISMAN *et al.* 1998].

7.4.4. *WALL PRESSURE PULSES AT CAVITY CLOSURE*

It is also possible to record the pressure pulses due to the vapor structures which are shed in the wake of partial cavities and collapse on the wall. From the point of view of erosion, pressure pulse height spectra are often used to characterize the aggressiveness of partial cavitating flows [DE *et al.* 1982, FRY 1989, IWAI *et al.* 1991].

The experimental determination of pressure pulse height spectra raises some difficulties:

- The pressure transducers must have a high natural frequency to reproduce as reliably as possible the sudden rise in pressure, typically of a few microseconds or less in duration. If not, the signal height is underestimated.
- The sensitive surface must be very small and in theory smaller than the size of the impacted area. If not, the measured pressure is actually the equivalent mean pressure which would give the same output if it were uniformly applied on the whole sensitive surface. This condition is practically impossible to meet so that the output signal is to be interpreted as the total force applied to the transducer rather than a true pressure.
- The transducer must obviously be sufficiently resistant not to be damaged.

- A dynamic calibration of the transducer needs to be carried out with a device capable of generating pressure pulses of rise time and amplitude comparable to the ones to be measured, say a few hundreds of MPa for a time of the order of the microsecond.

As an example, some typical results obtained by LE *et al.* (1993b) on the plano-circular foil already considered in sections 7.1.1 and 7.4.1 are presented here. The measurements consisted in counting the peaks whose amplitude is greater than a given threshold, which gives cumulative data.

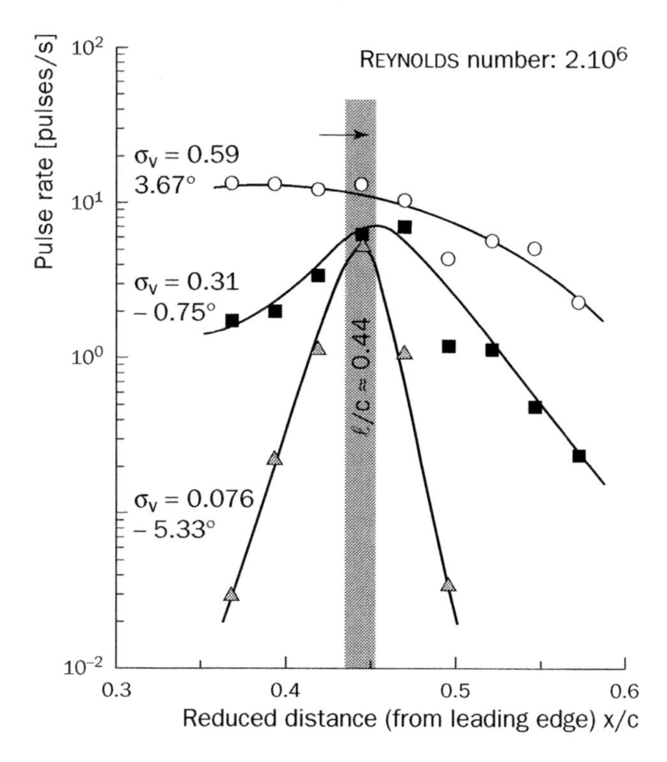

7.11 - Influence of partial cavitation patterns on the pressure pulse distribution around cavity closure for a constant cavity length

The ordinate is the number of pulses per second detected by a piezoelectric ceramic (sensitive surface area 0.64 mm², natural frequency 1.8 MHz). The height threshold is 0.5 V which corresponds approximately to a force of 0.32 N or an equivalent pressure of 0.5 MPa (if uniformly applied on the whole sensitive surface) [from LE et al., 1993b].

The region of maximum aggressiveness of a thin and stable cavity (case $\sigma_v = 0.076$ of figure 7.11) appears to be centered on the cavity closure. This proves that the production of bubbles by a stable partial cavity is concentrated around cavity closure and that the bubbles resorb relatively quickly because of a rapid increase in mean pressure.

As the cavitation number increases and the cavity becomes more unsteady, the pressure pulses are more widely distributed around closure. In the limiting case of cloud cavitation ($\sigma_v = 0.59$), there is no clear maximum in the distribution of pressure pulses and the dangerous zone, from an erosion viewpoint, may extend from the cavity leading edge to far downstream of closure. This spread accompanies a significant increase in the number of pressure pulses and also in their amplitude. This conclusion corroborates the well-known high erosive potential of cloud cavitation.

7.4.5. SCALING OF PULSE SPECTRA

In order to estimate the effect of either the liquid velocity or the length scale of the flow on pressure pulse height spectra, the following scaling procedure can be used. The approach is valid for geometrically similar flows, which requires not only similarity in the geometrical boundaries, but also equality of cavitation numbers to ensure equal relative cavity lengths.

It is assumed that each pressure pulse results from a shock wave mechanism occurring at the end of the bubble collapse, so that the acoustic impedance ρc (where ρ stands for the liquid density and c for the sound velocity) is considered as the major relevant physical property of the liquid. All other physical properties such as surface tension for instance are assumed secondary.

Let us denote by $\dot{n}$ the rate at which pulses, whose height is greater than a given pressure height Δp, occur per unit time and unit surface area. A simplified phenomenological analysis leads to the conclusion that the pulse rate is a function of the variables Δp, V, L and ρc:

$$\dot{n} = \text{Function}\,(\Delta p, V, L, \rho c) \tag{7.8}$$

where V is the flow velocity and L a characteristic length scale of the flow.

The VASCHY-BUCKINGHAM theorem allows us to write the previous relation in the following non-dimensional form:

$$\frac{\dot{n}L^3}{V} = \varphi\left(\frac{\Delta p}{\rho c V}\right) \tag{7.9}$$

where φ is a universal function for the considered family of similar flows.

Figure 7.12 validates to some extent the previous scaling law where the length scale L is kept constant (i.e. for a fixed geometry), with different flow velocities and a cavitation parameter assumed constant throughout. By dividing both the pulse rate $\dot{n}$ and the pulse height Δp by the velocity and plotting $\dot{n}/V$ as a function of $\Delta p/V$, a reasonable regrouping of the data is observed.

Equation (7.9) also shows the influence of the length scale L on the pulse rate $\dot{n}$. It appears that $\dot{n}$ varies like L^{-3}. In other words, short cavities have high rates of vapor bubble production. This principle is actually used in practice to increase the nuclei content of water using cavitating injectors of small-size.

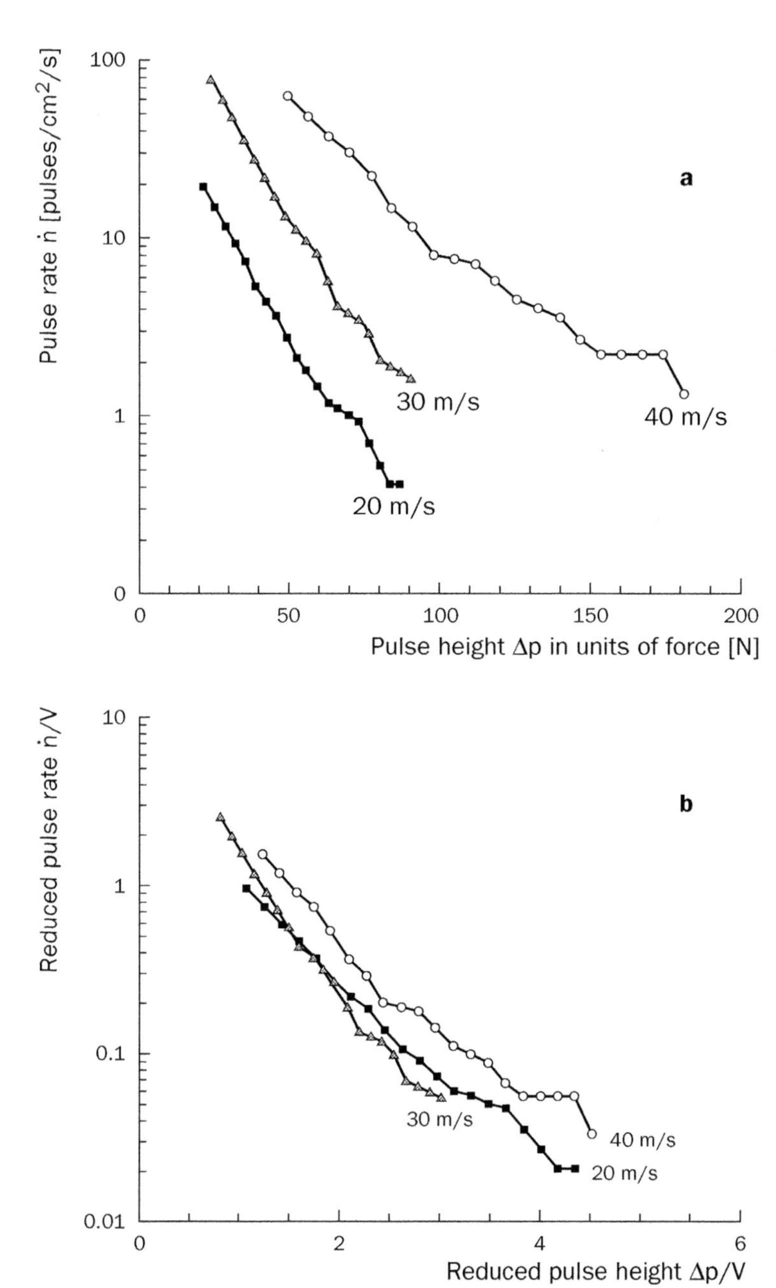

7.12 - Pressure pulse height spectra downstream of a partial cavity in a Venturi for three different water velocities and a constant cavitation number

Figure 7.12-a presents the original dimensional spectra and figure 7.12-b the reduced ones obtained by dividing both pulse rates and pulse heights by the flow velocity (unpublished results from the authors).

The scaling law represented by equation (7.9) can be used, to first approximation, to estimate the influence of the velocity, the length scale and the fluid properties on the aggressiveness of similar cavitating flows, taking into account inertia and acoustic impedance only, all other effects being assumed of minor importance. This question will be treated in more detail in chapter 12, where cavitation erosion is considered.

7.4.6. MAIN FEATURES OF THE NOISE EMITTED BY PARTIAL CAVITIES

When collapsing, an isolated bubble emits a noise very rich in high frequencies because of its sphericity and generally weak damping (see § 5.2). The situation is different in the case of partial cavities since the major source of noise is the near wake of the cavity where different kinds of vapor structures (often far from spherical bubbles) are present and where dissipation is high because of turbulence.

Besides the noise generated by the collapse itself, an additional component is due to the impact of interfaces on the solid walls. The result is a noise spectrum which extends in a broad band, from high to low, audible frequencies.

The noise emitted by a cavitating source is often difficult to measure in a cavitation tunnel because of interference due to reflection, transmission and absorption of acoustic waves by the tunnel walls. Despite these difficulties, global noise measurements are often conducted in order to know, at least qualitatively, how cavitation noise changes with the flow parameters.

When the cavitation number σ_v is reduced below non-cavitating values, the flow noise suddenly increases at cavitation inception. It continues to increase but more gradually and finally decreases when the cavity becomes fully developed. For a foil in a cavitation tunnel at a velocity around 8 m/s for instance, the noise level may increase by typically 30 dB over the whole spectrum at cavitation inception and its maximum may reach 150 to 160 dB. For traveling bubble cavitation, the noise increases continuously when σ_v decreases, as long as the bubbles remain separated and provided the air content is low so that damping remains small. The noise spectrum at high frequencies is usually higher for traveling bubble cavitation than for partial cavitation.

An increase in flow velocity at constant σ_v (i.e. for similar extents of partial cavitation) commonly leads to an increase in cavitation noise as already mentioned for pulse height spectra (see fig. 7.12).

As for the air content, it generally does not significantly affect the noise of developed cavities. For traveling bubble cavitation, the effect of air content is not obvious *a priori*. On one hand, a high air content in the water will result in a high air content inside the bubbles, and then in a damping effect. On the other hand, a high air content will generate a large concentration of air nuclei which represent as many sources of noise. The overall effect of air content on the noise of traveling bubble cavitation is therefore not at all straightforward.

The relation between noise and erosion has been studied since the fifties (see e.g. VARGA & SEBESTYEN 1972). KATO, YE and MAEDA (1988) proved that cavitation noise and pitting rate are strongly correlated on two-dimensional foils. This demonstrates that both phenomena basically depend upon the rate of production of cavitating structures. However, such a correlation cannot be used directly for the monitoring of cavitation erosion in hydraulic machinery, since there is no major difference in the noise signatures between erosive and non-erosive cavitation.

7.5. *THERMAL EFFECTS IN PARTIAL CAVITATION*

In the previous sections, heat transfer associated with phase change was disregarded. In many practical situations and particularly in cold water, this simplification is valid because the ratio of the vapor and liquid densities ρ_v / ρ_ℓ is very small (about 1/60,000 for cold water). If so, the mass flowrate required to fill the cavity with vapor is negligible and heat transfer associated with phase change is a secondary phenomenon.

Referring to the case of a spherical bubble (see § 5.3.1), this assumption is valid far from the critical point, which is the case of water at room temperature. When the temperature increases and becomes close to the critical temperature, the ratio ρ_v / ρ_ℓ increases and tends to unity. Heat transfer can no longer be ignored. This delays the development of cavitation.

Due to thermal effects, the temperature inside the cavity is lower than in the liquid. The vapor pressure is smaller and the cavitation parameter σ_v is greater, resulting in reduced cavitation. This occurs, for example, in the case of cavitating inducers in rocket engines using liquid hydrogen.

For partial cavitation, heat and mass transfer through the turbulent cavity interfaces are complex phenomena. The situation is complicated by the re-entrant jet and its unsteady character, and also by the possible two-phase nature of the cavity content as mentioned in section 7.3.2. The problem is generally approached empirically, the thermal effect usually taken as a correction to the mechanical problem which is considered as the driving force in cavitation.

7.5.1. *THE STEPANOFF B-FACTOR*

The thermal delay in pumps was first investigated by STEPANOFF (1961).[1] As mentioned in section 5.3.1, STEPANOFF defined the non-dimensional B-factor as the inverse ratio of the volume of liquid $\mathcal{V}_\ell$ to be cooled to supply the heat necessary for the production of a given volume $\mathcal{V}_v$ of vapor (see eq. 5.52):

1. See the reference in chapter 5.

$$B = \frac{\mathcal{V}_v}{\mathcal{V}_\ell} = \frac{\Delta T}{\frac{\rho_v L}{\rho_\ell c_{p\ell}}} \cong \frac{\Delta p_v}{\frac{\rho_v L^2}{\rho_\ell c_{p\ell} T}} \tag{7.10}$$

Here ΔT stands for the difference between the liquid bulk temperature T_∞ and the temperature T_c inside the cavity, while Δp_v is the corresponding variation in vapor pressure. Evaluation of the B-factor allows us to estimate the pressure drop and the σ_v increase.

A direct determination of the B-factor in liquid hydrogen and refrigerant 114 was made by MOORE and RUGGERI (1968) using similar Venturi nozzles with different length scales (see tab. 7.1). All data were compared to a given reference case. The authors proposed the following exponents for the effects of flow velocity, length scale and cavity length on the B-factor:

$$\frac{B}{B_{ref}} = \frac{\alpha_{ref}}{\alpha}\left(\frac{V}{V_{ref}}\right)^{0.8}\left(\frac{D}{D_{ref}}\right)^{0.2}\left[\frac{\ell/D}{(\ell/D)_{ref}}\right]^{0.3} \tag{7.11}$$

This correlation worked well for water, liquid nitrogen, liquid hydrogen and refrigerant 114. As an example, the measured vapor pressure difference due to thermal effects lies in the range 0.42 to 1.22 bars in liquid hydrogen at 20.3 K.

	Liquid hydrogen	**Refrigerant 114**
Length scale	scale 0.7	scales 0.7 and 1
Throat diameter D (mm)	24.5 mm	24.5 and 35 mm
Temperature (K)	$20.3 < T < 23.3$ K	$255 < T < 303$ K
Velocity (m/s)	$33.5 < V < 62.5$ m/s	$6 < V < 15$ m/s
Cavity length	$1.0 < \ell_c < 2.6$ D	$0.2 < \ell_c < 2.6$ D

Table 7.1 - Experimental conditions

The reference flow corresponds to the following conditions: refrigerant 114, scale 1, $T_{ref} = 299$ K, $V_{ref} = 6.0$ m/s, $\ell / D_{ref} = 0.72$, $\alpha_{ref} = 1.41\ 10^{-4}$ m²/s, $B_{ref} = 2.6$ [from MOORE & RUGGERI, 1968].

7.5.2. THE ENTRAINMENT METHOD

This method was derived by HOLL *et al.* (1975). The authors originally worked on axisymmetric flows around ogives, but their model is also suitable to partial cavities which develop on two-dimensional lifting foils or pump blades.

Principle

The flow is supposed steady. As vapor is released at the rear of the cavity, an equivalent volumetric flowrate q_v must be taken from the liquid and vaporized through the cavity interface such that the cavity volume remains constant on average (fig. 7.13). Because of heat transfer, a thermal boundary layer develops along the cavity interface, through which the temperature falls from T_∞ to T_c. At equilibrium, the cavity temperature T_c is such that the vapor production flowrate exactly balances the evacuated flowrate. The heat balance gives:

$$\rho_v q_v L = h \Delta T \ell \tag{7.12}$$

where h is the mean heat transfer coefficient over the cavity of length ℓ. Then:

$$\Delta T = \frac{\rho_v q_v L}{h\ell} \tag{7.13}$$

This relation is made non-dimensional by introducing the following parameters:

- the NUSSELT number (based on the cavity length ℓ): $Nu_\ell = \dfrac{h\ell}{\lambda}$,
- the REYNOLDS number (also based on ℓ): $Re_\ell = \dfrac{V\ell}{\nu}$,
- the liquid PRANDTL number: $Pr = \dfrac{\nu}{a}$ (a is the thermal diffusivity),
- the vapor flowrate coefficient: $C_{Qv} = \dfrac{q_v}{\ell V}$.

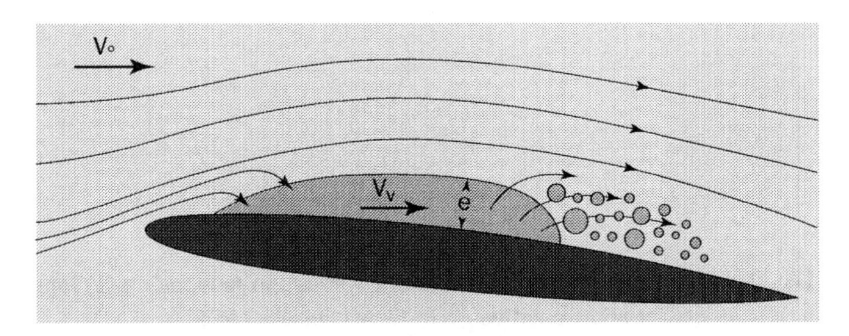

7.13 - Schematic illustration of vaporization in the front part of a cavity and vapor entrainment by the outer liquid flow and re-condensation in the aft part

The temperature difference is then expressed by the following non-dimensional equation:

$$B = C_{Qv} \frac{Re_\ell Pr}{Nu_\ell} \tag{7.14}$$

The calculation of ΔT or of its non-dimensional form B requires evaluation of two parameters:

- the flowrate coefficient C_{Qv}, which characterizes the entrainment of vapor at the rear of the cavity;
- the NUSSELT number Nu_{ℓ}, which characterizes heat transfer at the cavity interface.

Determination of the flowrate coefficient

Direct evaluation of the vapor flowrate coefficient is difficult. A rough estimate can be obtained by using ventilation. An artificial cavity is formed in the original non-cavitating flow by air injection. The procedure consists in adjusting the air flowrate so that the ventilated cavity has the same apparent length as the vapor cavity. Of course, those cavities are not quite similar since air is effectively incondensable. The cavity ends are different, and because of air compressibility effects, the behavior of the ventilated cavity depends to some extent upon the ambient pressure (see chap. 9). Thus the method can only give an order of magnitude.

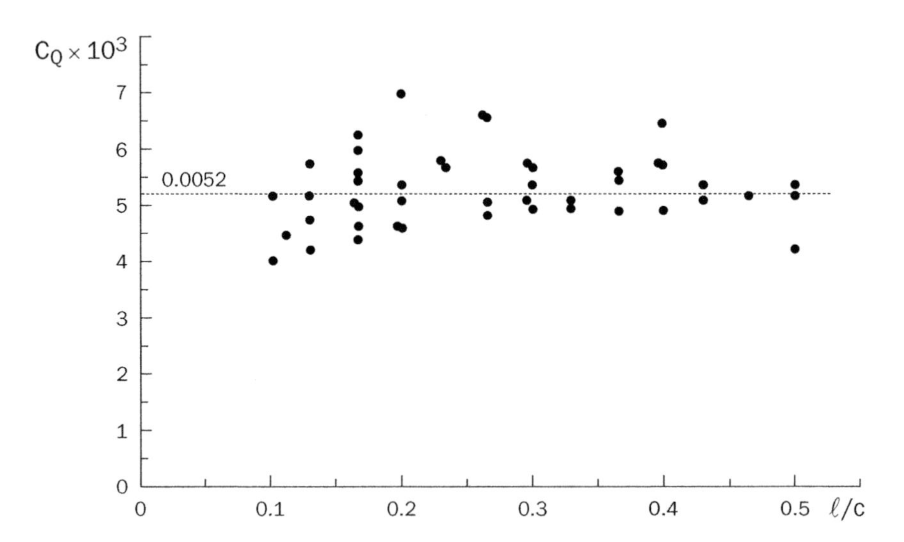

7.14 - Variation of the flowrate coefficient *versus* the reduced cavity length for short cavities *[from FRUMAN et al., 1991]*

FRUMAN *et al.* (1991a) carried out such measurements for stable cavities (fig. 7.14), i.e. for cavities whose length is smaller than half the foil chord length. Their results are somewhat scattered but do not show any dependence on the cavity length. The mean C_{Qv}-value is 0.0052. KAMONO *et al.* (1993) confirmed this trend for short cavities ($\ell/c < 0.6$) and showed that the flowrate coefficient increases significantly with the cavity length as the cavity begins to pulsate ($0.6 < \ell/c < 1$).

The vapor flowrate q_v can be expressed as $q_v = eV_v$ where e is the cavity thickness and V_v the mean velocity of the vapor inside the cavity. C_{Qv} can be then written as:

$$C_{Qv} = \frac{e}{\ell}\frac{V_v}{V} \tag{7.15}$$

For example, with $e/\ell = 5\%$ and $C_{Qv} = 0.0052$, V_v/V is equal to 0.1. The velocity of the vapor inside the cavity is about an order of magnitude smaller than that of the surrounding liquid.

Determination of the NUSSELT number

With respect to the NUSSELT number, the thermal boundary layer at the liquid-vapor interface is hardly characterized. One possible way to approach heat transfer there is to consider the interface as a solid wall and to estimate a value based on classical convective heat transfer correlations.

One typical method of determination of the heat transfer coefficient is based upon the REYNOLDS-COLBURN analogy between momentum and heat transfer. This gives the following classical relation (see e.g. HOLMAN 1997):

$$Nu = \frac{1}{2} C_f \, Re \, Pr^{1/3} \tag{7.16}$$

where C_f is the usual friction coefficient. Thus, the heat transfer is estimated on the basis of the well-known friction coefficient.

Equation (7.14) becomes:

$$B = \frac{2C_{Qv}}{C_f} Pr^{2/3} \tag{7.17}$$

and the value of the friction coefficient C_f is taken from empirical formulae.

If the interface is considered as a smooth flat plate in a turbulent flow, the following correlation can be used:

$$C_f = \frac{0.074}{Re_\ell^{0.2}} \tag{7.18}$$

It is often necessary to take into account the roughness of the liquid-vapor interface, which significantly enhances heat transfer in comparison with the smooth case [FRUMAN & BEUZELIN 1991b]. By introducing a roughness parameter ε and assuming that it is larger than the viscous boundary layer thickness, the mean friction coefficient can be considered as independent of the REYNOLDS number and given by the following correlation:

$$C_f = \left[1.89 + 1.62 \log \frac{\ell}{\varepsilon}\right]^{-2.5} \tag{7.19}$$

which is valid in the range $10^2 < \ell/\varepsilon < 10^6$.

Application

As an example, the above procedure is applied to tests carried out by HORD (1972) on partial cavities attached to foils, in liquid hydrogen and nitrogen. Temperature measurements were made at five points inside the cavity. Table 7.2 presents a comparison between calculated and measured data.

		Liquid hydrogen	Liquid nitrogen
Liquid temperature T_∞ (K)		20.70	77.71
Cavity length ℓ (mm)		44.4	50.8
Flow velocity V (m / s)		59.4	23.9
PRANDTL number		1.28	2.29
REYNOLDS number Re_ℓ		1.45 10^7	6.49 10^6
Friction coefficient C_f	smooth wall	0.0027	0.0032
	$\varepsilon = 0.05$ mm	0.0087	0.0084
	$\varepsilon = 0.5$ mm	0.0175	0.0167
Vapor flowrate coefficient C_{Qv}		0.0052	
$\rho_v L / \rho_\ell c_{p\ell}$ (K)		0.95	0.57
Calculated B-factor	smooth wall	4.53	5.63
	$\varepsilon = 0.05$ mm	1.41	2.14
	$\varepsilon = 0.5$ mm	0.69	1.09
Calculated ΔT (K)	smooth wall	4.30	3.21
	$\varepsilon = 0.05$ mm	1.34	1.22
	$\varepsilon = 0.5$ mm	0.66	0.62
Measured ΔT (K) (x_l is the distance, from the leading edge, of the station where cavity temperature was measured)	$x_l = 6$ mm	1.8	2.3
	$x_l = 11$ mm	2.0	2.3
	$x_l = 17.3$ mm	2.3	2.3
	$x_l = 26.8$ mm	2.4	2.2
	$x_l = 36.5$ mm	1.9	2.0

Table 7.2

In the absence of values for the flowrate coefficient C_{Qv}, use is made of the above-mentioned value 0.0052. Three different assumptions are considered for the estimation of the friction coefficient C_f:

- the smooth wall,
- wall roughness $\varepsilon = 0.05$ mm,
- wall roughness $\varepsilon = 0.5$ mm.

The friction coefficient increases with roughness, so that the temperature shift decreases. As seen from the table, the experimental values are within the range covered by the three assumptions.

In the above approach, the cavity is supposed filled with vapor with heat transfer occurring through the cavity interface. In fact, the void fraction inside the cavity may be smaller than unity [STUTZ *et al.* 1998] and the cavity may contain droplets whose area should contribute to the total interfacial area. If so, this model is no longer entirely appropriate. However, in the framework of the present approach, the effect of any increase in the interfacial area can be taken into account empirically by adjustment of the roughness ε which remains a free parameter of the model [FRUMAN & BEUZELIN 1991b].

7.6. SYSTEM INSTABILITY

In section 7.3, the cloud cavitation instability was discussed. This instability is intrinsic, in so far as it results from the dynamics of the cavity proper. Unlike the case of cloud cavitation, other instabilities which result from a coupling of the cavity dynamics with the rest of the hydraulic circuit can develop. Cavitation surge is a typical example of such a system instability.

To illustrate it, consider a cavitating hydrofoil in a duct of area A (fig. 7.15). A simplified one-dimensional model is used to analyse this configuration [WATANABE *et al.* 1998]. For simplicity, the downstream length is supposed infinite, so that the mass flowrate m [kg/s] downstream remains constant due to infinite inertia.

In steady state conditions, the cavity volume $\mathcal{V}$, the mass flowrate m and the pressure p in any cross-section are constant if head losses are neglected.

Under unsteady conditions, the fluctuation of mass flowrate in the upstream line is denoted m'(t) with p'(t) as the pressure fluctuation at the location of the hydrofoil. The inlet pressure is supposed constant. The momentum balance in the upstream duct is written:

$$-\frac{L}{A}\frac{dm'}{dt} = p' \qquad (7.20)$$

As for the mass balance, it simply states that the variations in cavity volume are balanced by the fluctuations in the incoming flowrate:

$$m' = -\rho\frac{d\mathcal{V}}{dt} \qquad (7.21)$$

Introduction of the classical cavitation compliance K allows us to state that variations in cavity volume are linearly related to the variations in pressure:

$$\rho\frac{d\mathcal{V}}{dt} = -K\frac{dp'}{dt} \qquad (7.22)$$

In other words, the cavitation compliance expresses how the cavity volume changes with the local pressure:

$$K = -\rho \frac{d\mathcal{V}}{dp'} \tag{7.23}$$

Cavitation compliance results from purely quasi-static considerations and do not involve any dynamic feature of the cavity.

By combining equations (7.20), (7.21) and (7.22), we get the following equation for the pressure fluctuation:

$$\frac{d^2 p'}{dt^2} + \frac{A}{LK} p' = 0 \tag{7.24}$$

The type of solution depends upon the sign of the cavitation compliance K. If K is positive, all the variables exhibit periodic oscillations at the frequency:

$$\omega = \sqrt{\frac{A}{LK}} \tag{7.25}$$

which is the natural frequency of the system made up of the cavity and the inlet line.

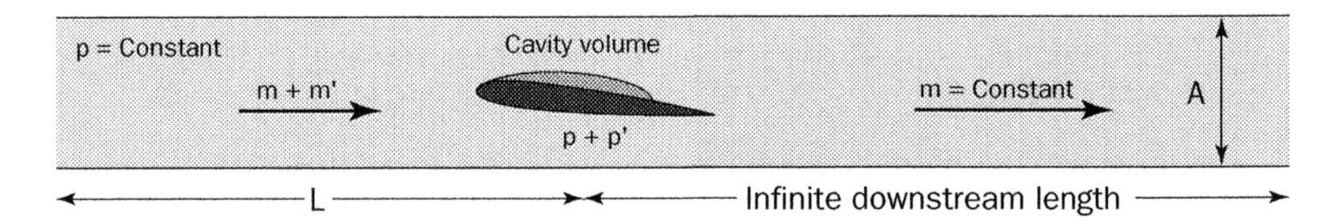

7.15 - The case of an isolated cavitating foil in a duct
For simplicity, the downstream length is supposed infinite.

This regime is typical of cavitation surge. The physical reason for oscillations is easy to understand. If it is supposed that the pressure at the location of the hydrofoil decreases, the cavity volume increases because of a positive value of the cavitation compliance. By continuity, the incoming flowrate decreases and the inertia of the upstream column makes the pressure at the location of the hydrofoil increase. Hence, the situation is stable and auto-oscillations develop.

From equation (7.25), it appears that the frequency of the pulsations results from a coupling between the cavity, characterized by its compliance, and the inlet line, characterized by its length L and area A. This behavior is typical of a system instability.

If the cavitation compliance were negative, the solution would not be periodic but would increase exponentially and the behavior would be unstable. In fact, the length of a cavity and hence its volume normally increases when the pressure decreases (see fig. 7.3), so that the cavitation compliance is usually positive. Hence, such an unstable behavior connected to negative values of the cavitation compliance is rather unusual.

The above covers the basic case of cavitation surge for an isolated hydrofoil in a duct. The principle of the analysis remains applicable to more complicated cases. DUTTWEILER and BRENNEN (2002) examined surge instability on a cavitating propeller tested in a water tunnel using the theoretical approach developed by BRENNEN and ACOSTA (1973). The procedure the authors developed illustrates the main steps to follow, in order to analyse a cavitation surge instability. The cavitation dynamics and the facility dynamics are considered as part of a coupled system. Each of them has to be characterized in order to predict the whole system dynamics. The main steps of the prediction are:

- The quasi-steady state response of the cavitation volume to changes in inlet conditions is characterized in terms of a cavitation compliance K and a mass flow gain factor M. The cavitation compliance describes the variation in cavity volume with pressure (see eq. 7.23), whereas the mass flow gain factor describes the variation in cavity volume with the angle of attack. The mass flow gain factor is needed for the modeling of cavitation instabilities in turbomachinery, because a change in inlet flowrate results in a change in angle of attack, as can be seen from consideration of the velocity triangle.
- The second step consists in modeling the dynamics of the facility by dividing it into elementary components (pipes, tanks...) of specified "resistance", "capacitance" and "inertance".
- Finally, the facility dynamics and the cavitation dynamics have to be coupled in order to identify potentially unstable behavior.

By applying the previous procedure to the case of a propeller in a cavitation tunnel, DUTTWEILER (2001) succeeded in predicting the characteristic frequency of the observed instability. This procedure can be followed for the analysis of any partial cavity instability of surge type.

7.7. PARTIAL CAVITY FLOW MODELING

The modeling of partial cavities is very similar to the case of supercavities, concerning the conditions for detachment from the solid wall and the boundary condition at the interface. However, the cavity termination and the reattachment of the flow to the wall bring additional difficulties, as well as the unsteady behavior of the cavity. Ad hoc methods (and even artifices) were used to reach agreement with the experimental results and particularly for the dependence of the cavity length or the drag and lift coefficients on the cavitation parameter. An abundant literature has been devoted to this topic since the mid-seventies.

As for the steady state, both closed and open models have been developed. For closed models, the closure condition can be imposed right at the cavity termination or along a transition zone, whose length is adjusted for best representation of the flow. The reattachment of the flow to the wall gives a curvature to the interface which is directed towards the liquid and generates an overpressure near the cavity closure. Generally, such closed models do not result in a one-to-one relation $\ell(\sigma)$ (see e.g. NISHIYAMA & ITO 1977, LEMONNIER & ROWE 1988). They are more appropriate to the modeling of short and thin cavities, which only weakly disturb the flow.

Thick cavities, however, are best represented by open models in which the cavity is followed by a wake whose displacement thickness is constant or increases downstream. Such a wake can be generated by sources blowing at the cavity end or through a permeable wall [YAMAGUCHI & KATO 1983, ITO 1986, ROWE & BLOTTIAUX 1993]. It induces an additional drag which simulates the dissipation associated with thick cavities.

As for unsteady cavities, a detailed presentation of the main difficulties in the modeling of partial cavity flows based on the classical non-linear inviscid theory (with the effects of viscosity on cavity detachment and cavity wake included) can be found in KINNAS (1998), together with a comprehensive review of the main works devoted to 2D and 3D cavity flow modeling.

Other approaches have been developed to model unsteady cavitation on the basis of either the EULER or NAVIER-STOKES equations completed with a cavitation model. A homogeneous fluid of variable density is usually considered to represent the water-vapor mixture. As an example, the barotropic model [Reboud & Delannoy 1994] introduces a constitutive law $\rho(p)$ for this equivalent fluid whose density is assumed to change continuously from the vapor density to the liquid density when the pressure curve crosses the vapor pressure value. There is also the two-phase bubble flow model [KUBOTA *et al.* 1992] wherein the cavity is considered as a homogeneous cluster of spherical bubbles. The local void fraction is computed from the resolution of a RAYLEIGH-PLESSET equation for the bubble cluster. Such a model, which requires the stipulation of a nuclei density, satisfactorily simulated the cyclic behavior of partial cavities.

REFERENCES

BRENNEN C.E. & ACOSTA A.J. –1973– Theoretical quasi-static analyses of cavitation compliance in turbopumps. *J. Spacecraft and Rockets* **10**(3), 175-180.

CALLENAERE M. –1999– Étude physique des poches de cavitation partielle en écoulement interne. *Thesis*, Grenoble University (France).

CALLENAERE M., FRANC J.P. & MICHEL J.M. –2001– The cavitation instability induced by the development of a re-entrant jet. *J. Fluid Mech.* **444**, 223-256.

DE M.K. & HAMMITT F.G. –1982– New method for monitoring and correlating cavitation noise to erosion capability. *J. Fluids Eng.* **104**, 434-442.

DE LANGE D.F. –1996– Observation and modelling of cloud formation behind a sheet cavity. *PhD Thesis*, Twente University (the Netherlands).

DE LANGE D.F. & DE BRUIN G.J. –1998– Sheet cavitation and cloud cavitation, re-entrant jet and three-dimensionality. *Appl. Sci. Res.* **58**, 91-114.

DRIVER D.M. & SEEGMILLER H.L. –1985– Features of a reattaching turbulent shear layer in divergent channel flow. *AIAA Journal* **23**(2), 163-171.

DUTTWEILER M.E. –2001– Surge instability on a cavitating propeller. *PhD Thesis, Cal. Inst. Techn.*, Pasadena (USA).

DUTTWEILER M.E. & BRENNEN C.E. –1998– Partial cavity instabilities. *Proc. US-Japan Seminar: Abnormal Flow Phenomen in Turbomachines*, Osaka (Japan).

DUTTWEILER M.E. & BRENNEN C.E. –2002– Surge instability on a cavitating propeller. *J. Fluid Mech.* **458**, 133-152.

FRUMAN D.H., BENMANSOUR I. & SERY R. –1991a– Estimation of the thermal effects on cavitation of cryogenic liquids. *Proc. ASME Cavitation and Multiphase Flow Forum*, Portland (USA).

FRUMAN D.H. & BEUZELIN F. –1991b– Effets thermiques dans la cavitation des fluides cryogéniques. *La Houille Blanche* **7/8**, 557-561.

FRY S.A. –1989– The damage capacity of cavitating flow from pulse height analysis. *J. Fluids Eng.* **111**, 502-509.

FURNESS R.A. & HUTTON S.P. –1975– Experimental and theoretical studies of two-dimensional fixed-type cavities. *J. Fluids Eng.* **97**, 515-522.

HOLL J.W., BILLET M.L. & WEIR D.S. –1975– Thermodynamics effects on developed cavitation. *J. Fluids Eng.* **97**, 507-514.

HOLMAN J.P. –1997– Heat transfer. *McGraw-Hill Book Company Ed.*

HORD J. –1973– Cavitation in liquids cryogens – II. Hydrofoil. *NASA CR-2156*, 157 p.

ITO J. –1986– Calculation of partially cavitating thick hydrofoil and examination of a flow model at cavity termination. *Proc. Int. Symp. on Cavitation*, Sendai (Japan).

IWAI Y., OKADA T., NASHIYA N. & FUKUDA Y. –1991– Formation and progression of vibratory cavitation erosion. *Proc. 1st Joint ASME/JSME Fluids Eng. Conf.*, Portland (USA), June 23-27.

KAMONO H., KATO H., YAMAGUCHI H. & MIYANAGA M. –1993– Simulation of cavity flow by ventilated cavitation on a foil section. *ASME Cavitation and Multiphase Flow Forum*, FED **153**, 183-189.

KATO H. –1984– Thermodynamic effect on incipient and developed sheet cavitation. *Proc. Int. Symp. on Cavitation Inception*, New Orleans (USA).

KATO H., YE Y.P. & MAEDA M. –1989– Cavitation erosion and noise study on a foil section. *Proc. ASME Int. Symp. on Cavitation in Hydraulic Structures and Turbomachinery*, Albuquerque (USA).

KAWANAMI Y., KATO H., YAMAGUCHI H., TAGAYA Y. & TANIMURA M. –1997– Mechanism and control of cloud cavitation. *J. Fluids Eng.* **119**, 788-795.

KAWANAMI Y., KATO H. & YAMAGUCHI H. –1998– Three-dimensional characteristics of the cavities formed on a two-dimensional hydrofoil. *Proc. 3rd Int. Symp. on Cavitation*, vol. 1, Grenoble (France), 191-196.

KINNAS S.A. –1998– The prediction of unsteady sheet cavitation. *Proc. 3rd Int. Symp. on Cavitation*, vol. 1, Grenoble (France), 19-36.

KUBOTA S., KATO H., YAMAGUCHI H. & MAEDA M. –1987– Unsteady structure measurement of cloud cavitation on a foil section using conditional sampling technique. *Proc. Int. Symp. on Cavitation Research Facilities and Techniques*, Boston (USA), 161-168.

KUBOTA S., KATO H., YAMAGUCHI H. –1992– A new modelling of cavitating flows: a numerical study of unsteady cavitation on a hydrofoil section. *J. Fluid Mech.* **240**, 59-96.

LABERTEAUX K.R. & CECCIO S.L. –2001– Partial cavity flows. Part 1 – Cavities forming on models without spanwise variation. Part 2 – Cavities forming on test objects with spanwise variation. *J Fluid Mech.* **431**, 1-41 and 43-63.

LE Q., FRANC J.P. & MICHEL J.M. –1993a– Partial cavities: global behaviour and mean pressure distribution. *J. Fluids Eng.* **115**, 243-248.

LE Q., FRANC J.P. & MICHEL J.M. –1993b– Partial cavities: pressure pulse distribution around cavity closure. *J. Fluids Eng.* **115**, 249-254.

LEMONNIER H. & ROWE A. –1988– Another approach in modelling cavitating flows. *J. Fluid Mech.* **195**, 557-580.

MAEDA M. ,YAMAGUCHI H. & KATO H. –1991– Laser holography measurement of bubble population in cavitation cloud on a foil section. *Proc. 1st Joint ASME/JSME Fluids Eng. Conf.*, FED **116**, Portland (USA), 67-75.

MICHEL J.M. –1978– Demi-cavité formée entre une paroi solide et un jet plan de liquide quasi parallèle: approche théorique. *DRME Contract* **77/352**, Rpt 4.

MOORE R.D. & RUGGERI R.S. –1968– Prediction of thermodynamic effects of developed cavitation. *NASA*, Rpt TN D-4899, Washington DC (USA).

NISHIYAMA T. & ITO J. –1977– Calculation of partially cavitating flow by singularity method. *Trans. JSME* **43**(370), 2165-2174.

PO W.Y. & CECCIO S.L. –1997– Diffusion induced bubble populations downstream of a partial cavity. *J. Fluids Eng.* **119**, 782-787.

REBOUD J.L. & DELANNOY Y. –1994– Two-phase flow modelling of unsteady cavitation. *Proc. 2nd Int. Symp. on Cavitation*, Tokyo (Japan), 34-44.

REISMAN G.E., WANG Y.C. & BRENNEN C.E. –1998– Observations of shock waves in cloud cavitation. *J. Fluid Mech.* **355**, 255-283.

ROWE A. & BLOTTIAUX O. –1993– Aspects of modelling partially cavitating hydrofoils. *J. Ship Res.* **37**(1), 34-48.

SEBESTYEN G. & VARGA J.J. –1972– Determination of cavitation hydrodynamic intensity by noise measurements. *Proc. 2nd JSME Int. Symp. on Fluid Machinery and Fluidics*, Tokyo (Japan).

STUTZ B. & REBOUD J.L. –1997a– Two-phase flow structure of sheet cavitation. *Phys. Fluids* **9**(12), 3678-3686.

STUTZ B. & REBOUD J.L. –1997b– Experiments on unsteady cavitation. *Exp. Fluids* **22**, 191-198.

WATANABE S., TSUJIMOTO Y., FRANC J.P. & MICHEL J.M. –1998– Linear analyses of cavitation instabilities. *Proc. 3rd Int. Symp. on Cavitation*, vol. 1, 347-352, *J.M. MICHEL and H. KATO Ed.*

YAMAGUCHI H. & KATO H. –1983– Non-linear theory for partially cavitating hydrofoils. *Trans. JSNA* **152**, 117-124.

APPENDIX: SONIC VELOCITY IN A LIQUID/VAPOR MIXTURE WITH PHASE CHANGE

We consider a bubbly flow in thermodynamic equilibrium in which the void fraction is α. The bubbles are assumed to be filled with saturated vapor only. Surface tension is neglected and thus the pressure in the liquid is equal to the vapor pressure p_v. In addition, the mixture is treated as homogeneous, which requires that the wave length be much larger than the size of the dispersed phase, i.e. the bubble size.

A unit volume of mixture contains a volume α of vapor and $(1-\alpha)$ of liquid. The corresponding masses of vapor and liquid are $\alpha\rho_v$ and $(1-\alpha)\,\rho_\ell$ where ρ_v and ρ_ℓ are the vapor and liquid densities respectively. The density of the mixture is $\rho = \alpha\rho_v + (1-\alpha)\,\rho_\ell$.

Assuming this volume undergoes a pressure variation from p_v to $p_v + \delta p_v$, then in order to calculate the velocity of sound $c = \sqrt{\dfrac{\delta p_v}{\delta \rho}}$, we must evaluate $\delta\rho$.

The pressure variation induces a phase change. If δm is the mass of liquid changed into vapor, the mass of vapor is now $\alpha\rho_v + \delta m$ and the vapor density becomes $\rho_v + \dfrac{\partial \rho_v}{\partial p}\delta p_v = \rho_v + \dfrac{\delta p_v}{c_v^{\,2}}$. Thus the vapor volume becomes:

$$\frac{\alpha\rho_v + \delta m}{\rho_v + \dfrac{\delta p_v}{c_v^{\,2}}} \cong \alpha + \frac{\delta m}{\rho_v} - \alpha\frac{\delta p_v}{\rho_v\, c_v^{\,2}} \tag{7.26}$$

Calculations are similar for the liquid, with δm changing sign and α changed to $1-\alpha$. The recalculated liquid volume is then:

$$(1-\alpha)-\frac{\delta m}{\rho_\ell}-(1-\alpha)\frac{\delta p_v}{\rho_\ell c_\ell^2} \tag{7.27}$$

with density:

$$\rho' = \frac{\alpha\rho_v + (1-\alpha)\rho_\ell}{\left[\alpha + \frac{\delta m}{\rho_v} - \alpha\frac{\delta p_v}{\rho_v c_v^2}\right] + \left[(1-\alpha) - \frac{\delta m}{\rho_\ell} - \frac{(1-\alpha)\,\delta p_v}{\rho_\ell c_\ell^2}\right]} \tag{7.28}$$

or

$$\frac{\delta\rho}{\rho} \cong -\delta m\left[\frac{1}{\rho_v} - \frac{1}{\rho_\ell}\right] + \delta p_v\left[\frac{\alpha}{\rho_v c_v^2} + \frac{1-\alpha}{\rho_\ell c_\ell^2}\right] \tag{7.29}$$

Thus the sound velocity c is given by:

$$\frac{1}{\rho c^2} \cong \frac{\alpha}{\rho_v c_v^2} + \frac{1-\alpha}{\rho_\ell c_\ell^2} - \frac{\delta m}{\delta p_v}\left[\frac{1}{\rho_v} - \frac{1}{\rho_\ell}\right] \cong \frac{\alpha}{\rho_v c_v^2} + \frac{1-\alpha}{\rho_\ell c_\ell^2} - \frac{\delta m}{\rho_v\,\delta p_v} \tag{7.30}$$

The phase change δm depends on the nature of the thermodynamic transformation. If the transformation is adiabatic and the latent heat of vaporization is taken from the liquid whose temperature varies by δT, the heat balance gives:

$$\delta m\, L = -(1-\alpha)\,\rho_\ell\, c_{p\ell}\,\delta T \tag{7.31}$$

where L is the latent heat of vaporization and $c_{p\ell}$ the liquid heat capacity. The contribution of the vapor to the previous heat balance is negligible, except for void fractions close to unity. In the case of vaporization ($\delta m > 0$), the liquid temperature obviously decreases ($\delta T < 0$).

Equation (7.31) gives:

$$\frac{\delta m}{\delta p_v} = -\frac{(1-\alpha)\,\rho_\ell\, c_{p\ell}}{L}\,\frac{\delta T}{\delta p_v} \tag{7.32}$$

The term $\frac{\delta T}{\delta p_v}$ in the last equation is the inverse of the slope of the vaporization-condensation curve. It is calculated on the basis of the CLAPEYRON relation (5.45) and approximated by $\frac{T}{\rho_v L}$ according to equation (5.46).

Returning to equation (7.30) we obtain finally:

$$\frac{1}{\rho c^2} \cong \frac{\alpha}{\rho_v c_v^2} + \frac{1-\alpha}{\rho_\ell c_\ell^2} + \frac{(1-\alpha)\rho_\ell c_{p\ell} T}{(\rho_v L)^2} \tag{7.33}$$

In the case of water at room temperature, we have the following orders of magnitude (see § 5.3.1):

$$c_v \cong 380 \text{ m/s} \quad \text{and} \quad \rho_v c_v^2 \cong 2{,}500 \text{ kg/m/s}^2$$

$$c_\ell \cong 1{,}500 \text{ m/s} \quad \text{and} \quad \rho_\ell c_\ell^2 \cong 2.25\,10^9 \text{ kg/m/s}^2$$

$$\frac{(\rho_v L)^2}{\rho_\ell c_{p\ell} T} \cong 1.5 \text{ kg/m/s}^2$$

Except in very special cases and particularly for void fractions close to unity, the last term in equation (7.33) is predominant because of the phase change. It reduces the speed of sound dramatically compared to the case without phase change (for which it is well known that the velocity of sound is smaller than the values for each phase separately). For example, for a void fraction of 50%, equation (7.33) gives the speed of sound as 0.08 m/s whereas, ignoring the last term, it is about 3 m/s (i.e. without phase change).

8. BUBBLES AND CAVITIES ON TWO-DIMENSIONAL FOILS

This chapter is mainly concerned with developed cavitation on two-dimensional hydrofoils as it occurs on blades of rotating machinery or lifting hydrofoils. The mechanisms which govern inception and development of attached cavitation and traveling bubble cavitation are presented in sections 8.1 and 8.2 respectively. It is shown that attached cavities are strongly connected to the boundary layer, while bubble cavitation chiefly depends on the pressure distribution and the liquid nuclei content. The possible competition and interaction between both modes of cavitation is examined in section 8.3 while roughness effects are examined in section 8.4. From an experimental viewpoint, the control of cavitation patterns requires specific control of the water quality by means of equipment devoted to de-aeration, nuclei seeding and nuclei measurement.

8.1. ATTACHED CAVITATION

In chapters 6 and 7, the main features of attached cavities, partial cavities as well as supercavities, were reviewed. Two critical regions were pointed out: cavity detachment and cavity closure. The latter was treated in detail in chapter 7 for partial cavitation. Here the focus is on cavity detachment, whose location is, *a priori*, unknown in the case of a wall of continuous slope.

8.1.1. CAVITATION INCEPTION ON A CIRCULAR CYLINDER

The case of a circular cylinder or a sphere is particularly representative of the connection between cavitation inception and the boundary layer.

In non-cavitating conditions, the drag coefficient of a cylinder presents a sudden decrease for a critical value of the REYNOLDS number. This value depends to a small extent upon the operating conditions (such as wall roughness or turbulence level) but usually lies around 3.10^5. It is well-known that this drop corresponds to the transition to turbulence of the boundary layer flow.

Figure 8.1 shows the corresponding states of the boundary layer, as observed from visualizations, together with the variation of the drag coefficient versus the REYNOLDS number.

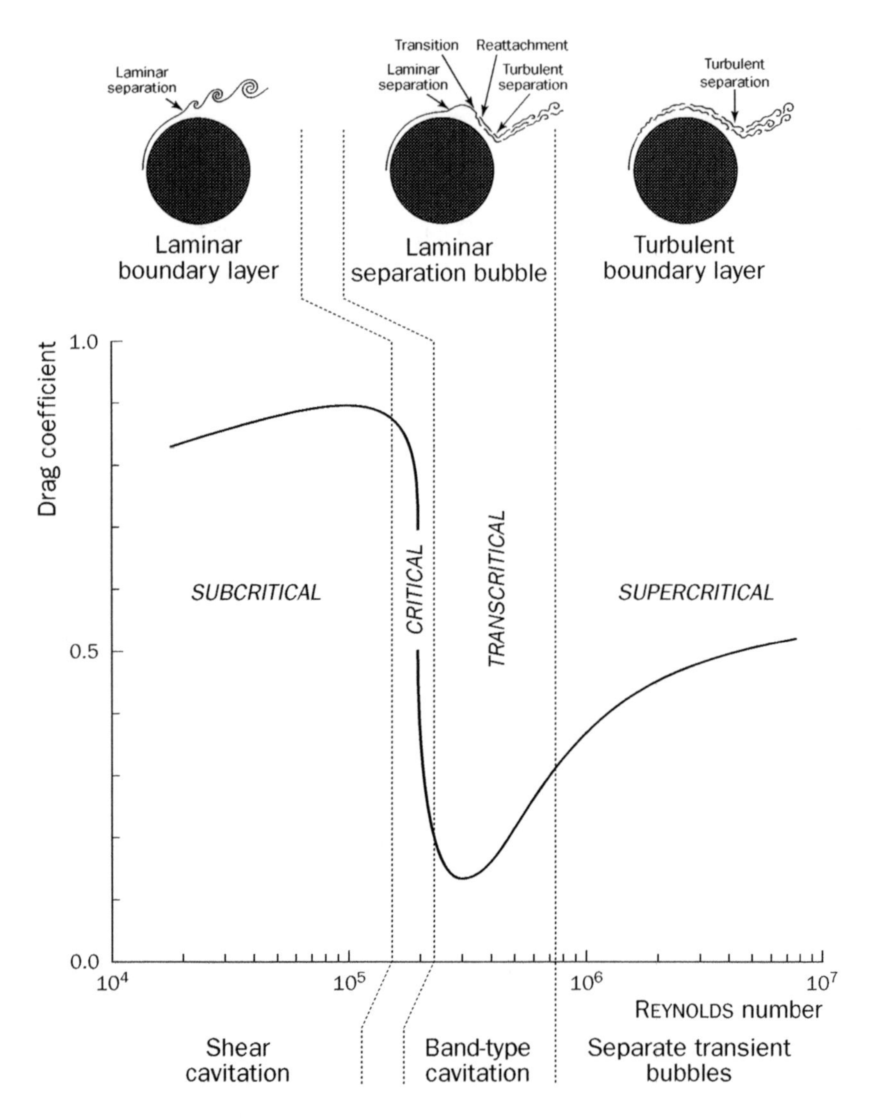

8.1 - Drag coefficient of a circular cylinder versus the REYNOLDS number and associated cavitation inception patterns and boundary layer state

Below the critical value of the REYNOLDS number, the boundary layer is laminar and separates from the body at an angle of between 70 and 80 degrees with respect to the stagnation point. The wake is thick and the drag is large. Transition to turbulence occurs away from the wall, in the free shear layer. It progressively approaches the separation point as the REYNOLDS number is increased.

In the transcritical region, the boundary layer is dramatically modified. Transition to turbulence is so close to the cylinder that the boundary layer reattaches on the wall. Laminar separation, transition to turbulence and turbulent reattachment are concentrated within a small region and form the so-called "separation bubble". The boundary layer, which has become turbulent, can withstand a stronger adverse pressure gradient than if it were laminar and separates later, at about 130 degrees. The wake becomes narrower and the drag smaller.

Photographs on figure 8.2 show the corresponding cavitation patterns at inception. Although the difference in REYNOLDS numbers is very small, the cavitation looks very different.

At Re = 270,000, cavitation is not attached to the body, but appears in the wake, in the form of vapor filaments which are characteristic of shear cavitation (see chap. 11).

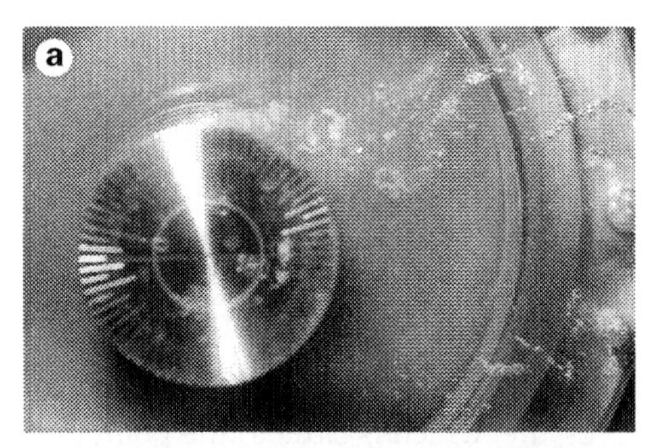

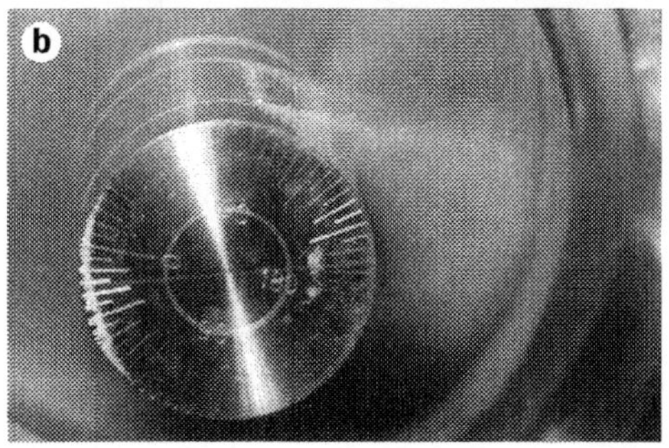

8.2
Cavitation patterns at inception in the vicinity of the critical REYNOLDS number for a circular cylinder
(a) Re = 270,000
(b) Re = 290,000

Conversely, for a slightly higher REYNOLDS number of 290,000, cavitation appears as a narrow sheet of vapor attached to the wall at an angle of about 115 degrees (visible on figure 8.2-b on the upper surface of the cylinder). It is generally referred to as band-type cavitation. It is initially made up of small bubbles, originating in nuclei trapped in the laminar separation bubble, where they grow slowly by diffusion of dissolved air. A further decrease in the ambient pressure results in the transformation of the band into a continuous attached vapor cavity. In this transcritical regime, the cavitation band, in which the pressure is close to the vapor pressure, develops downstream of the minimum pressure location. Hence, the liquid particles are in a metastable state while traveling through a region where the pressure is lower than the vapor pressure.

For large values of the REYNOLDS number, the flow regime is supercritical. Transition to turbulence moves upstream of the laminar separation and suppresses the laminar separation bubble. Cavitation appears as separate transient bubbles –on the condition that water contains nuclei which can actually be activated by the minimum pressure– or as three-dimensional vapor cones attached to isolated roughness elements.

8.1.2. CAVITY PATTERNS ON A TWO-DIMENSIONAL FOIL

In the case of hydrofoils, various cavity flow patterns can be observed according to the angle of attack and the cavitation number. For a proper observation of attached cavities in a hydrodynamic tunnel, it is essential to eliminate as much as possible traveling bubble cavitation in favor of attached cavities, and hence to use strongly deaerated water, so that almost no nucleus is activated.

Figure 8.3 gives a mapping of the various cavity flow patterns which have been observed on a NACA 16012 hydrofoil, for a fixed REYNOLDS number, when the incidence and the cavitation number (or simply the ambient pressure) are modified [FRANC & MICHEL 1985].

For small values of the cavitation number, supercavitation is observed for any angle of attack. The supercavity detaches itself from the rear part of the foil for an angle of attack of around zero (region 1), and the detachment point progressively moves upstream, towards the leading edge, as incidence increases. The detachment line is almost straight in the spanwise direction in regions 1 and 3, whereas it becomes strongly three-dimensional in the intermediate zone 2.

In region 1, the detachment point is far downstream of the point of minimum pressure, which is close to the point of maximum foil thickness. Thus, as for trans-critical flow around a cylinder, the liquid particles are in a metastable state in front of the cavity. This is why deaerated water is needed to observe this cavity flow regime.

For high values of the cavitation number, the pattern progressively evolves from a partial pure vapor cavity (region 3'), to a two-phase cavity (region 4) with a smaller void fraction and, finally, to cavitation in the shear layer bordering the recirculating zone due to stall at high angle of attack (region 5).

Within a very narrow domain of attack angles (around 4 degrees), the behavior of the cavity flow is rather unexpected. When the cavitation number is decreased from non-cavitating conditions, a leading edge cavity appears first. As the cavitation number is further decreased, the leading edge cavity completely disappears before a new cavity develops, at sufficiently low values of the cavitation number. The disappearance of the leading edge cavity is associated with an unexpected displacement of the detachment point, which moves downstream as σ_v decreases until the cavity disappears. This behavior, connected to the S-shaped limiting curve, results from the strong interaction between the cavity and the boundary layer. Similar behavior of the detachment point was also observed by MICHEL (1988) on a different foil, for velocities of up to 30 m/s, and approximately the same σ_v-values.

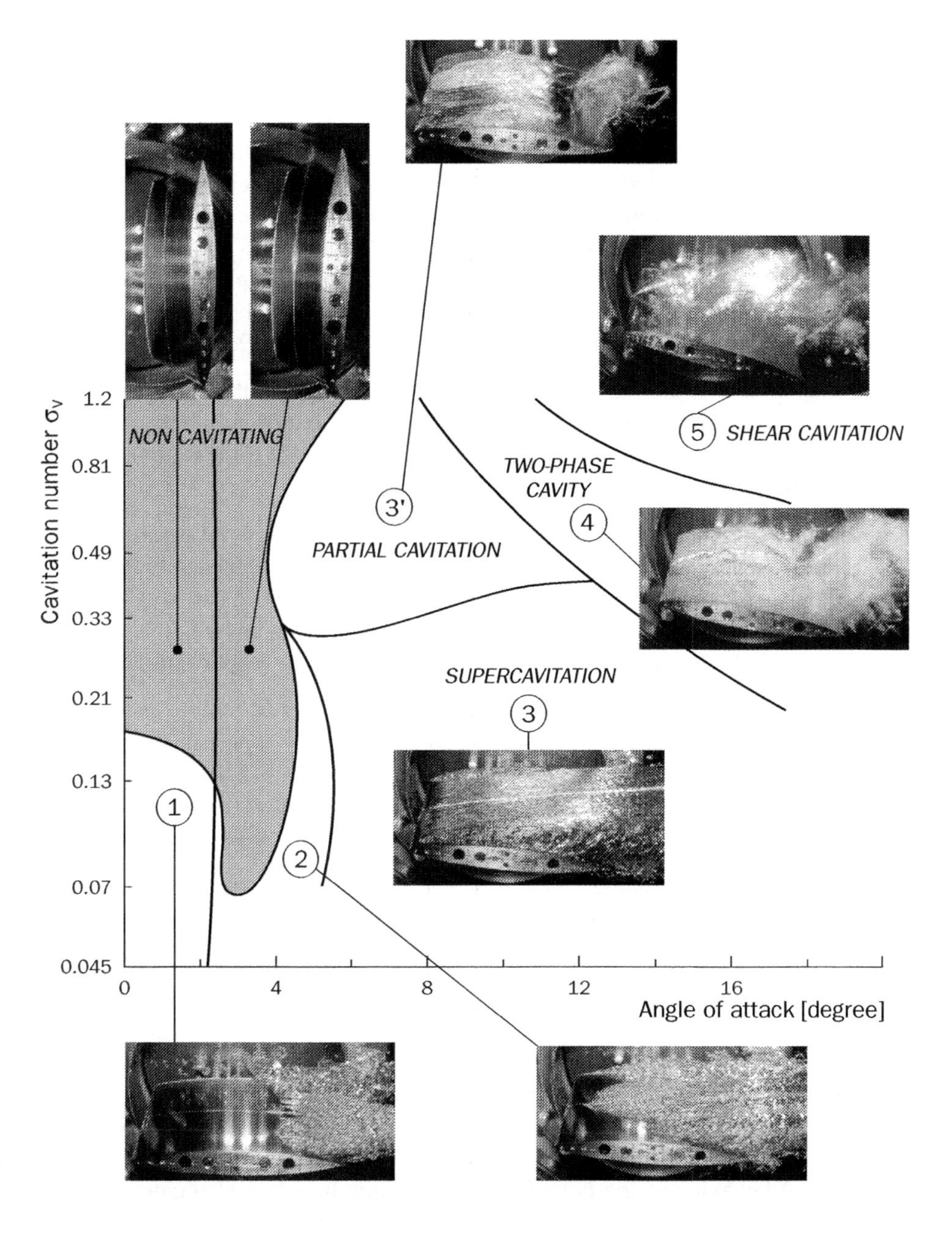

8.3 - Cavity patterns on a NACA 16012 hydrofoil at Re = 10^6 in strongly deaerated water

The foil is set at mid-height in the free surface channel of a hydrodynamic tunnel. The foil chord is 0.10 m and the channel height 0.40 m [from FRANC & MICHEL, 1985].

8.1.3. BOUNDARY LAYER FEATURES ON A SLENDER FOIL

A simple but efficient way to investigate the boundary layer on slender foils at small angles of attack consists in first computing the wall pressure distribution using an inviscid potential flow approach. The boundary layer is then computed using, for example, an integral method completed by a semi-empirical prediction of transition to turbulence (see e.g. ARNAL *et al.* 1984). There is usually no need to iterate as the modification of the initially calculated pressure field by the boundary layer is negligible for slender bodies. Predictions from such calculations agree fairly well with dye injection visualizations [FRANC & MICHEL 1985].

As an example, wall pressure distributions for two values of the attack angle and the corresponding calculated positions of laminar separation and transition to turbulence in the boundary layer are presented on figure 8.4 for the non-cavitating NACA 16012 foil considered in the previous section.

On the upper side and at an incidence of 2 degrees (see fig. 8.4-a), laminar separation, which should occur at the rear part of the foil, is prevented by transition to turbulence. At 5 degrees (see fig. 8.4-b), due to a strong adverse pressure gradient, laminar separation occurs close to the leading edge immediately followed by transition. Downstream of this separation bubble, the upper side boundary layer is fully turbulent and separates close to the trailing edge.

A summary of the nature of the boundary layer which develops on the upper side of the foil as a function of the angle of attack is given in figure 8.5. On slender foils, the position of laminar separation does not depend on the REYNOLDS number (nor on the turbulence level), contrary to turbulent transition or turbulent separation.

Comparing the non-cavitating boundary layer of figure 8.5 to the cavity patterns of figure 8.3, it appears that the location of cavity detachment corresponds fairly well to the location of laminar separation, whether it occurs at the rear of the foil or near its leading edge.

Moreover, the unexpected behavior of the cavitation discussed in the previous section and occurring for incidences between about 2 and 5 degrees, corresponds to the domain in which laminar separation and transition to turbulence jump from the rear to the front part of the foil. The rather unstable and three-dimensional behavior of the cavities observed in region 2 of figure 8.3, is related to the intermittent behavior of the boundary layer in this transitional region. Between two consecutive turbulent spots, the boundary layer is laminar and an attached cavity can develop locally for a short time, before being swept out by the next turbulent spot. This intermittent behavior of cavities is currently observed in experimental tests of propeller models.

Thus, the main features of inception and even of developed attached cavitation are strongly correlated to the behavior of the boundary layer in non-cavitating conditions.

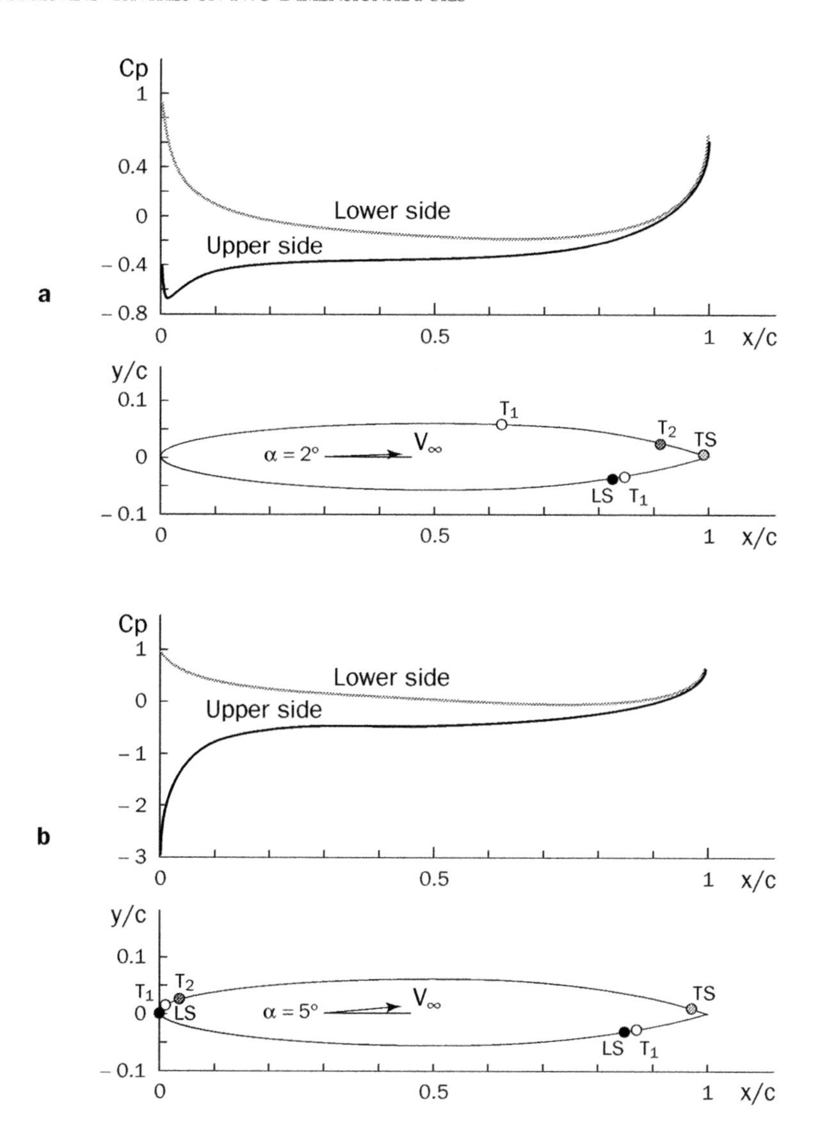

8.4 - Calculated pressure coefficient distribution and boundary layer behavior on a NACA 16012 hydrofoil in non-cavitating conditions

(chord length 0.10 m, flow velocity 12 m/s, turbulence intensity 0.165%)

LS: laminar separation; T_1: start of transition; T_2: end of transition; TS: turbulent separation. Between T_1 and T_2, intermittency of the turbulence grows from zero to one. Thus, T_2 also refers to the beginning of the fully turbulent boundary layer flow.

(a) $\alpha = 2°$, $C_L = 0.25$ - (b) $\alpha = 5°$, $C_L = 0.60$

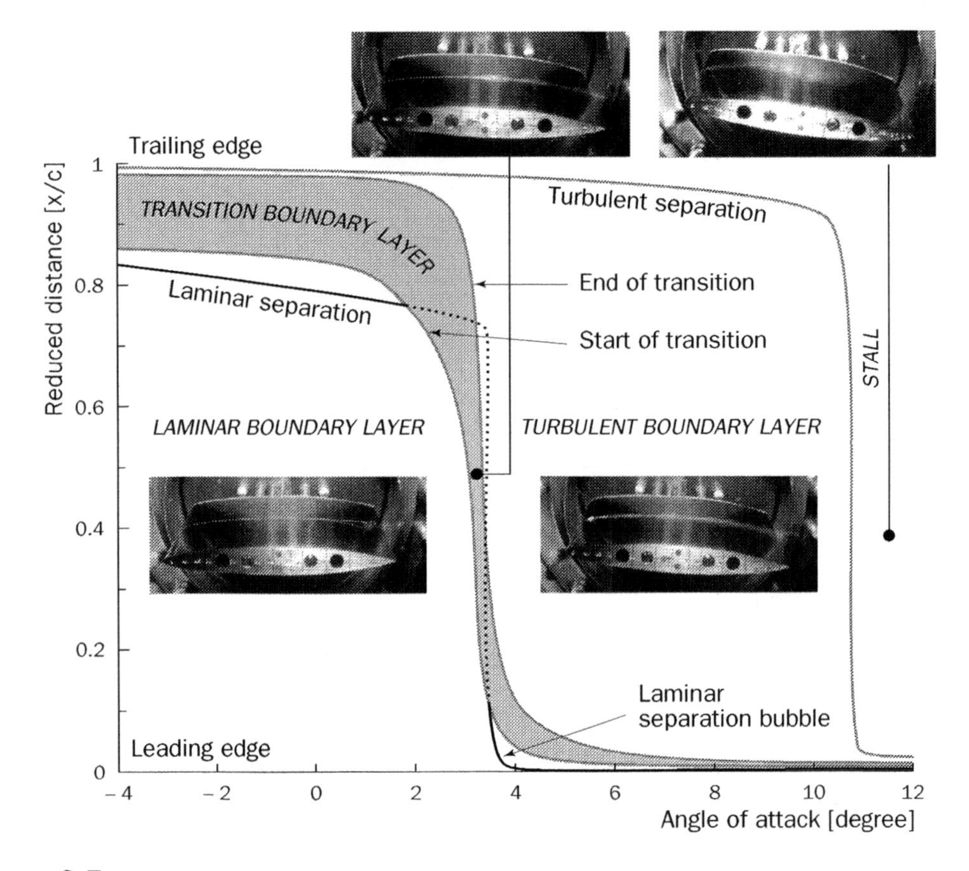

8.5 - Location of boundary layer characteristic points versus the angle of attack for the upper side of a NACA 16012 hydrofoil under non-cavitating conditions
(REYNOLDS number 6.10^5, turbulence level 0.14%)

8.1.4. THE CONNECTION BETWEEN LAMINAR SEPARATION AND DETACHMENT

Cavitation inception

The connection between laminar separation and attached cavitation was first put forward in 1968 by ALEXANDER in order to explain the difference between the minimum value of the pressure coefficient on the wall and the cavitation number at inception σ_{vi}. In 1973, ARAKERI and ACOSTA demonstrated experimentally, using schlieren and holographic methods, that the inception of attached cavitation on ogives occurs in the small recirculating bubble following laminar separation of the boundary layer, as it does for the transcritical flow around a circular cylinder.

Thus, at inception, the pressure at the laminar separation point is equal to the vapor pressure, so that the cavitation inception number is equal in magnitude to the pressure coefficient at laminar separation:

$$\sigma_{vi} = -Cp_{LS} \tag{8.1}$$

With reference to the minimum pressure coefficient, one can write:

$$\sigma_{vi} = -Cp_{min} - \Delta Cp \tag{8.2}$$

with

$$\Delta Cp = \frac{p_{LS} - p_{min}}{\frac{1}{2}\rho V^2} > 0 \tag{8.3}$$

Hence, there is a delay in cavitation inception with respect to the expected inception at the minimum pressure point. This delay corresponds to the additional decrease in pressure necessary for the pressure at the laminar separation point to reach the vapor pressure.

Developed attached cavities

The development of cavitation alters the non-cavitating pressure distribution. This generally causes the laminar separation point to move upstream. Thus, it is necessary to take into account the interaction between the boundary layer flow and the cavitating flow in order to predict the location of cavity detachment.

Considering the development of cavitation to be a small perturbation of the non-cavitating flow, ARAKERI (1975) suggested correlating the shift in the position of laminar separation due to the development of an attached cavity (subscript C) to the distance between laminar separation (subscript LS) and the minimum pressure point under non-cavitating conditions (subscript NC). On the basis of experimental data on ogives, ARAKERI proposed the following empirical relation:

$$s_{LS,C} - s_{LS,NC} = -2.37\left[1 + \frac{\sigma}{Cp_{LS,NC}}\right]\left[s_{LS,NC} - s_{Cp_{min,NC}}\right] \tag{8.4}$$

where s stands for the curvilinear distance.

In ARAKERI's approach, the distance λ between laminar separation and cavity detachment (see fig. 8.6) is treated in an empirical way. This can be considered analogous to the problem of a viscous liquid film spread over a wall by a wedge. This distance is related to the boundary layer thickness at separation and to the TAYLOR-SAFMAN parameter $\mu U/S$, which represents the effect of surface tension S on cavity detachment.

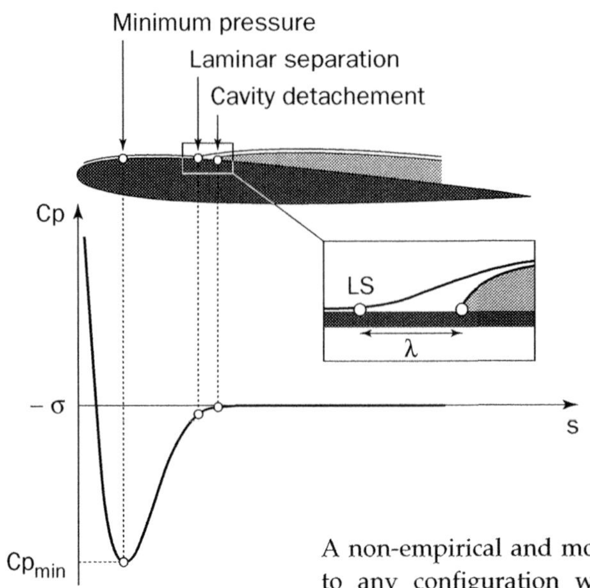

8.6 The connection between laminar separation and cavity detachment (ARAKERI's approach)

A non-empirical and more general method applicable to any configuration was proposed by FRANC and MICHEL in 1985. It is based upon an iterative procedure, and is applicable to largely developed cavities. Neglecting the effects of surface tension, FRANC and MICHEL made the assumption that a cavity detaches right at the point of separation of the boundary layer. In their approach, the distance λ between laminar separation and cavity detachment is assumed negligible (see fig. 8.6). If required, it can be estimated from ARAKERI's correlation.

The iterative scheme consists in using a potential method to compute the cavity flow with an initial guess of the position of cavity detachment. A boundary layer calculation then gives the position of laminar separation. The process is repeated until the distance between laminar separation and cavity detachment vanishes. The convergence is ensured by the fact that, when the detachment point is moved upstream, the adverse pressure gradient decreases and thus the laminar separation point approaches detachment. If the cavity detachment point is chosen too far upstream, no separation can occur since the adverse pressure gradient will be too small. If it is chosen too far downstream, the boundary layer separates prematurely, far from cavity detachment. After convergence, this detachment criterion guarantees that laminar separation occurs exactly where the vapor pressure is reached.

Good agreement was found between prediction and experiment in the case of a two-dimensional hydrofoil, for very different values of incidence and cavitation number and so for various locations of cavity detachment between leading edge and trailing edge [FRANC & MICHEL 1985].

As in the case of cavitation inception, the cavity, which is at vapor pressure, is downstream of a region where the pressure is lower than the vapor pressure. This is possible only in the absence of nuclei which otherwise would be destabilized and give rise to bubble cavitation (see § 8.2).

From a physical viewpoint, the boundary layer separation presents a relatively dead region for the cavity to be sheltered from the oncoming flow. If the boundary layer flow does not separate, the cavity cannot attach to the wall and is swept away. This is the case when transition to turbulence occurs. If so, the turbulent boundary layer can withstand the adverse pressure gradient without separating and no cavity can attach to the wall. This situation illustrates the destructive effect of turbulence on attached cavitation. It is sometimes used to prevent the development of attached cavitation.

If it is assumed that the cavity detaches at the point of minimum pressure, then the boundary layer would not separate, due to the absence of an adverse pressure gradient. Consequently, the condition of mechanical equilibrium of the cavity could not be fulfilled (see also § 6.1.2). However, in the case of leading edge cavities at high angle of attack, laminar separation is very close to the point of minimum pressure, such that this latter can usually be considered as a good approximation of the position of cavity detachment.

In the small region between laminar separation and cavity detachment, the flow is far from being two-dimensional and detachment often presents a three dimensional structure at small-scales. Furthermore, vortices originating in the shear layer are shed into the downstream flow near the cavity interface [AVELLAN & DUPONT 1988].

As a final point, it should be mentioned that the connection between laminar separation and cavity detachment still holds in unsteady conditions [FRANC & MICHEL 1988].

8.2. *TRAVELING BUBBLE CAVITATION*

Cavitation patterns can depend strongly on water quality. Attached cavitation generally occurs in deaerated water, whereas traveling bubble cavitation is favored by the presence of microbubbles. This latter is due to the growth of nuclei as they travel through low pressure regions, before collapsing further downstream where the pressure recovers [KODAMA *et al.* 1981]. In cavitation test facilities, artificial nuclei seeding is used to change the water quality in order to examine the various types of cavitation and their interactions. In practice, traveling bubble cavitation can be seen moving over the blades of hydraulic machinery for low values of the cavitation number, either at the outlet of FRANCIS turbine models or at the inlet of centrifugal pumps near their design point.

8.2.1. *THE EFFECT OF WATER QUALITY AND NUCLEI SEEDING*

By way of example, the influence of nuclei seeding on the cavitation patterns observed on a non-symmetrical NACA 16209 foil in a hydrodynamic tunnel [BRIANÇON-MARJOLLET *et al.* 1990] is presented here.

With deaerated water, the mapping presented in figure 8.7 appears to be very comparable qualitatively to the one presented in figure 8.3 for a different foil. In particular, the S-shaped curve related to cavitation inception is confirmed. At low incidences and σ_v-values, the cavitation inception number for attached cavitation on the rear part of the foil is very close in magnitude to the pressure coefficient at laminar separation as expressed in equation (8.1).

Nuclei seeding strongly affects the cavitation patterns as can be seen from a comparison between figure 8.7 and figure 8.8. The non-cavitating domain is seriously reduced. Traveling bubble cavitation replaces trailing edge super-cavitation and extends far inside the non-cavitating domain. The limiting curve between cavitating and non-cavitating regimes, under nuclei injection, is close to the $-Cp_{min}$ curve. The difference between σ_{vi} and $-Cp_{min}$ is connected to the susceptibility pressure p_s of the liquid and is given by (see § 2.4.2):

$$\sigma_{vi} - (-Cp_{min}) = -\frac{p_v - p_s}{\frac{1}{2}\rho V^2} \tag{8.5}$$

For a given water quality, the higher the flow velocity, the smaller this difference.

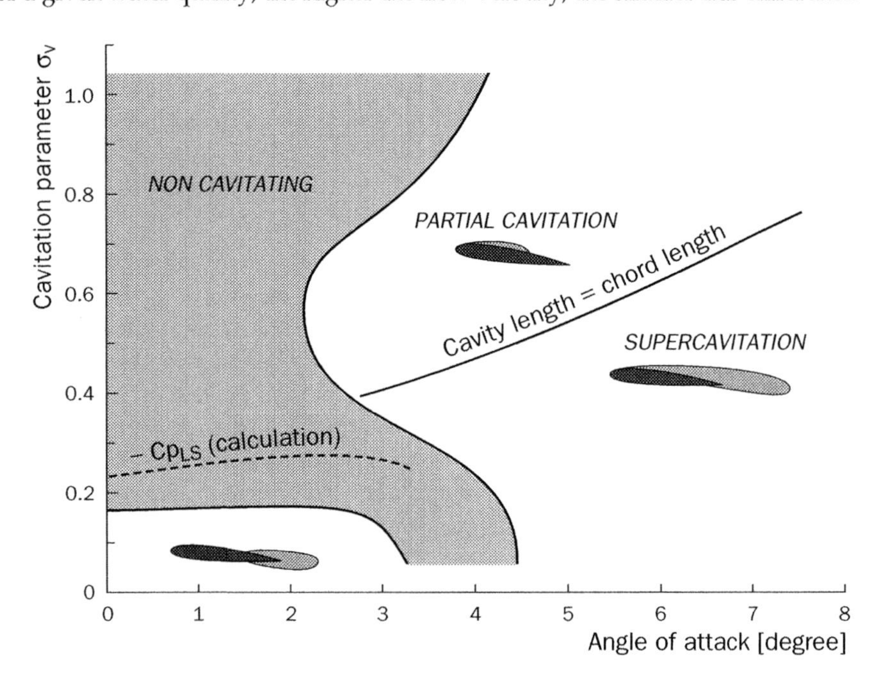

8.7 – Attached cavitation patterns on a NACA 16209 hydrofoil in deaerated water at a REYNOLDS number of 10^6
[from BRIANÇON-MARJOLLET et al., 1990]

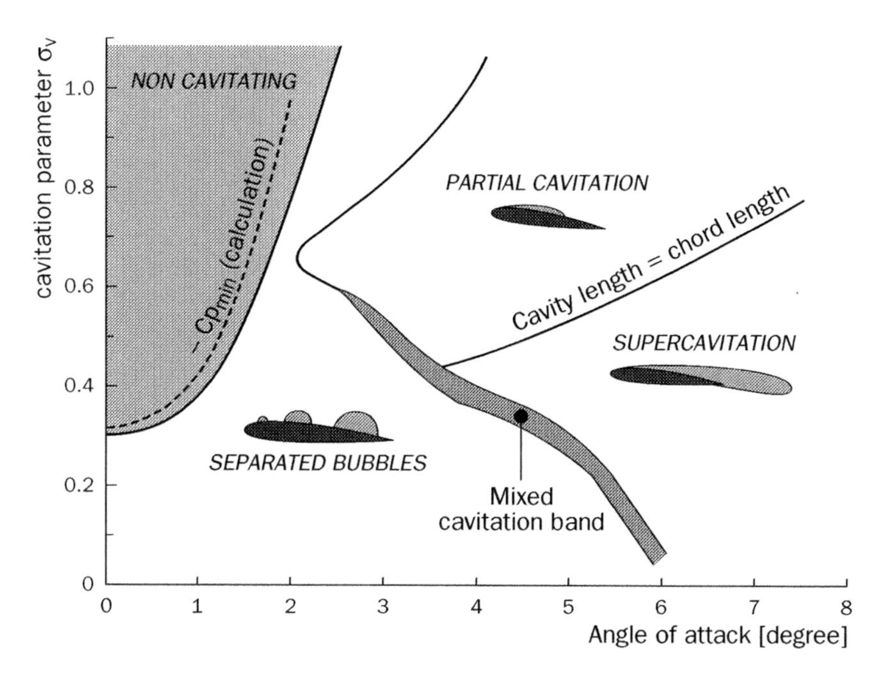

8.8 - Cavitation patterns on a NACA 16209 hydrofoil with nuclei seeding at a REYNOLDS number of 10^6

The experimental conditions correspond to a strong seeding, around 5 nuclei/cm^3. [from BRIANÇON-MARJOLLET et al., 1990]

Bubbles are not very sensitive to the boundary layer features such as laminar separation or transition to turbulence. Their development is principally controlled by the pressure distribution and the minimum pressure.

At high enough angles of attack, leading edge partial cavities as well as super-cavities are not affected by nuclei seeding. A region of transition (named mixed cavitation band in figure 8.8-b) between traveling bubble cavitation and leading edge cavitation is observed in which both cavitation patterns alternate in space and time. The extent of this transition region depends upon the nuclei injection rate. As this increases, the separated bubble region expands at the expense of the mixed region.

In case of traveling bubble cavitation, the pressure on the non-wetted parts of the foil, inside the different bubbles, is equal to the vapor pressure. The bubble growth limits the pressure drop and consequently the lift compared with the non-cavitating flow. This effect is visible on figure 8.9 which shows the decrease in lift with development of either a trailing edge cavity or traveling bubble cavitation at zero incidence. Similar trends are observed for the evolution of pressure head or of the efficiency versus the cavitation number, in rotating machines.

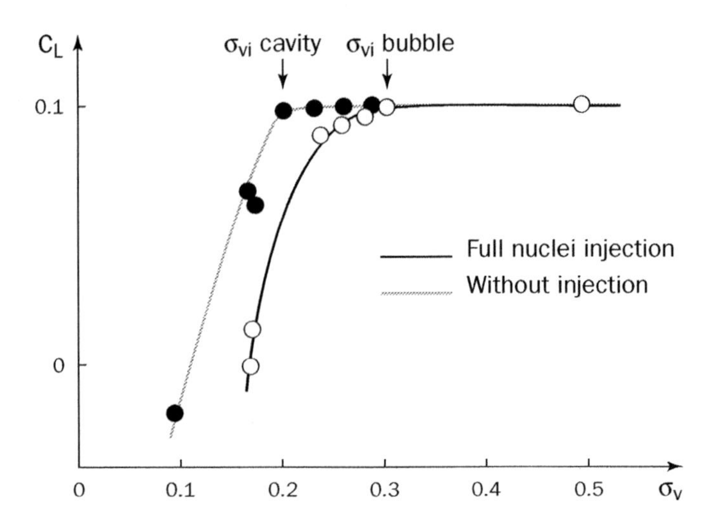

8.9 - Lift coefficient as a function of σ_v and nuclei injection ($\alpha = 0°$)

Usually, when bubbles explode on the upper side of a foil, they remain almost spherical. If the initial nucleus is very close to the wall, they are hemispherical. There is hardly any slip between the bubbles and the liquid and the difference between the bubble and liquid velocities does not generally exceed 10%. Ultra rapid movies show that the growth in radius is effectively proportional to the distance from the leading edge and in reasonable agreement with equation (3.59).

8.2.2. SCALING LAW FOR DEVELOPED BUBBLE CAVITATION

Consider the case of an isolated bubble exploding in the vicinity of a foil. It is assumed to be spherical or hemispherical depending on its proximity to the surface. The RAYLEIGH-PLESSET equation is used to predict the evolution of the bubble radius. Effects of gas pressure and surface tension are negligible as soon as the microbubble becomes a macroscopic cavitation bubble.

By transforming the time derivatives into space derivatives, via the relation :

$$ds = V(s)\,dt \tag{8.6}$$

where V(s) is the local velocity at the curvilinear distance s, the RAYLEIGH-PLESSET equation becomes:

$$R\ddot{R} + \frac{3}{2}\dot{R}^2 - \frac{1}{\rho V^2} R\dot{R}\dot{p} = \frac{p_v - p(s)}{\rho V^2} - \frac{4\nu \dot{R}}{VR} \tag{8.7}$$

Usually, the local pressure $p(s)$ on the foil is taken to be that at infinity for the bubble. In obtaining equation (8.7), the BERNOULLI equation for the homogeneous liquid flow was used. Initial conditions are that the nucleus is generally assumed to be in equilibrium with a zero initial radial velocity. As for the initial bubble size, this has no practical importance and can be disregarded, as mentioned in section 3.6.4.

The bubble radius R as well as the distance s along the foil are non-dimensionalised using the chord c of the foil. The non-dimensional form for equation (8.7) is then:

$$\overline{R}\,\ddot{\overline{R}} + \frac{3}{2}\dot{\overline{R}}^2 - \frac{\dot{Cp}}{2(1-Cp)}\overline{R}\,\dot{\overline{R}} = -\frac{Cp-\sigma_v}{2(1-Cp)} - \frac{4}{Re\sqrt{1-Cp}}\frac{\dot{\overline{R}}}{\overline{R}} \qquad (8.8)$$

Re is the REYNOLDS number based on the foil chord and the velocity at infinity.

Now we refer to the full scale flow as the prototype flow and the geometrically similar smaller scale flow as the model. Consequently, equation (8.8) shows that the growth of bubbles is similar for equal values of the cavitation number σ_v and of the REYNOLDS number. In fact, viscous effects are often weak, particularly for large bubbles, and it can be assumed that the previous statement is valid even when the REYNOLDS scaling law is not fulfilled. Hence, the non-dimensional radius $\overline{R}(\overline{s})$ is the same for both model and prototype, i.e. the radius is proportional to the length scale at the same relative distance value $\overline{s}$. That result is an extension of equation (3.59).

Consider the case of several bubbles exploding on the upper side of the model and the prototype. If inertia and pressure forces are again assumed to be dominant, the scaling law between both flows requires that the exploding bubbles are at the same locations relative to the foil. In practice, this condition needs only be satisfied statistically, which requires that the number of bubbles in similar volumes are identical. In other words, given a prototype flow and a geometrically similar model, the concentrations of active nuclei must satisfy the scaling law:

$$\frac{N_m}{N_p} = \left(\frac{L_p}{L_m}\right)^3 \qquad (8.9)$$

This relation, in which L_p and L_m stand for characteristic length scales of the prototype and the model respectively, was given first by HOLL and WISLICENUS (1961). It implies that the nuclei concentration for model tests must be higher than for the prototype. This is important when testing models of turbomachines, as it requires an additional scaling of the nuclei concentration between model and prototype to ensure the similarity of traveling bubble cavitation patterns. However, this condition need not be fulfilled at high enough nuclei concentrations, when saturation occurs (see next section).

8.2.3. *SATURATION*

Partial saturation

In the upstream region of the cavitating zone, the exploding bubbles are generally separate. If the nuclei concentration is large enough, they merge at a given distance x_s from the leading edge. The downstream part of the cavitating zone is said to be saturated. The higher the nuclei concentration, the earlier saturation occurs (fig. 8.10).

If this happens at a bubble radius R_S, the bubble surface density per unit wall surface area is of the order of $1/4R_S^2$. Assuming that these cavitation bubbles originate in nuclei contained in a layer of thickness δ, the nuclei concentration required to reach saturation with bubbles of radius R_S can be estimated via the expression:

$$N_S \cong \frac{1}{4\delta R_S^2} \tag{8.10}$$

The bubble radius R_S can be calculated using the RAYLEIGH-PLESSET equation (8.8). In the special case of an approximately constant pressure distribution on the foil, R_S can be estimated using equation (3.59), where s is replaced by x_s:

$$R_S \cong \sqrt{\frac{-Cp_{min} - \sigma_v}{3(1 - Cp_{min})}}\, x_s \tag{8.11}$$

As for the thickness δ, a rough estimate can be obtained using EULER's equation:

$$\frac{dp}{dn} = \rho \frac{V^2}{r} \tag{8.12}$$

This equation is applied in the vicinity of the minimum pressure point, where the radius of curvature of the wall is denoted by r. Moving away from the wall, the pressure progressively increases and less and less nuclei are activated. A threshold value of 10% of the pressure difference $\Delta p = p_v - p_{min}$ is chosen to define the thickness δ. In other words, it is supposed that the active nuclei are contained in a layer of fluid where the pressure does not differ from the minimum pressure by more than 10% of Δp. This threshold is somewhat arbitrary but was chosen to give a reasonable account of experimental results [BRIANÇON-MARJOLLET *et al.* 1990]. Thus, the order of magnitude of this liquid layer thickness $\delta_{10\%}$ is:

$$\delta_{10\%} \cong \frac{0.1(p_v - p_{min})}{\left|\frac{dp}{dn}\right|} \cong \frac{0.1(p_v - p_{min})}{\rho \frac{V_{max}^2}{r}} \cong \frac{-Cp_{min} - \sigma_v}{20(1 - Cp_{min})} r \tag{8.13}$$

Substitution of equations (8.11) and (8.13) into equation (8.10), leads to:

$$N_s \cong 15\left[\frac{1-Cp_{min}}{-Cp_{min}-\sigma_v}\right]^2 \frac{1}{r x_s^2} \tag{8.14}$$

This equation gives the order of magnitude of the nuclei concentration required to reach saturation at a distance x_s from the leading edge of a foil. It depends on

- its minimum pressure coefficient Cp_{min}, and
- the mean radius of curvature r of the wall.

This concentration is infinite in the limiting case of inception corresponding to $\sigma_v = -Cp_{min}$ and progressively decreases as σ_v diminishes. Although the present analysis is approximate for many reasons including the fact that interactions between bubbles have been disregarded, it exhibits the key parameters controlling saturation.

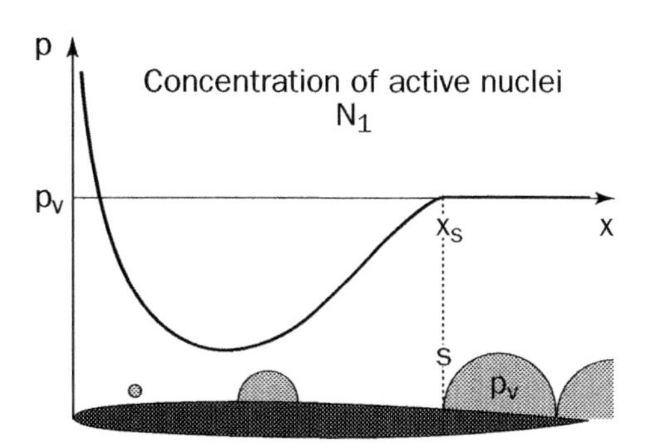

8.10
Evolution of the traveling bubble cavitation and particularly of the point of saturation S for an increasing nuclei concentration

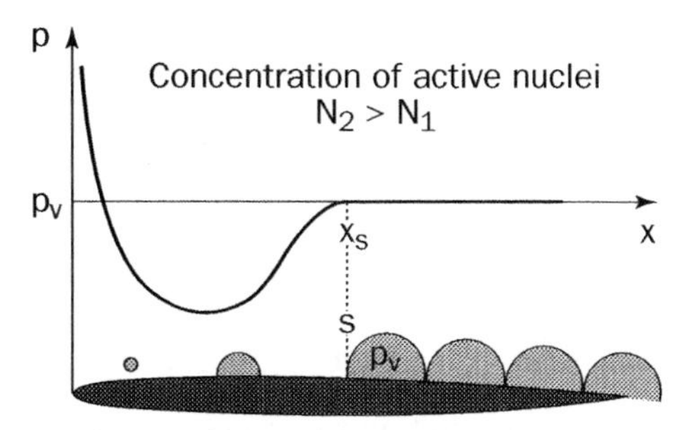

It must be noted that downstream of the saturation point, the bubbles form a kind of continuous cavity in which the pressure is almost equal to the vapor pressure (fig. 8.10).

Full saturation

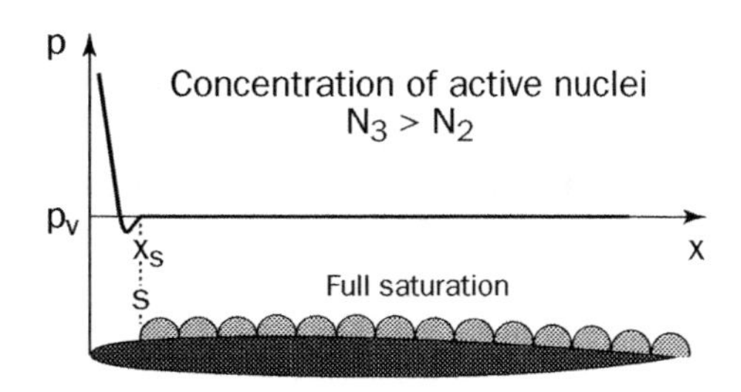

If the nuclei concentration is increased (see fig. 8.10), the saturation distance x_S decreases, i.e. the point of saturation moves upstream. The region of constant pressure equal to vapor pressure extends upstream at the expense of the region in which p drops below p_v. The saturation distance x_s tends to a limit corresponding to a pressure distribution that is constant and equal to the vapor pressure on most of the upper side of the foil, leaving only a very small zone where pressure is below p_v. The foil is said to be fully saturated in so far as the cavitating configuration is unchanged by further increase in the nuclei concentration.

8.3. INTERACTION BETWEEN BUBBLES AND CAVITIES

The low pressure region which precedes an attached cavity can activate oncoming cavitation nuclei and generate traveling bubble cavitation which will inevitably interact with the original attached cavity. For example, an attached cavity can be replaced by traveling bubble cavitation, if the nuclei concentration is high enough.

8.3.1. EFFECT OF EXPLODING BUBBLES ON A CAVITY

When a hemispherical bubble explodes on the upper side of a foil, the pressure distribution is locally and temporarily altered and a pressure gradient appears in the vicinity of the bubble (fig. 8.11).

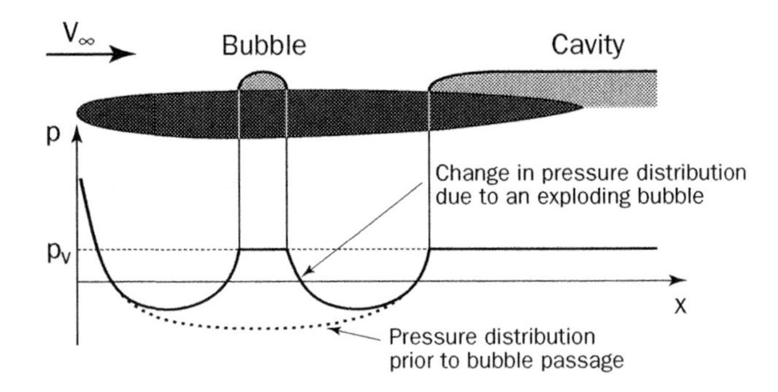

8.11 - Schematic representation of the change in pressure distribution on a foil due to an exploding bubble

Upstream of the bubble, this gradient can give rise to different effects (fig. 8.12):

- If the cavitation parameter is low enough, the bubble is accompanied by a kind of transient cavity, which reveals the existence of a local separation of the boundary layer moving with the bubble (fig. 8.12-a).

- In other cases, it has been observed [BRIANÇON-MARJOLLET *et al.* 1990] via dye injections, that the exploding bubble triggers transition to turbulence. A turbulent spot grows and moves along behind the bubble (fig. 8.12-b).

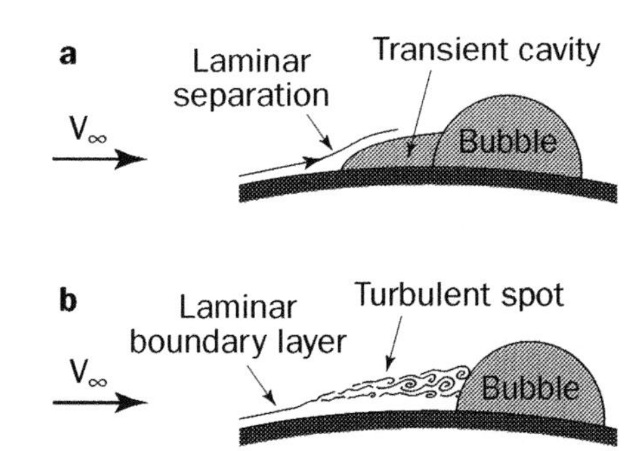

8.12
The two main effects of an exploding bubble on the boundary layer

(a) generation of a local laminar separation and of a subsequent transient cavity
(b) generation of a turbulent spot moving along behind the bubble

Consider the case of a cavity attached to the rear part of a foil. When a hemispherical bubble reaches the cavity detachment line, it passes over the cavity as if it were a solid wall. However, the turbulent spot which follows the bubble causes the boundary layer flow to re-attach to the wall and consequently destroys the mechanical equilibrium of the cavity. A part of the cavity, with a spanwise width of the order of the bubble diameter, is swept outwards and carried away by the flow. It can be said that the bubble does not push the cavity, but rather pulls it via the action of its associated turbulent spot.

After the transit of the bubble and its turbulent spot, conditions favorable to laminar separation and attached cavitation are recovered. If the nuclei density is high enough such that a new bubble arrives before attached cavitation is re-established, traveling bubble cavitation will definitively replace attached cavitation.

8.3.2. CRITICAL NUCLEI CONCENTRATION FOR TRANSITION BETWEEN ATTACHED CAVITATION AND TRAVELING BUBBLE CAVITATION

It is difficult to predict the critical nuclei concentration which will lead to the disappearance of attached cavitation by this mechanism since the conditions for destabilization of the boundary layer by an exploding bubble are unclear. However, it is possible to estimate an order of magnitude as follows.

Attached cavitation induces a constant pressure downstream of the point of cavity detachment and a lower pressure value upstream. As a consequence, the adverse pressure gradient forces the boundary layer to separate just upstream of cavity detachment. Similarly, for traveling bubble cavitation, the pressure beyond the saturation point can be considered constant with a lower value upstream which allows the bubble to grow.

If it is assumed that both regimes, traveling bubble cavitation and an attached cavity, coexist, then it is necessary to compare the relative positions of both cavity detachment and saturation. If saturation occurs upstream of detachment, the traveling bubble cavitation will impose a constant pressure field there. The adverse pressure gradient will be reduced and become insufficient to make the boundary layer separate. In this way, no cavity can actually attach to the wall. As a consequence, if the nuclei concentration is high enough, the change in mean pressure distribution due to the bubbles will no longer be compatible with attached cavitation. This is different from the previous mechanism of destabilization of the boundary layer by transient turbulent spots but gives an upper limit for the transition from attached cavitation to traveling bubble cavitation.

The critical nuclei concentration can therefore be estimated using equation (8.14) in which the saturation distance x_s is replaced by the distance to cavity detachment x_D. Considering the previous case of a NACA 16209 foil with chord $c = 100$ mm, $r = 450$ mm, $\sigma_v = 0.056$, $x_s = 75$ mm and $Cp_{min} = -0.18$, the following estimates are found, using the relationships of section 8.2.3:

$$R_s = 14 \text{ mm}$$
$$\delta_{10\%} = 2.4 \text{ mm}$$
$$N_s = 0.54 \text{ nuclei}/\text{cm}^3$$

Those orders of magnitude, and more especially the critical value of the nuclei concentration for transition between attached and traveling bubble cavitation, are consistent with experimental values [BRIANÇON *et al.* 1990].

From this approach, it can easily be understood why a leading edge cavity is almost insensitive to nuclei injection. In practice, the seeding systems currently used in hydrodynamic tunnels only lead to saturation for foils at small angles of attack. For large angles, it is almost impossible to reach saturation near the leading edge. The radius of curvature r and the distance x_s are so small near the leading edge that the concentration at saturation given by equation (8.14) is huge and far beyond the capabilities of usual seeding devices. In such cases, only leading edge cavities are possible. Similarly, current experience in turbomachines shows that attached cavities are much more likely to occur on the blades when operating off design point.

8.3.3. THE PREDICTION OF CAVITATION PATTERNS

The following predictive rules result from the previous considerations.

- In water without active nuclei, i.e. in which the existing nuclei are not activated by the minimum pressure, attached cavitation develops in the region of laminar separation, if the boundary layer actually separates from the wall. Equation (8.1), which correlates the inception cavitation number to the magnitude of the pressure coefficient at laminar separation:

  $$\sigma_{vi} = -Cp_{LS}$$

  must then be applied.

For developed cavities, the link between cavity detachment and laminar separation is still valid. However, the prediction of the location of detachment has to take into account the modification of the pressure distribution due to the development of the cavity.

- Conversely, if water contains a large number of nuclei which are destabilized and grow in the region of minimum pressure, equation (8.5), which correlates the inception cavitation number to the magnitude of the minimum pressure coefficient via the liquid susceptibility pressure p_s:

$$\sigma_{vi} = -Cp_{min} - \frac{p_v - p_s}{\frac{1}{2}\rho V^2}$$

must be applied. Traveling bubble cavitation is more often observed for low values of the angle of attack and of the cavitation parameter.

More complicated situations can occur because of the interaction between either an attached cavity and the boundary layer, or between the attached cavity and the traveling bubbles. Typical examples of this are:

- In the small range of moderate attack angles for which the boundary layer undergoes transition to turbulence, intermittency can lead to the formation of highly unsteady and three-dimensional attached cavities.
- If the nuclei density is moderate, traveling bubbles and attached cavities can coexist in a transitional region, the extent of which diminishes as the nuclei density increases.

8.4. *ROUGHNESS AND CAVITATION INCEPTION*

In the previous sections, it was implicitly assumed that the walls were smooth enough so that roughness effects were negligible. In practice, surface roughness can affect the onset of cavitation (e.g. on the walls of concrete spillways or on the blades of turbomachines or ship propellers) depending on their surface finish. Isolated roughness usually generate small, attached cavities, which release a high number of tiny bubbles. In turbomachines, such cavitation may be erosive, without any detectable change in the global performance.

Little is known from a fundamental viewpoint and practically all data have been obtained empirically. The problem has mainly been studied by HOLL (1960), ARNDT and IPPEN (1968) and ARNDT (1979, 1981) who proposed a two step predictive method. The influence of surface roughness on cavitation inception is first analysed on a flat plate (i.e. without a pressure gradient). The results are then applied to the case of roughness elements placed on an originally smooth body.

For an isolated surface irregularity on a flat plate, the incipient cavitation number σ_{i0} depends on the shape of the roughness element, on its height h relative to the local boundary layer thickness δ and on the REYNOLDS number $Re_\delta = V\delta / \nu$ (V is the external velocity). The critical cavitation number in this case may be written as [HOLL 1960]:

$$\sigma_{i0} = C\left(\frac{h}{\delta}\right)^a (Re_\delta)^b \tag{8.15}$$

In this analysis, the influence of the shape factor H of the boundary layer is ignored due to the lack of available data. According to HOLL (1960), the critical cavitation number at inception decreases slightly when the shape factor increases for given values of the velocity V and of the relative protuberance height h/δ.

The values of the exponents a and b and the coefficient C can be found in ARNDT's review (1981) for various 2D and 3D small-sized roughness elements. For example, for a 2D triangular protuberance, $a = 0.361$, $b = 0.196$, $C = 0.152$, and the critical cavitation number is of the order of 1. An asymptotic value for σ_{i0} is expected for high values of the REYNOLDS number and values larger than 1 of the ratio h/δ.

While an isolated protuberance produces only a local perturbation of the boundary layer flow, the major effect of uniformly distributed roughness is an increase in turbulence intensity. Cavitation occurs then in the core of large eddies of the boundary layer. For distributed roughness, ARNDT and IPPEN (1968) showed that the cavitation inception number could be linearly correlated to the friction coefficient C_f:

$$\sigma_{i0} = 16 C_f \tag{8.16}$$

Let us consider a smooth body and assume that an isolated irregularity or distributed roughness is placed at a given point A on its surface. The problem is to determine the incipient cavitation number of the rough body σ_{iR} from

- the incipient cavitation number of the smooth one σ_{iS}, and
- the characteristics of the roughness, in particular its critical cavitation number σ_{i0} determined by the appropriate equation (8.15) or (8.16).

The minimum pressure coefficient on the rough body at the location of the roughness is:

$$Cp_R = \frac{p_R - p_\infty}{\frac{1}{2}\rho V_\infty^2} = Cp_S + \frac{p_R - p_S}{\frac{1}{2}\rho V^2}\frac{V^2}{V_\infty^2} \tag{8.17}$$

p_R is the minimum pressure at point A due to the roughness, p_S the original pressure at the same location on the smooth body and Cp_R (Cp_S respectively) the corresponding pressure coefficients.

The flow velocity V at the location of the irregularity is estimated from BERNOULLI's equation:

$$V \cong V_\infty \sqrt{1 - Cp_S} \tag{8.18}$$

so that equation (8.17) becomes:

$$Cp_R = Cp_S + \frac{p_R - p_S}{\frac{1}{2}\rho V^2}(1 - Cp_S) \tag{8.19}$$

At cavitation inception on the roughness element, $p_R \cong p_v$ and $\frac{p_R - p_S}{\frac{1}{2}\rho V^2}$ is identified with $-\sigma_{i0}$. The pressure coefficient is then decreased due to roughness as follows:

$$Cp_R = Cp_S - \sigma_{i0}(1 - Cp_S) \tag{8.20}$$

The effect of the roughness is stronger when it is located near the minimum pressure point where the local velocity is the highest. If so, the incipient cavitation number on the rough body (considered here as the negative of the minimum pressure coefficient) is then:

$$\sigma_{iR} = \sigma_{iS} + \sigma_{i0}(1 + \sigma_{iS}) \tag{8.21}$$

This equation gives the increase in the value of the incipient cavitation number due to the roughness.

For example, consider a triangular 2D singularity of height $h = 0.1$ mm, located at the point of minimum pressure on a streamlined body whose minimum pressure coefficient is $Cp_{min\,S} = -\sigma_{iS} = -0.5$. If the boundary layer thickness is $\delta = 1$ mm with a flow velocity of 12 m/s, equation (8.15) gives $\sigma_{i0} \cong 0.42$ with $\sigma_{iR} \cong 1.13$. If the height of the roughness element is $h = 1$ mm, $\sigma_{i0} \cong 0.96$ and $\sigma_{iR} \cong 1.94$. These values of σ_{iR} for the rough body are much higher than that of the smooth one ($\sigma_{iS} = 0.5$). The effect of roughness on cavitation inception may then be quite important.

REFERENCES

ALEXANDER A.J. –1968– An investigation of the relationship between flow separation and cavitation. *NPL Unpublished Report.*

ARAKERI V.H. –1975– Viscous effects on the position of cavitation separation from smooth bodies. *J. Fluid Mech.* **68**, 779-799.

ARAKERI V.H. & ACOSTA A.J. –1973– Viscous effects in the inception of cavitation on axisymmetric bodies. *Trans. ASME I – J. Fluids Eng.* **95**, 519-527.

ARNAL D., HABIBALLAH M. & COUSTOLS E. –1984– Théorie de l'instabilité laminaire et critère de transition en écoulement bi et tri-dimensionnel. *La Recherche Aérospatiale* **2**.

ARNDT R.E.A. –1981– Cavitation in fluid machinery and hydraulic structures. *Ann. Rev. Fluid Mech.* **13**, 273-328.

ARNDT R.E.A. & IPPEN A.T. –1968– Rough effects on cavitation inception. *J. Basic Eng.* **90**, 249-261.

ARNDT R.E.A., HOLL J.W., BOHN J.C. & BECHTEL W.T. –1979– The influence of surface irregularities on cavitation performances. *J. Ship Res.* **23**, 157-170.

AVELLAN F. & DUPONT P. –1988– Étude d'un sillage d'une poche de cavitation partielle se développant sur un profil bi-dimensionnel. *La Houille Blanche* **7/8**, 507-516.

BRIANÇON-MARJOLLET L., FRANC J.P. & MICHEL J.M. –1990– Transient bubbles interacting with an attached cavity and the boundary layer. *J. Fluid Mech.* **218**, 355-376.

FRANC J.P. & MICHEL J.M. –1985– Attached cavitation and the boundary layer: experimental and numerical treatment. *J. Fluid Mech.* **154**, 63-90.

FRANC J.P. & MICHEL J.M. –1988– Unsteady attached cavitation on an oscillating hydrofoil. *J. Fluid Mech.* **193**, 171-189.

HOLL J.W. –1960– The inception of cavitation on isolated surface singularities. *J. Basic Eng.* **82**, 169-183.

KODAMA Y., TAKE N., TAMIYA S. & KATO H. –1981– The effect of nuclei on the inception of bubble and sheet cavitation on axisymmetric bodies. *J. Fluids Eng.* **103**, 557-563.

MICHEL J.M. –1988– Recherches récentes sur la cavitation à l'Institut de Mécanique de Grenoble. *La Houille Blanche* **7/8**, 517-526.

9. VENTILATED SUPERCAVITIES

By injecting gas into the low pressure regions of liquid flows, it is possible to obtain artificial cavities globally similar to the vapor cavities generated by natural cavitation.

Ventilation is currently used to limit the risk of erosion in spillways of large concrete dams, especially in tropical zones where a great amount of water has to be discharged in a short time [CHANSON 1999]. Other applications can be found in the field of ventilated hydrofoils or propellers and also supercavitating torpedoes for which the presence of gas in the cavity can be due to the propulsion device or to the entrainment of atmospheric air at their entry into water.

Ventilation decreases the relative cavity underpressure σ_c by increasing the cavity pressure, so that large artificial cavities comparable to natural supercavities can be obtained. However, two main differences exist between artificial and natural supercavities. Firstly, the non-condensable character of the gas causes different behavior in the aft part of the cavity. This furthermore depends upon the ambient pressure due to the effects of gas compressibility. Secondly, gravitational effects can be expected, as large cavities are obtained even for small velocities. One key parameter in all of this is the gas flowrate to be injected into the wake in order to obtain a given cavity length, drag or lift.

In section 9.1, we consider the case of two-dimensional ventilated flows behind a wedge or behind a step, the latter (called "half-cavity") being representative of ventilated flows in spillways. Typical features of these flows effected by gravity and cavity pulsations are presented. Section 9.2 is devoted to the axisymmetric case of a horizontal flow around a disc or a cone, with special attention given to gravitational effects. Finally, the theoretical analysis by PARISHEV (1978) of the stability and the pulsations of ventilated cavities is presented in section 9.3.

9.1. TWO-DIMENSIONAL VENTILATED CAVITIES

9.1.1. VENTILATED HYDROFOILS

The concept of the ventilated hydrofoil was proposed, about forty years ago, for the lift of rapid hydrofoil boats as well as for their propulsion by ventilated propellers.

The typical shape of base-vented hydrofoils is shown on figure 9.1. The thickness increases downstream parabolically, with some camber of the midline. The aft part of the foil is truncated. Its upper side is usually shorter than its lower side.

9.1
Typical shape of a truncated foil with a ventilated base

Ventilation is operated artificially from an air compressor on board. Air passes through the legs of the boat or through the propeller hub and is injected at the base of the foils or the propeller blades. Thus, the base pressure is increased and the drag is lowered.

When the angle of attack increases, air tends to invade the foil upper side, which results in a decrease of the lift. Such a regime must be avoided, so that the operating range of incidences is usually rather small [ROWE 1979].

This kind of foil is suited to boat speeds in the range 40-80 knots, for which classical non-cavitating foils are inoperative. For larger speeds, only supercavitating foils with a non-wetted upper side can be used.

From an experimental viewpoint, it is essential that the hydrodynamic tunnel in which ventilation tests are conducted is able to eliminate the injected air, so that the water entering the test section is free of any traces of air after recirculation in the tunnel loop. If the flow is unsteady, as is the case for pulsating cavities, the test facility must ensure a good decoupling between the test section and the rest of the loop in order to avoid any influence of the facility on the results.

In the case of three-dimensional foils of finite span, additional air cavities develop in the tip vortices. They coexist with the two-dimensional air cavities whose behavior is not significantly modified by them, except for very small values of the wing aspect ratio [VERRON & MICHEL 1984].

9.1.2. THE MAIN PARAMETERS

The global features of ventilated flows such as cavity length, lift or drag, depend on the relative cavity underpressure σ_c (see § 1.4.3):

$$\sigma_c = \frac{p_r - p_c}{\frac{1}{2}\rho V^2} \tag{9.1}$$

p_r is a reference pressure and p_c the cavity pressure.

The pressure inside the cavity is the sum of the vapor pressure (as the cavity is usually saturated with water vapor) and the pressure of the injected air:

$$p_c = p_v + p_{air} \tag{9.2}$$

In the case of substantial cavity pressure fluctuations, p_{air} is the mean value of the air pressure inside the cavity. For ventilated flows, p_v is generally negligible compared to p_{air}.

The air flowrate is normally imposed and the resulting air pressure inside the cavity is *a priori* unknown. It depends upon the mode of evacuation of air at the rear of the cavity, which, in turn, depends upon the global flow geometry, particularly the cavity length and the circulation, so that all parameters are strongly coupled.

Analysis of the experimental results [MICHEL 1971, 1984] suggests that the relative underpressure be split into two terms:

- the classical cavitation parameter σ_v, which characterizes the ambient reference pressure p_r:

$$\sigma_v = \frac{p_r - p_v}{\frac{1}{2}\rho V^2} \tag{9.3}$$

- and the relative mean pressure of the air inside the cavity, which takes into account its elastic behavior due to compressibility:

$$\sigma_a = \frac{p_{air}}{\frac{1}{2}\rho V^2} \tag{9.4}$$

Thus, the relative cavity underpressure σ_c is given by:

$$\sigma_c = \sigma_v - \sigma_a \tag{9.5}$$

As σ_c is usually small, the two parameters σ_v and σ_a are close in value.

In the case of pulsating cavities, the regimes of pulsation are fairly well identified via the ratio σ_c / σ_a, or equivalently σ_c / σ_v (see SILBERMAN & SONG, 1961, and SONG, 1962, who first described the pulsation regimes).

As for the pulsation frequency f, the experimental results are satisfactorily correlated by introducing the following non-dimensional frequency:

$$\varphi = \frac{f}{\sqrt{\frac{p_{air}}{\rho d \ell}}} \tag{9.6}$$

where ℓ is the mean cavity length and d a fixed reference length, typically the chord of the forebody.

As for the mass flowrate of air Q_m, two non-dimensional flowrate coefficients can be considered. These are:

- the mass flowrate coefficient:

$$C_{Qm} = \frac{Q_m}{\rho V S} \tag{9.7}$$

- and the volumetric flowrate coefficient:

$$C_{Qv} = \frac{Q_m}{\rho_{air} V S} \tag{9.8}$$

Here, S is a reference surface area. The mean air density ρ_{air} can be determined from the measurement of the cavity pressure p_c combined with the BOYLE-MARIOTTE law. In the last equation, the quantity Q_m / ρ_{air} represents the volumetric flowrate of air.

If we introduce the mean velocity of the air inside the cavity V_{air} and take an area close to the cavity cross-section for S, we have approximately:

$$Q_m \cong \rho_{air} V_{air} S \tag{9.9}$$

so that the volumetric air flow coefficient C_{Qv} gives the order of magnitude of the ratio of the air velocity to the water velocity:

$$C_{Qv} \cong \frac{V_{air}}{V} \tag{9.10}$$

The values of C_{Qv} typically lie in the range 0.05 to 0.5.

Among the other classical non-dimensional parameters to be considered, there is the FROUDE number:

$$Fr = \frac{V}{\sqrt{gd}} \tag{9.11}$$

which is often determined using the cavity length ℓ in place of the reference length d.

9.1.3. *CAVITY LENGTH*

Experimental tests conducted on various configurations have shown that the relative mean cavity length ℓ/d, as well as the global lift and drag coefficients C_L and C_D, depend primarily on the relative cavity underpressure σ_c and, to a smaller extent, on the FROUDE number, i.e.:

$$\frac{\ell}{d}(\sigma_c, Fr) \tag{9.12}$$

In the case of a ventilated foil, the incidence α should be included.

The influence of the relative cavity underpressure on cavity length is shown in figure 9.2 in the case of ventilated cavities behind a wedge.

The experimental points correspond to various values of the cavitation number σ_v, ranging from 0.4 to 10. Thus, the ambient pressure and consequently the different air flow regimes (see § 9.1.5) do not significantly influence the mean cavity length.

The variation of ℓ/c with σ_c suggests a power law (see chap. 6):

$$\frac{\ell}{c} \cong A\,\sigma_c^{\;-n} \tag{9.13}$$

The table in figure 9.2 gives the values of A and n for three different submersion depths. These values are close to those for vapor supercavities. The theoretical asymptotic value of n for an infinite medium is 2 (see chap. 6), while n approaches unity for small values of the submersion depth.

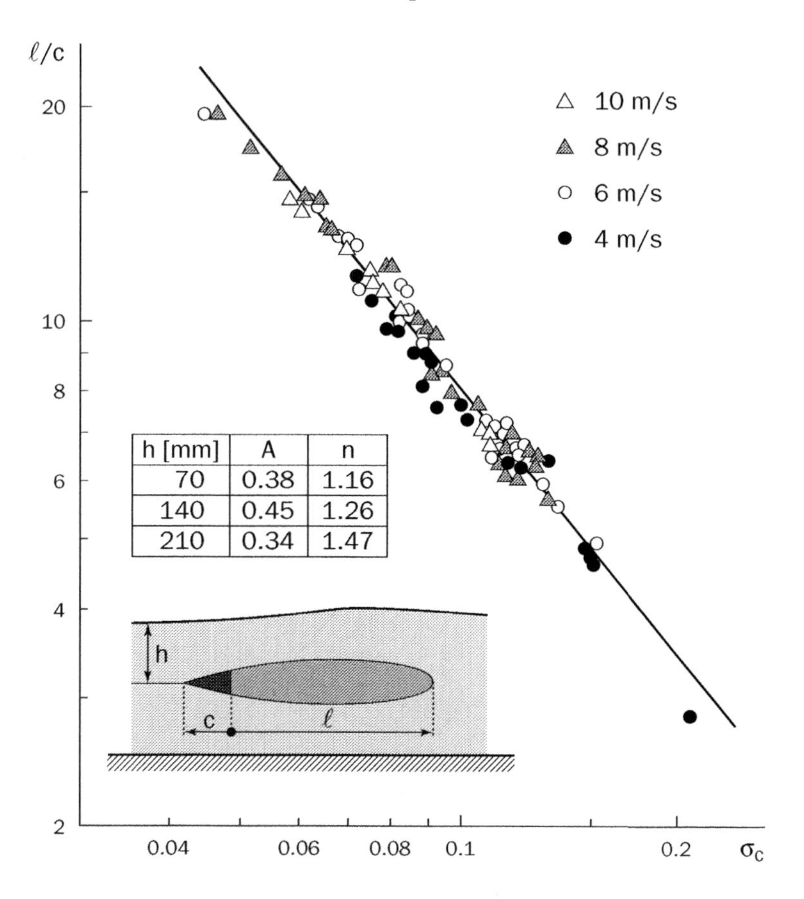

h [mm]	A	n
70	0.38	1.16
140	0.45	1.26
210	0.34	1.47

9.2 - Variation of the non-dimensional mean cavity length versus the relative underpressure σ_c for ventilated cavities behind a two-dimensional wedge (chord 60.5 mm, base 17 mm) for four different values of the water velocity

The wedge is at mid-height of a free surface channel of 280 mm in depth. Values of A and n for two other immersion depths are given in the table [from MICHEL, 1971].

In the range of velocities considered here, the FROUDE number has only negligible influence on the cavity length, except for the smallest values of σ_c where a slight deviation with respect to the power law becomes visible on figure 9.2. If the FROUDE number is calculated using the cavity length, as suggested in section 6.2.2:

$$Fr_\ell = \frac{V}{\sqrt{g\ell}} \tag{9.14}$$

gravitational effects are expected to be negligible if Fr_ℓ is much larger than one. For the smallest values of σ_c considered above, this condition is not fulfilled as the FROUDE number is relatively small and close to one (typically $Fr_\ell = 2.4$ for $\ell / c = 18$, $\ell = 1.09$ m and $V = 8$ m/s), so that gravitational effects become more significant.

The dependence of the cavity length on the cavity underpressure σ_c in the half-cavity configuration presented in figure 9.3 is qualitatively very similar to the case of a wedge [LAALI & MICHEL 1984]. This configuration is representative of the flow in spillways aerators. It is simpler than the hydrofoil since the effects of incidence and circulation are ruled out.

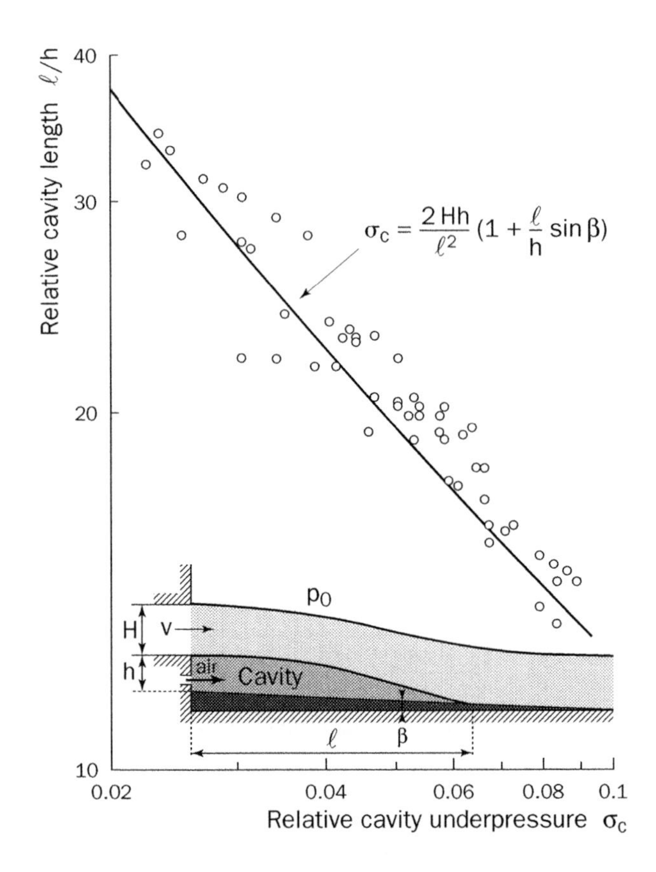

9.3 - Cavity length versus σ_c in the case of the half-cavity

The cavity is formed between an initially horizontal plane water jet of height $H = 138.9$ mm and a wall with a small slope $\beta = 3.1°$. The step height is $h = 30$ mm. The reference pressure p_r used in the definition of σ_c and σ_v is the pressure p_0 on the upper free surface of the jet. The different points correspond to various values of σ_v and of the velocity (between 4 and 12 m/s) [from LAALI & MICHEL, 1984].

9.1.4. *AIR FLOWRATE AND CAVITY PRESSURE*

The cavity pressure p_c depends mainly upon the air flowrate and to a smaller extent upon the ambient pressure. This dependency can be written non-dimensionally as follows:

$$\sigma_c(C_{Qv}, \sigma_v) \tag{9.15}$$

A schematic representation of the previous relationship is given on figure 9.4. It shows the complex behavior of ventilated cavities with hysteresis effects. The experimental procedure used to obtain the graph in figure 9.4 involves gradually increasing the air flowrate up to high values before decreasing it.

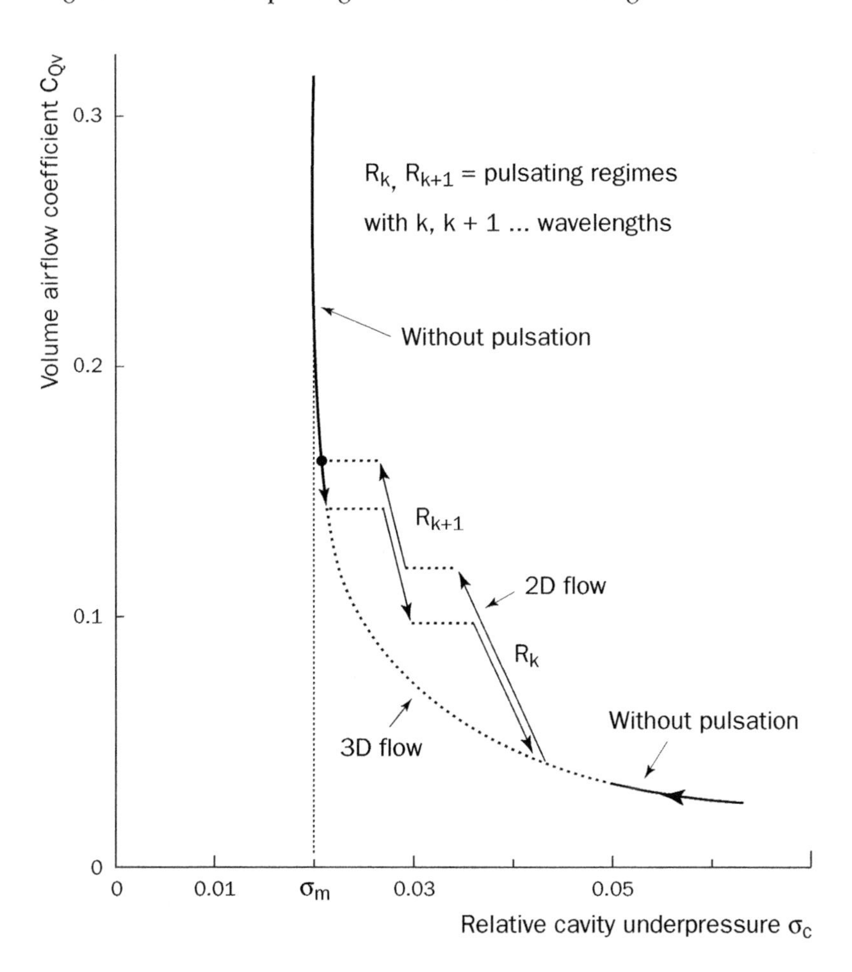

9.4 - Typical evolution of the relative cavity underpressure for an increasing, then decreasing, air flowrate (σ_v is constant)

The ambient pressure, and therefore the cavitation parameter σ_v, is large enough so that no vapor cavity preexists before air injection. In fact, if such a vapor cavity were present, the phenomena would practically be unchanged except for small σ_v-values (lower than 0.2 typically) for which a very small rate of air injection would result in a large increase in cavity length, disclosing a kind of global flow instability.

Roughly speaking, the curve of figure 9.4 is L-shaped. For very small injection rates, the air forms separate bubbles which are entrapped in the alternate BÉNARD-KÁRMÁN vortices of the body wake. For larger air flowrates, a continuous cavity appears and it becomes possible to measure the cavity pressure and to determine the corresponding σ_c parameter. This condition corresponds to the starting point on the right-hand side of the curve.

A subsequent small increase in the air flowrate produces a large increase in the air pressure and in the cavity length. This is due to the difficulty for air to escape from the cavity. On the contrary, on the vertical branch of the curve, air is evacuated via big bubbles, which break from the cavity on its upper interface upstream of its closure point due to gravity. This mode of air evacuation is very efficient, so that the cavity pressure generally reaches a maximum. Hence, the parameter σ_c tends to a minimum (σ_m) and the cavity length to a maximum.

The connecting region between the horizontal and the vertical branch of the curve is characterized by periodic pulsations of the air cavity, which will be considered more extensively in next section.

Examples of experimental curves $C_{Qv}(\sigma_c)$, corresponding to three water velocities in the half-cavity configuration are shown in figure 9.5.

The minimum value σ_m of the relative cavity underpressure σ_c depends mainly upon the FROUDE number $Fr = V/\sqrt{gH}$, as shown on figure 9.6. It decreases with the FROUDE number, i.e. with the flow velocity for a given jet height. The parameter σ_m is also an increasing function of σ_v at constant FROUDE number, as seen in figure 9.6, where several experimental points are indexed with the corresponding σ_v-values.

For reference purposes, figure 9.6 shows the curve $\sigma_c = 2/Fr^2$ corresponding to the case of a jet with equal values of pressure on both sides. The pressure on the free surface is assumed equal to the pressure inside the cavity and the jet is subject only to gravity. When σ_m is smaller than $2/Fr^2$, which occurs for small values of the FROUDE number, the pressure in the cavity is greater than the ambient pressure. The water jet lies on a cushion of pressurized air and the transverse pressure gradient promotes the rise of air bubbles. An increase in the air flow compels the air to escape the cavity from its upper interface by separate bubbles, without any increase in cavity length.

However, for larger values of the FROUDE number, the transverse pressure gradient diminishes and tends to be reversed, which makes the rise of air bubbles more difficult, so that a greater value of the cavity length, or a smaller value of σ_m, is required for the appropriate air evacuation.

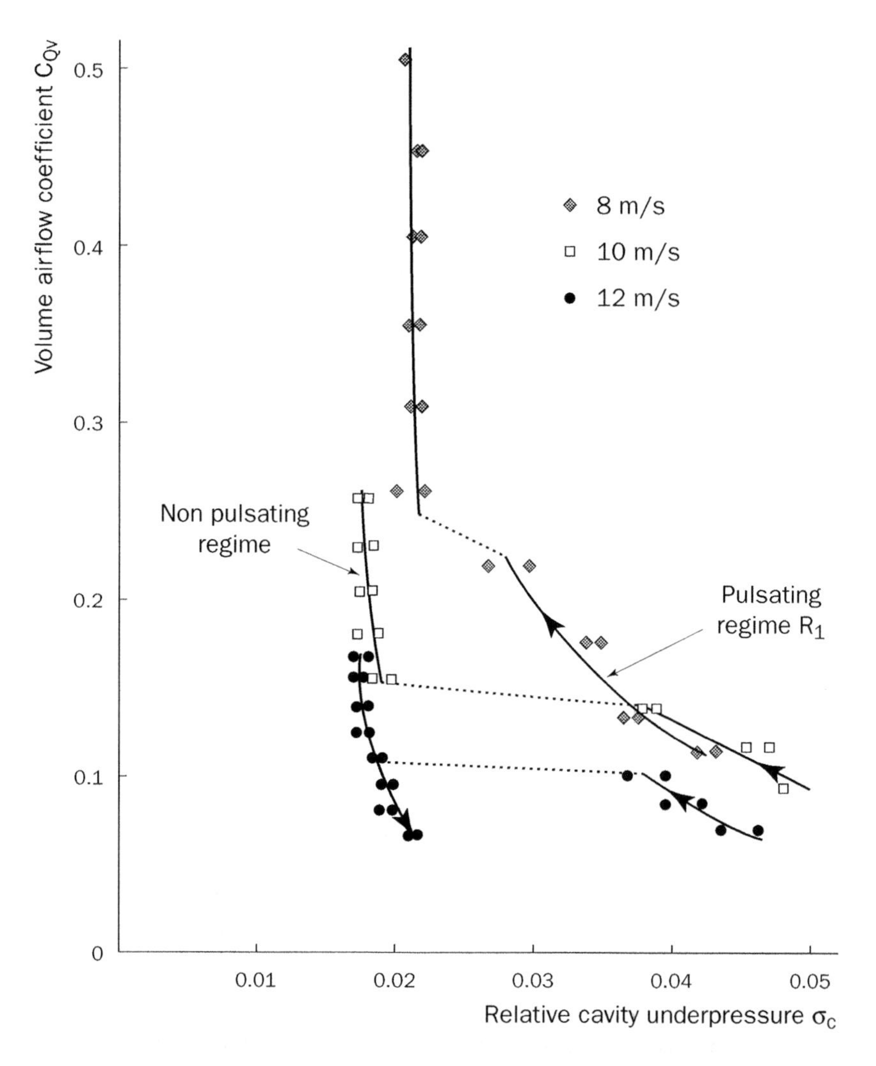

9.5 - Evolution of the cavity pressure σ_c with the air flowrate C_{Qv} at constant ambient pressure $\sigma_v = 0.44$ in the case of the half-cavity configuration

($H = 100.3$ mm, $h = 49.2$ mm)

The area S in the air flow coefficient is based on the height h and the width of the test section (see fig. 9.3) [from LAALI & MICHEL, 1984].

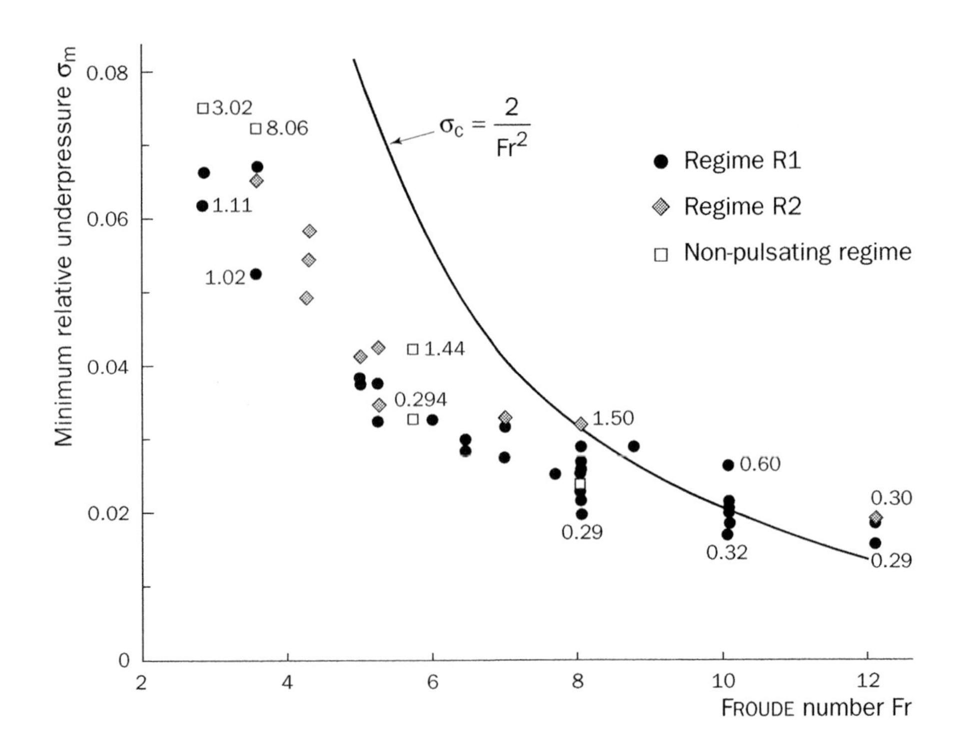

9.6 - Variation of the minimum relative underpressure σ_m with the FROUDE number for the half-cavity configuration

Several experimental points are indexed with the corresponding σ_v-values.

[from LAALI & MICHEL, 1984]

9.1.5. PULSATION REGIMES

The intermediate region in figure 9.4 is characterized by periodic pulsations of the air cavity. In this case, undulations of increasing amplitude are convected on both free frontiers of the ventilated cavity at a speed equal to the liquid velocity on the interfaces. They join at the rear part of the cavity, allowing air to escape periodically under the form of separate air pockets, as shown on figure 9.7.

The air pressure undergoes periodic and almost sinusoidal fluctuations in time. The minimum pressure occurs when an air pocket leaves the cavity. At the same time, undulations of the free surface arise at the two trailing edges of the wedge.

The cavity, as it reaches its minimum length, has the shape of one or several spindles. The corresponding flow regimes are named R_1, R_2, ..., R_k according to the number k of spindles. Regime R_2 is presented in figure 9.12. In general, high order regimes are obtained for large values of the ambient pressure or, more precisely, for small values of the parameter σ_c / σ_a.

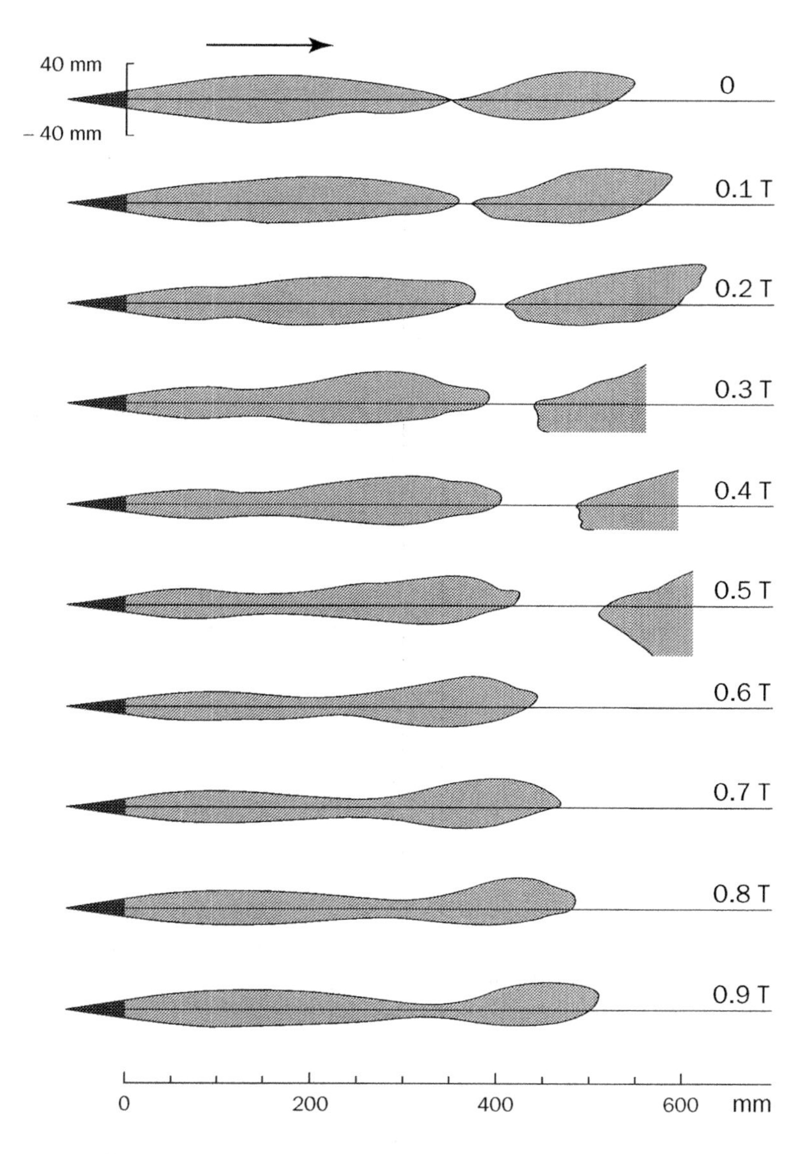

9.7 - Evolution of a one-wave cavity during one period of pulsation for V = 4 m/s, $\sigma_c = 0.111$, $\sigma_c/\sigma_a = 0.137$

The pulsation frequency is $f = 11.1$ Hz for this R_1 regime. A slight gravitational effect is visible [from MICHEL, 1971].

The passage from one regime R_k to the next R_{k+1} occurs for critical values of the volumetric air flow coefficient. It is accompanied by a jump in both the cavity pressure and the cavity length. The critical values of C_{Qv} are different for increasing or decreasing injection rates (see fig. 9.4), so that the curve $C_{Qv}(\sigma_c)$ presents hysteresis loops.

The wake of the cavity is made up of separate and almost periodic air pockets, which are emitted during successive periods. Further investigations have shown that air is not present as a continuous medium in the air pocket but rather as small bubbles organized in alternate vortices.

When gravitational effects are negligible, the transient re-entrant jet and the air pockets are symmetrical with respect to the flow direction. Conversely, when the FROUDE number is small, i.e. for small water velocities, the re-entrant jet is more or less deflected upwards and the air pockets are no longer symmetrical (fig. 9.8).

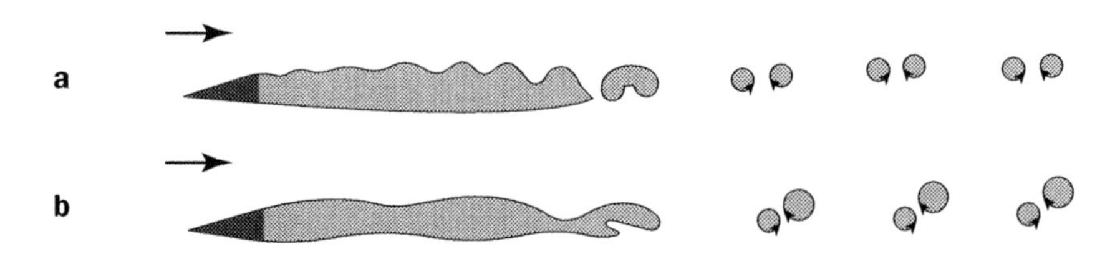

9.8 - Influence of gravity on the re-entrant jet direction and on the arrangement of the vortices in the wake
Case (a) corresponds to a smaller velocity than case (b) and hence to larger gravitational effects.

The basic reason for the discontinuous release of air at the rear part of a ventilated cavity is connected to its geometry. Since the cavity is a low pressure region, the curvature of its frontiers tends to be directed inwards. Thus, in steady two-dimensional or axisymmetric flows, the free streamlines tend to join and to produce a re-entrant jet which prevents the outflow of air.

This blockage effect can be illustrated, *a contrario,* experimentally. If two parallel plates are placed downstream of a ventilated pulsating cavity in such a way that the curvature of the free streamlines is locally reversed (see fig. 9.9), the flow of air is no longer discrete but continuous. Cavity pulsations disappear and the flow is globally steady [MICHEL 1984].

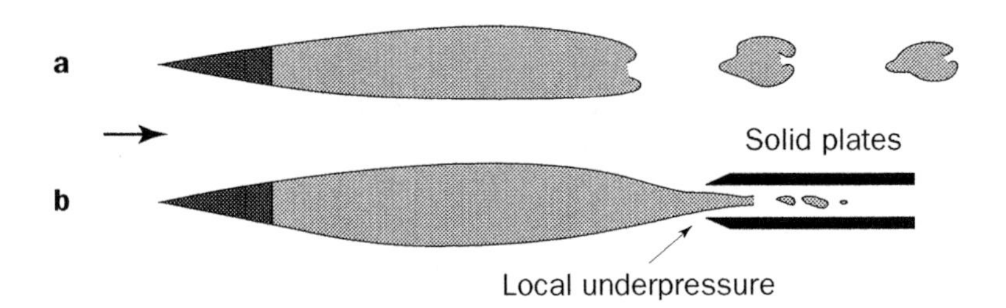

9.9 - Influence of the curvature of the free streamlines on the mechanism of evacuation of air into the wake
(a) discontinuous - (b) continuous (without pulsation)

Another mechanism of air release can be observed in the case of ventilated three-dimensional hydrofoils of small aspect ratio. Part of the injected air is continuously evacuated through the tip vortices, so that the cavity pulsations become weaker and may even be suppressed [VERRON 1977, VERRON & MICHEL 1984].

9.1.6. *PULSATION FREQUENCY*

For each regime R_k, the pulsation frequency varies approximately as the square root of the mean pressure of the air inside the cavity $\sqrt{p_{air}}$ (see fig. 9.10 for regime R_1). This leads to the non-dimensional frequency φ defined by relation (9.6). This parameter particularly depends upon the ratio σ_c / σ_a:

$$\varphi\left[\frac{\sigma_c}{\sigma_a}\right] \tag{9.16}$$

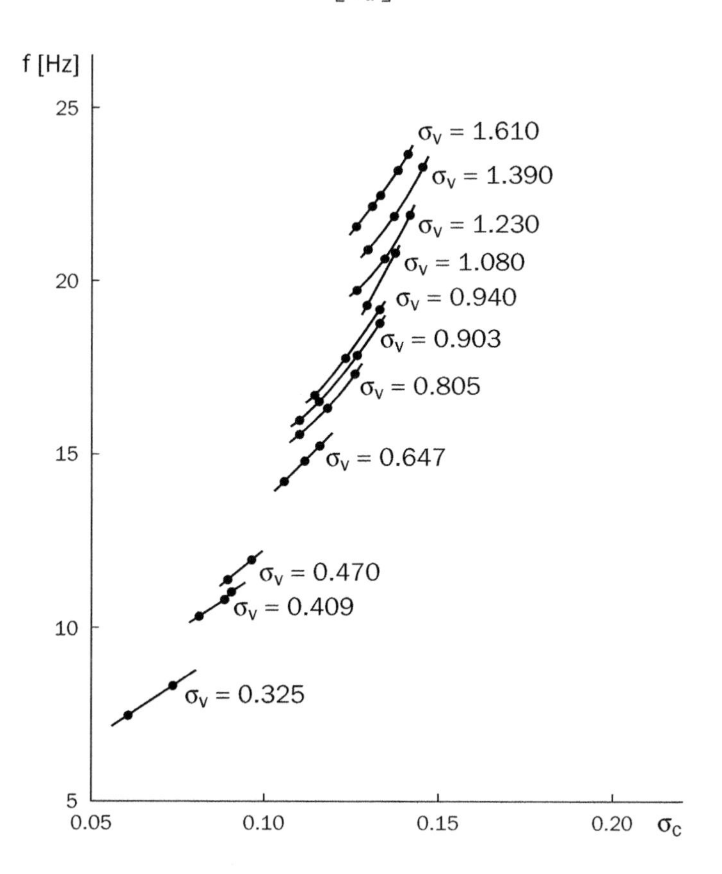

9.10 - Variations of the pulsation frequency with the relative cavity underpressure σ_c for increasing values of σ_v in the case of a wedge
(regime R_1, immersion depth 21 cm, $V = 8$ m/s) [from MICHEL, 1984]

A typical example of this relationship is presented in figure 9.11. It is remarkable that the various pulsating regimes correspond to different ranges of values of the ratio σ_c / σ_a. In particular, the one-wave regime corresponds approximately to the range 0.06-0.25, for a closed channel. It is the same for a free surface channel. The higher the order of the regime, the narrower the range of σ_c / σ_a. For each regime, the dependence $\varphi(\sigma_c / \sigma_a)$ is roughly linear.

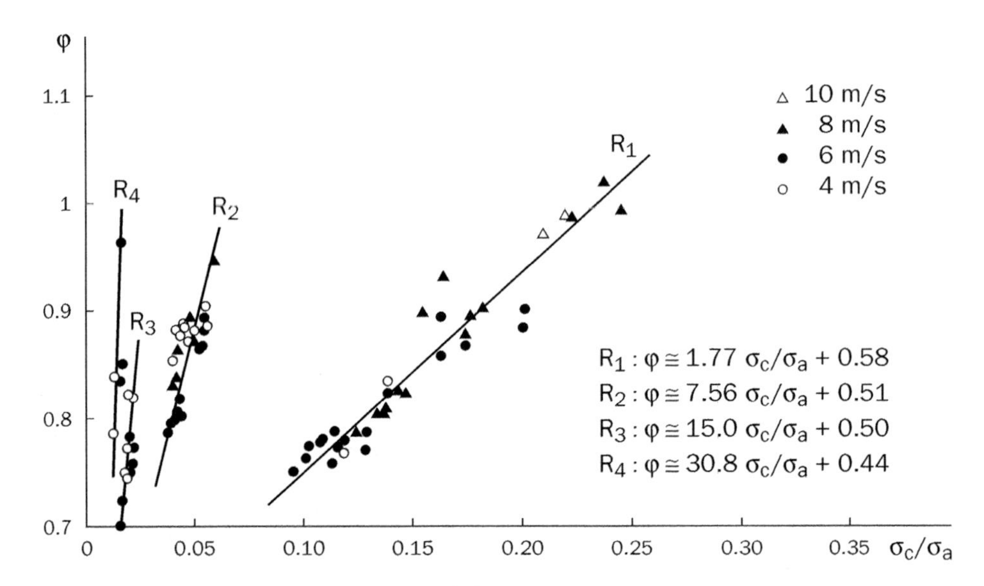

9.11 - Non-dimensional frequency versus σ_c/σ_a in the case of a wedge
(immersion depth 7 cm) [from MICHEL, 1984]

On the whole, the non-dimensional frequency φ lies in a rather limited range, between about 0.7 and 1. The limits are slightly smaller in the case of a larger submersion depth. If the non-dimensional frequency φ is assumed roughly constant, it follows that the pulsation frequency can be considered as approximately proportional to $\ell^{-1/2}$. Similar results have been obtained by LAALI and MICHEL (1984) in the half-cavity configuration.

In the case of lifting foils, the circulation around the cavity can also influence the mode of air evacuation. Only slight variations of the angle of attack, all other parameters being kept constant (in particular σ_v and the mass air flowrate Q_m), may result in a change of regime and then of cavity pressure, lift and drag [MICHEL 1984].

9.1.7. CONCERNING THE PULSATION MECHANISM

Figure 9.12 presents typical variations of several variables during the pulsation of a ventilated cavity: the pressure inside the cavity p_c, inside the shed air pocket p_p and at cavity closure p_f as well as the velocity of the shed air pocket V_p and of the cavity closure point V_f.

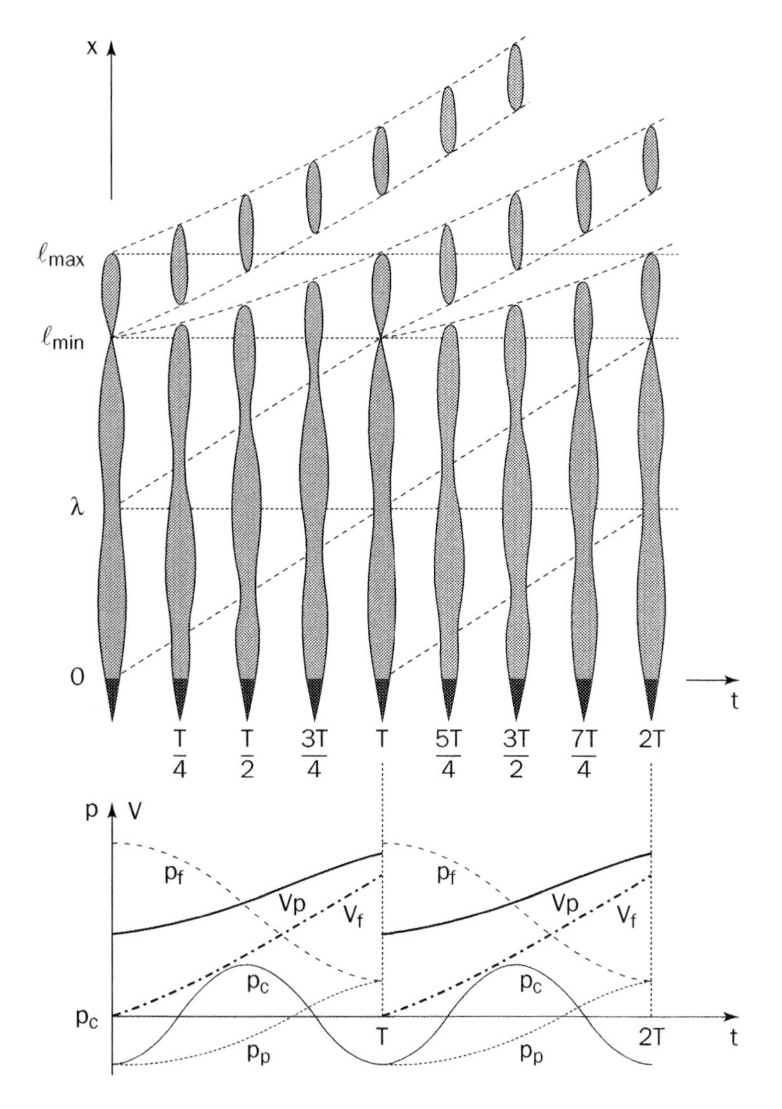

9.12 - Schematic evolution of an R2-cavity behind a ventilated wedge
[from MICHEL, 1984]

All these parameters undergo consistent variations during the cycle.

- The cavity pressure p_c effectively oscillates sinusoidally in time and is minimum when an air pocket is shed.
- When both cavity interfaces meet, the pressure at cavity closure p_f is maximum whereas the pressure p_p in the air pocket is minimum and equal to the cavity pressure. The difference between both pressures gradually decreases as the air pocket moves away from the cavity. As the pressure inside the shed pocket increases, its volume decreases, while it is accelerated in the cavity wake.

- The velocity V_f of the cavity closure point is close to zero at the instant of shedding and then increases and tends to reach the velocity of the air pocket previously shed.

Since the cavity is considered closed at any time, the mass flowrate Q_m of the injected air can be identified with the instantaneous increase in the mass of air enclosed in the cavity. The latter varies as the product of the cavity pressure p_c and the cavity volume. In addition, the mass of air enclosed in each pocket which is shed is given by Q_m/f.

Assuming that the undulations of the cavity interface, which are responsible for the pulsations, travel at the fluid velocity V_c on the interface:

$$V_c = V\sqrt{1+\sigma_c} \tag{9.17}$$

the wavelength λ is given by:

$$\lambda = \frac{V_c}{f} \tag{9.18}$$

In the R_k regime, experiments show that the minimum cavity length ℓ_{min} is given by:

$$\ell_{min} = k\lambda = k\frac{V_c}{f} \tag{9.19}$$

This relation is consistent with the simultaneous occurrence of

- the shedding of an air pocket, and
- the birth of undulations of the free surfaces at the wedge trailing edges.

By introducing the non-dimensional pulsation ω_k of regime k defined by:

$$\omega_k = \frac{\pi f \ell_{min}}{V_c} \tag{9.20}$$

the experimental relation (9.19) takes the following non-dimensional form:

$$\omega_k = k\pi \tag{9.21}$$

WOODS (1964, 1966, see references in chapter 6) developed a model of pulsating ventilated cavities, assuming that the total volume of the cavity and its wake remains constant in order to avoid the pressure singularity at infinity which occurs in a two-dimensional flow field. This condition is equivalent to assuming that the variations in cavity volume are balanced by the variations in the volume of the first air pocket of the wake. It is partly confirmed by experimental observations of pulsating cavities in free surfaces channels, which show that, in most cases, the free surface is not affected by the cavity pulsations. The energy is mainly transferred along the streamwise direction, but not along the transverse one. This allows us to better understand how cavity pulsations also occur in closed channels.

For the non-dimensional pulsation ω_k, WOODS obtains the following theoretical relation:

$$\omega_k = k\pi - 1.17 \tag{9.22}$$

which has to be compared to equation (9.21). The higher the order of the pulsating regime, the smaller the relative difference.

Finally, several features of two-dimensional ventilated cavities are very similar to the case of axisymmetric cavities such as those theoretically described by PARISHEV (1978) (see § 9.3).

9.2. *AXISYMMETRIC VENTILATED SUPERCAVITIES*

9.2.1. *DIFFERENT REGIMES OF VENTILATED CAVITIES*

For axisymmetric ventilated supercavities in horizontal flows, two main problems have to be considered, the mode of evacuation of air at the rear of the cavity and the deformation of the cavity under gravity [SEMENENKO 2001, SAVCHENKO 2001].

Three modes of air leakage have been observed:

- When the FROUDE number $Fr = V_\infty/\sqrt{gd}$ and the relative cavity underpressure σ_c are large enough, i.e. for large velocities and short cavities, gravitational effects are small and the cavity is axisymmetric. Its tail is filled with foam which is periodically rejected in the form of toroidal vortices (§ 9.2.2).
- For moderate values of the FROUDE number and long enough cavities (i.e. small enough values of σ_c), gravity is important. The tail end of the cavity has two hollow vortex tubes by which air is evacuated into the wake (§ 9.2.4).
- The third regime of gas leakage corresponds to pulsating cavities. It occurs for high values of the flowrate coefficient. The behavior of axisymmetric pulsating cavities is close to the 2D case described in section 9.1. It depends upon the parameter σ_c/σ_v which plays a role similar to σ_c/σ_a introduced in section 9.1.2.

CAMPBELL *et al.* (1958) suggested from empirical considerations that gravitational effects would be important if:

$$\sigma_c Fr < 1 \tag{9.23}$$

Values of $\sigma_c Fr$ lower than 1 lead to the second regime of gas leakage by hollow vortex tubes.

A different criterion, obtained theoretically by BUYVOL (1980), covers the experimental data better:

$$\sigma_c^{3/2} Fr^2 < 1.5 \tag{9.24}$$

Effectively, gravity is negligible for $\sigma_c^{3/2} Fr^2$ larger than 10.

9.2.2. GAS EVACUATION BY TOROIDAL VORTICES

In the case of negligible gravitational effects, the cavity is closed by an unstable re-entrant jet which produces a foam at the rear of the cavity. The external part of the foam is reentrained by the outer flow, while the central part moves as a counterflow. The result is the formation of almost periodic ring vortices which evacuate the foam (fig. 9.13). The STROUHAL number $S = f\ell / V_\infty$ (where ℓ is the mean cavity length) is around 0.31. This order of magnitude is comparable to the one which is found for partial vapor cavities (see chap. 7).

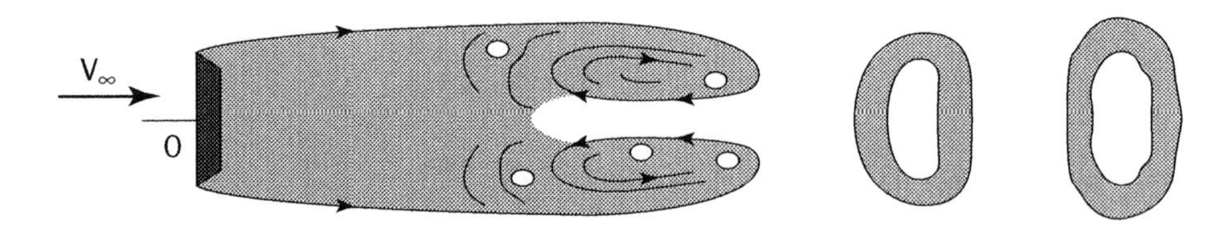

9.13 - Gas leakage by toroidal vortices

In the absence of gravitational effects, LOGVINOVICH (1973) proposed the following semi-empirical formula for the entrained air flowrate:

$$Q = \gamma V_\infty S_C \left(\frac{\sigma_V}{\sigma_C} - 1 \right) \tag{9.25}$$

in which γ lies in the range 0.01-0.02 and S_C is the area of the cavity mid-section. The flowrate vanishes for vapor cavities since $\sigma_V = \sigma_C$.

9.2.3. DEFORMATION OF THE CAVITY AXIS BY GRAVITY

For small enough FROUDE numbers, gravity tends to curve the cavity axis upwards and deform the cavity cross-section.

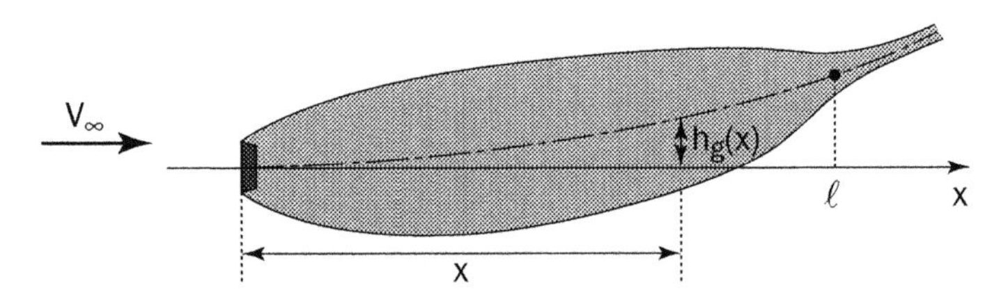

9.14 - Cavity deformation due to gravity

Assuming that, for each transverse slice of the cavity, the vertical momentum is balanced by the buoyancy force, the upwards drift of the cavity can be estimated by the following formula [SEMENENKO 2001]:

$$\frac{h_g}{\ell} \cong \frac{4(1+\sigma_c)}{3 Fr_\ell^{\,2}} \left(\frac{x}{\ell}\right)^2 \tag{9.26}$$

where Fr_ℓ is the FROUDE number based on cavity length given by equation (9.14). This relationship agrees with experimental results in the range $0.05 \le \sigma_c \le 0.12$ and $2.0 \le Fr_\ell \le 3.5$.

9.2.4. *GAS EVACUATION BY TWO HOLLOW TUBE VORTICES*

When submitted to strong gravitational effects, an originally axisymmetric cavity gives rise to two tube vortices by which air is evacuated continuously into the wake (fig. 9.15). This phenomenon was described first by COX and CLAYDEN (1957) and later by EPSHTEIN (1970, 1971).

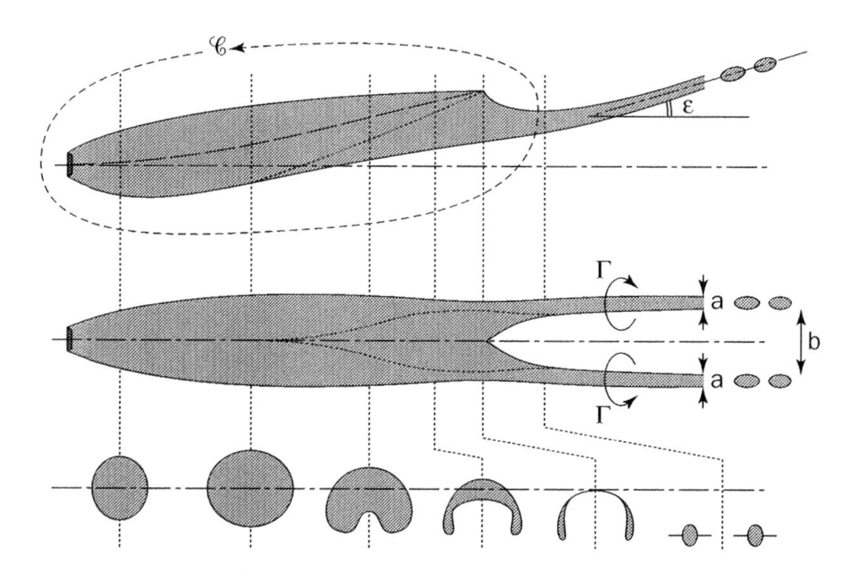

9.15 - Gas evacuation into the wake of a supercavitating disc through two hollow tube vortices

There is a phenomenological model based on the approach of COX and CLAYDEN. Experiments were conducted using discs of various diameter d (1.27-1.91 and 2.54 cm) and for σ_c-values in the range 0.11 to 0.23. Flow velocities were lower than 2.5 m/s, so that the FROUDE number $Fr = V_\infty / \sqrt{gd}$ was below 4.9.

Because of gravity, the velocity on the cavity is not constant, contrary to the cavity pressure p_c which is assumed constant. The velocity on the upper side of the cavity V_+ is smaller than that on the lower side V_-. The BERNOULLI equation:

$$p_c + \frac{1}{2}\rho V_\pm^{\,2} + \rho g z_\pm = p_\infty + \frac{1}{2}\rho V_\infty^{\,2} \tag{9.27}$$

allows us to determine the velocities on the cavity interface:

$$V_{\pm} = V_{\infty} \sqrt{1 - \sigma_c - \frac{2 g z_{\pm}}{V_{\infty}^2}} \tag{9.28}$$

Assuming σ_c negligible and keeping only the first order term for gravity, this equation is approximated by:

$$V_{\pm} \cong V_{\infty} - \frac{g z_{\pm}}{V_{\infty}} \tag{9.29}$$

The velocity difference results in a circulation Γ. Considering a path $\mathscr{C}$ around the cavity and the body, and situated in the plane of symmetry, the circulation along this path is:

$$\Gamma = \oint_{\mathscr{C}} \vec{V} . d\vec{s} \cong \int_0^{\ell} (V_- - V_+)\, dx \cong \int_0^{\ell} \frac{g}{V_{\infty}} (z_+ - z_-)\, dx \cong \frac{gA}{V_{\infty}} \tag{9.30}$$

where A is the intersectional area. If the cavity is considered as an axisymmetric ellipsoid of maximum diameter d_c and length ℓ, we have for the area A and for the volume $\mathscr{V}$ of the cavity:

$$\mathscr{V} \cong \frac{4\pi}{3} \frac{d_c^{\,2} \ell}{8}$$
$$A \cong \frac{\pi d_c \ell}{4} \tag{9.31}$$

The circulation Γ is discharged in the two counter rotating trailing vortices of the cavity wake. It is responsible of a lift force on the cavity which is directed downwards and which balances the buoyancy force on the cavity $\rho g \mathscr{V}$. The lift can be roughly estimated on the basis of the KUTTA-JOUKOWSKI theorem for two-dimensional flows around lifting foils, taking for the span length the distance b between the two tube vortices. Thus, the vertical force balance is:

$$\rho \Gamma V_{\infty} b \cong \rho g \mathscr{V} \tag{9.32}$$

This equation gives the order of magnitude of b:

$$b = \frac{\mathscr{V}}{A} \cong \frac{2}{3} d_c \tag{9.33}$$

It shows that the distance between the two vortices is of the order of the maximum cavity diameter.

Because of its rotation, each trailing vortex induces an upwards velocity $\Gamma / 2\pi b$ on the other which causes it to incline at an angle ε to the horizontal. The expression for ε is:

$$\tan\varepsilon \cong \frac{\Gamma}{2\pi b V_\infty} \cong \frac{3}{16}\frac{g\ell}{V_\infty^2} = \frac{3}{16}\frac{1}{Fr_\ell^2} \tag{9.34}$$

For a given tube vortex of vapor core diameter a, the actual velocity on the tube interface is given by the sum of the flow velocity V_∞ and the tangential velocity $\Gamma/\pi a$. The BERNOULLI equation is written:

$$p_c + \frac{1}{2}\rho\left(V_\infty^2 + \frac{\Gamma^2}{\pi^2 a^2}\right) = p_\infty + \frac{1}{2}\rho V_\infty^2 \tag{9.35}$$

The order of magnitude of the tube diameter is therefore:

$$\frac{a}{d_c} \cong \frac{1}{4Fr_\ell^2\sqrt{\sigma_c}} \tag{9.36}$$

Thus, the main geometrical characteristics of the flow can be determined when the disc diameter and the relative cavity underpressure are given.

To evaluate the gas flowrate Q_v, the velocity of air in the tubes, V_m, is required. In practice, the ratio of the air velocity V_m to the liquid flow velocity V_∞ usually varies between 1 and 5. The gas flowrate evacuated by the two vortices is then given by:

$$Q_v \cong 2\frac{\pi a^2}{4} V_m \tag{9.37}$$

and the corresponding volumetric flowrate coefficient, defined by:

$$C_{Qv} = \frac{Q_v}{V_\infty d_c^2} \tag{9.38}$$

is

$$C_{Qv} \cong \frac{\pi}{32 Fr_\ell^4 \sigma_c}\frac{V_m}{V_\infty} \tag{9.39}$$

This formula accounts for experimental results in the ranges $0.8 < Fr_\ell \sigma_c^{1/4} < 1.5$ and $0.08 < C_{Qv} < 20$. Another, more general, formula applicable to disks and cones of diameter d was given by EPSHTEIN (1970):

$$C_{Qv} = \frac{Q_v}{V_\infty d^2} = \frac{0.42 C_{D0}^2}{\sigma_c\left[\sigma_c^3 Fr_\ell^4 - 2.5 C_{D0}\right]} \tag{9.40}$$

Here, $C_{D0} = \dfrac{D_0}{\frac{1}{2}\rho V_\infty^2 \frac{\pi d^2}{4}}$ is the drag coefficient of the cavitator for $\sigma_c = 0$. In the case of a disk, $C_{D0} = 0.82$. The length scales used in the definition of the flowrate coefficient in equations (9.39) and (9.40) are not the same. In the first case it is the maximum cavity diameter d_c and in the second the cavitator diameter d. It is interesting to note that if a reference size defined by $d_{ref} = d\sqrt{C_{D0}}$ is taken instead of ℓ and d in the definitions of the FROUDE number and the flowrate coefficient, the semi-empirical formula (9.40) takes a universal form which does not depend upon the drag coefficient C_{D0}.

EPSHTEIN's formula (9.40) shows that the minimum value of σ_c diminishes when the FROUDE number increases, according to the law:

$$\sigma_{min} \approx 1.36 \sqrt[3]{\frac{C_{D0}}{Fr^4}} \qquad (9.41)$$

At a fixed FROUDE number, when the limiting value is approached, there is a change in the air flow regime and cavity pulsations occur for high air flowrates.

9.3. *ANALYSIS OF PULSATING VENTILATED CAVITIES*

PARISHEV's approach (1978) is relevant to the field of system dynamics and helps explain the main features of pulsating ventilated cavities. Although the original configuration considered by PARISHEV was axisymmetric, several theoretical results are quite close to the experimental data obtained by SILBERMAN and SONG (1961) and MICHEL (1971, 1984) on two-dimensional plane flows. Hence, the mechanisms of cavity pulsation can be considered as basically similar in 2D and 3D situations.

9.3.1. *BASIC EQUATIONS*

It is assumed that the cavity contains a constant mass of gas, i.e. that there is no gas supply inside the cavity nor is there entrainment at the rear. Otherwise, the rate of gas supply as well as the rate of gas leakage at the rear of the cavity must be specified and included in the mass balance of the gas contained in the cavity. An empirical formula such as equation (9.25) can be used to express the rate of gas entrained.

It is also assumed that the pressure is uniform along the axis, i.e. that no acoustic waves travel inside the cavity. This hypothesis is valid if the cavity length is small with respect to the sound wavelength at the pulsation frequency. Such a condition was met in the laboratory experiments of SONG (1961) and MICHEL (1971), but it might be necessary to re-examine it in the case of very long cavities.

The cavity pulsations are assumed to take place around a mean static shape and the cavity pressure to fluctuate with an amplitude $\tilde{p}_c$ around a constant mean value p_{c0}:

$$p_c(t) = p_{c0} + \tilde{p}_c(t) \qquad (9.42)$$

The time-dependent evolution of the cavity pressure $p_c(t)$, cavity volume $\mathcal{V}(t)$, and cross-sectional area $S(x,t)$ at each station x along the axis are governed by the following equations:

– equation (6.68) for the evolution of the cross-sectional area of the cavity. This is a statement of the LOGVINOVICH independence principle of cavity expansion:

$$\left[\frac{\partial}{\partial t}+V_{\infty}\frac{\partial}{\partial x}\right]^2 S=-\kappa\,\frac{p_{\infty}-p_c(t)}{\rho} \tag{9.43}$$

where the dimensionless parameter κ is given by equation (6.70):

$$\kappa \cong \frac{4\pi C_D d^2}{\sigma_c^2 \ell_0^2} \tag{9.44}$$

C_D is the drag coefficient, d the cavitator diameter and ℓ_0 the mean cavity length;

– the equation for the volume of air inside the cavity (it is assumed there is no solid body there):

$$\mathcal{V}(t)=\int_{x=0}^{x=\ell(t)} S(x,t)\,dx \tag{9.45}$$

– the thermodynamic law of evolution of the gas inside the cavity:

$$p_c(t)\,\mathcal{V}(t)^{\gamma}=C \tag{9.46}$$

with $\gamma=1$ or 1.4 according to the kind of transformation considered, isothermal or adiabatic. The value of the constant C is known from the mean conditions.

To understand the full details of the coupling between equations (9.43), (9.45) and (9.46), let us consider a step by step procedure of computation. If it is assumed that all variables are known at a given time step, the new shape of the cavity $S(x,t)$ at the next time step can be deduced from equation (9.43). Then, equation (9.45) allows us to compute the new cavity volume, and equation (9.46) the new cavity pressure at the next time step, and so on... Hence, the cavity dynamics can be fully predicted from the above set of equations.

The analytical developments conducted by PARISHEV are based on a Lagrangian approach. The integrals are expressed with a time lag term linked to the time required for a fluid particle to travel the length of the cavity. After transformation and linearisation of the equations, PARISHEV obtains a differential equation for the fluctuations of the cavity pressure. Two different time scales are introduced:

– a time scale T which characterizes the air cavity behavior with regard to its possible unsteadiness:

$$T=\sqrt{\frac{\mathcal{V}_0}{\gamma p_{c0}}\,\frac{\rho}{\kappa \ell_0}} \tag{9.47}$$

– the classical transit time τ_0 required for a fluid particle to travel the cavity length at a velocity V:

$$\tau_0=\frac{\ell_0}{V} \tag{9.48}$$

In all previous expressions, the index 0 refers to the mean static conditions.

Using these characteristic time scales, PARISHEV introduces the following non-dimensional times:

$$\bar{t} = \frac{t}{T} \tag{9.49}$$

$$\bar{\tau}_0 = \frac{\tau_0}{T} \tag{9.50}$$

The differential equation for the pressure fluctuation amplitude is then:

$$\frac{d^3 \tilde{p}_c}{d\bar{t}^3}(\bar{t}) + \left[\frac{d\tilde{p}_c}{d\bar{t}}(\bar{t}) + \frac{d\tilde{p}_c}{d\bar{t}}(\bar{t} - \bar{\tau}_0)\right] - \frac{2}{\bar{\tau}_0}\left[\tilde{p}_c(\bar{t}) - \tilde{p}_c(\bar{t} - \bar{\tau}_0)\right] = 0 \tag{9.51}$$

If the air flowrate is taken into account, a second order term of the form $\alpha \dfrac{d^2 \tilde{p}_c}{d\bar{t}^2}$, in which α is a semi-empirical constant, has to be added to the left-hand side of equation (9.51). In the case of a body inside the cavity, the air volume would be diminished, which would change the characteristic time T and the parameter $\bar{\tau}_0$.

An important conclusion of the above analysis is that the pressure fluctuations depend only on the non-dimensional time $\bar{\tau}_0$. In the case of an ellipsoidal cavity of maximum radius R_0 and length ℓ_0, this parameter can be computed using the arguments presented in sections 6.4.3 and 6.4.4. Combining equations (6.52), (6.55) and (6.69), the cavity slenderness is given by:

$$\frac{2R_0}{\ell_0} = \sqrt{\frac{\sigma_c \kappa}{4\pi}} \tag{9.52}$$

so that the mean cavity volume is:

$$\mathcal{V}_0 = \frac{4}{3}\pi R_0^{\ 2}\frac{\ell_0}{2} = \frac{1}{24}\sigma_c \kappa \ell_0^{\ 3} \tag{9.53}$$

Hence, the non-dimensional time $\bar{\tau}_0$ is expressed as:

$$\bar{\tau}_0 = \sqrt{\frac{24\gamma}{\sigma_c}\frac{p_{c0}}{\rho V^2}} \tag{9.54}$$

Using the definitions given in section 9.1.2, we have:

$$\sigma_a = \sigma_v - \sigma_c = \frac{p_{c0} - p_v}{\frac{1}{2}\rho V^2} \tag{9.55}$$

Finally, by neglecting the vapor pressure in comparison with the air pressure in the cavity, we obtain:

$$\bar{\tau}_0 \cong \sqrt{12\gamma \frac{\sigma_a}{\sigma_c}} \tag{9.56}$$

In conclusion, the theoretical analysis corroborates the experimental finding according to which the cavity pulsation frequency is controlled by the parameter σ_c / σ_a only.

9.3.2. ANALYSIS OF THE PRESSURE FLUCTUATION EQUATION

To analyze the stability of ventilated cavities, PARISHEV looked for exponential solutions of the pressure fluctuation equation of the form:

$$\tilde{p}_c(\bar{t}) = e^{\alpha \bar{t}} \tag{9.57}$$

with

$$\alpha = \mu + i\omega \tag{9.58}$$

The behavior is stable if μ is negative and unstable if μ is positive.

Figure 9.16 presents the nine first solutions $\alpha_j = \mu_j + i\omega_j$ $(j = 1, 2...)$ as functions of the single parameter $\bar{\tau}_0$. For each solution #j, the parameter μ_j is negative for $\bar{\tau}_0 < j\pi\sqrt{2}$ and positive otherwise. Thus, from the viewpoint of linear stability, the solution is stable only for $\bar{\tau}_0$-values smaller than $\pi\sqrt{2}$. Beyond that limit, there is at least one positive value of $\bar{\mu}$, which leads to instability.

In the linear case, the oscillations at the angular non-dimensional frequency $\bar{\omega}$ would have an unbounded amplitude. This is not possible in real non-linear systems with the result that self-oscillations with a finite amplitude and a frequency close to the linear frequency will actually take place.

Thus, for a given value of $\bar{\tau}_0$, the most probable regime is the one which corresponds to the maximum value of $\bar{\mu}$. Figure 9.16 shows that each solution holds within a limited interval of values for $\bar{\tau}_0$, so that different regimes of oscillations develop according to that value. Each regime #j (bounded by the vertical lines) is characterized by its corresponding frequency $\bar{\omega}_j$ and discontinuities in $\bar{\omega}_j$ occur at the limits of the $\bar{\tau}_0$ intervals, i.e. at regime change.

It is clear from figure 9.16 that the non-dimensional frequency $\bar{\omega}$ takes a value close to one and that its variation within each regime decreases when the regime number j (or $\bar{\tau}_0$) increases. This is corroborated by experiments in 2D configurations. More especially, it can be shown that the frequency for regime #j is approximately such that:

$$\bar{\omega}_j \bar{\tau}_0 \cong j \times 2\pi \tag{9.59}$$

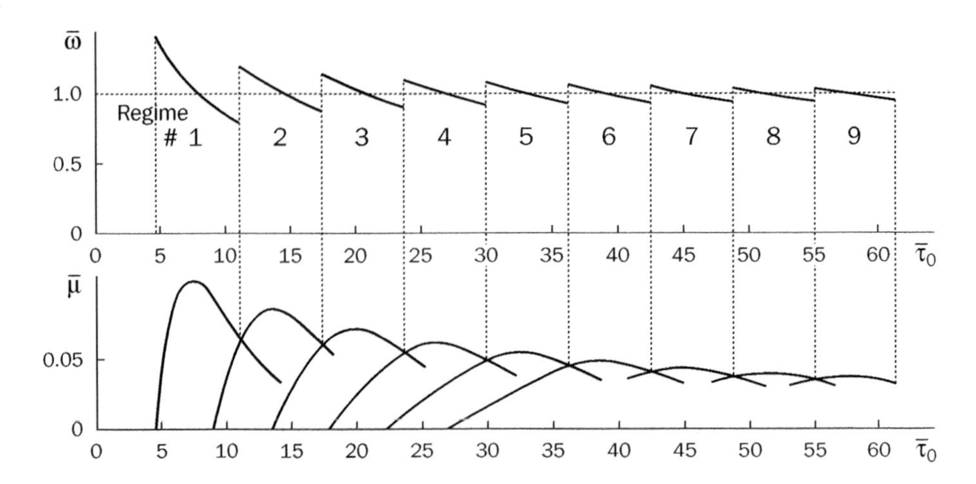

9.16 - The nine first regimes of cavity pulsation *[from PARISHEV, 1978]*

Assuming, as in section 9.1.7, that the undulations of the cavity interface in regime #j have a wavelength λ_j given by:

$$\lambda_j = \frac{2\pi V}{\omega_j} \tag{9.60}$$

the wavelength is then:

$$\lambda_j = \frac{2\pi \ell_0}{\tau_0\, \omega_j} = \frac{2\pi \ell_0}{\bar{\tau}_0\, \bar{\omega}_j} \tag{9.61}$$

From equation (9.59), it follows that $\ell_0 \cong j \times \lambda_j$. In other words, regime #j corresponds to j undulations of the cavity interface, as observed experimentally.

The interval size for the $\bar{\tau}_0$-value is approximately 2π. This means that the #j regime corresponds to the following domain of variation of $\bar{\tau}_0$:

$$\pi\sqrt{2} + (j-1) \times 2\pi < \bar{\tau}_0 < \pi\sqrt{2} + j \times 2\pi \tag{9.62}$$

9.3.3. COMPARISON WITH EXPERIMENTS

A comparison of PARISHEV's theory with the experimental results obtained in the two-dimensional plane configuration presented in section 9.1.6 shows good agreement between them when referring to the dependence of the pressure fluctuations on σ_c / σ_a.

Relation (9.62) which gives the domain of variation of the parameter $\bar{\tau}_0$ corresponding to regime #j can be transposed in terms of the parameter σ_c / σ_a using equation (9.56). Table 9.1 gives the domain of variation of σ_c / σ_a corresponding to the first four regimes of cavity pulsation in the case of an isothermal transformation of the gas

$(\gamma = 1)$. The predicted ranges of σ_c / σ_a-values for each regime are quite close to the experimental ones, which are given on figure 9.11.

Regime #	$\overline{\tau}_0$	σ_c / σ_a
1	$\pi\sqrt{2} < \overline{\tau}_0 < \pi\left(\sqrt{2}+2\right)$	$0.10 < \sigma_c / \sigma_a < 0.61$
2	$\pi\left(\sqrt{2}+2\right) < \overline{\tau}_0 < \pi\left(\sqrt{2}+4\right)$	$0.041 < \sigma_c / \sigma_a < 0.10$
3	$\pi\left(\sqrt{2}+4\right) < \overline{\tau}_0 < \pi\left(\sqrt{2}+6\right)$	$0.022 < \sigma_c / \sigma_a < 0.041$
4	$\pi\left(\sqrt{2}+6\right) < \overline{\tau}_0 < \pi\left(\sqrt{2}+8\right)$	$0.014 < \sigma_c / \sigma_a < 0.022$

Table 9.1 - Domain of variation of $\overline{\tau}_0$ and σ_c/σ_a for the first four regimes of cavity pulsation

As for the non-dimensional pulsation frequency φ defined by equation (9.6):

$$\varphi = \frac{\overline{\omega}}{2\pi T \sqrt{\dfrac{p_{c0}}{\rho d \ell_0}}} \tag{9.63}$$

it can be written, using equations (9.55) and (9.56):

$$\varphi = \frac{\overline{\omega}\sqrt{6\gamma}}{\pi} \sqrt{\frac{d}{\sigma_c \ell_0}} \tag{9.64}$$

The length d corresponds to the diameter of the cavitator. Taking into account the power law given in the table of figure 9.2 for the variation of the cavity length with the cavitation parameter, it can easily be shown that the ratio $d / \sigma_c \ell_0$ is of the order of unity. Referring to figure 9.16, $\overline{\omega}$ is also of the order of unity, so that the non-dimensional pulsation frequency φ is predicted to be of the order of $\sqrt{6\gamma / \pi}$. According to the type of transformation, this gives 0.78 (isothermal case) or 0.92 (adiabatic case). Finally, the non-dimensional pulsation frequency φ appears to be of the order of unity, as are the experimental values given in figure 9.11.

On the whole, PARISHEV's theory agrees fairly well with the experimental results although the flow of air inside the cavity is not taken into account. There are some disparities such as the fact that the linear stability condition $\overline{\tau}_0 < \pi\sqrt{2}$ has no counterpart in the experimental field or that the global instability mentioned in section 9.1.4 for low σ_v-values is not predicted since σ_c / σ_a is the only relevant parameter.

In conclusion, the pulsation frequency of a ventilated cavity can be considered as a kind of natural frequency of the cavity. It is fixed by the σ_c / σ_a parameter and basically depends on the mean characteristics of the cavity though not on the flow of air inside. The ventilation process adapts itself to this frequency and strongly influences global parameters such as cavity length and pressure, which, in turn, control the pulsation frequency.

REFERENCES

BUYVOL V.N. –1980– Slender cavities in flows with perturbations (in Russian). *Nauvoka Dunka Ed.*, Kiev (Ukraine).

CAMPBELL I.J. & HILBORNE D.V. –1958– Air entrainment behind artificially inflated cavities. *Proc. 2nd Int. Symp. on Naval Hydrodynamics*, Washington DC (USA).

CHANSON H. –1999– The hydraulics of open channel flow – An introduction. *Butterworth-Heineman Publishers*, Oxford (England), 544 p.

COX R.N. & CLAYDEN W.A. –1957– Air entrainment at the rear of a steady cavity. *Symp. on Cavitation in Hydrodynamics.* NPL, Teddington (England).

EPSHTEIN L.A. –1970– Theoretical methods of similarity in problems of ship hydromechanics. *Sudostroenie Publishing House,* Leningrad (Russia).

EPSHTEIN L.A. –1971– Characteristics of ventilated cavities and some scale effects. *Proc. IUTAM Symp. on Rapid Non-Steady Liquid Flows*, Leningrad, 173-185.

LOGVINOVICH G.V. –1973– Hydrodynamics of flows with free boundaries. *Halsted Press,* 215 p.

LOGVINOVICH G.V. –1976– Problems of the theory of axixymmetrical cavities. *Tsagi* **1797**.

LAALI A.R. & MICHEL J.M. –1984– Air entrainment in ventilated cavities: case of the fully developed "half-cavity". *J. Fluids Eng.* **106**(3), 327-335.

MICHEL J.M. –1971– Ventilated cavities: a contribution to the study of pulsation mechanism. *Proc. IUTAM Symp. on Rapid Non-Steady Liquid Flows*, Leningrad (Russia), 343-360.

MICHEL J.M. –1984– Some features of water flows with ventilated cavities. *J. Fluids Eng.* **106**(3), 319-326.

PARISHEV E.V. –1978– Systems of non-linear differential equations with time-lag for the description of non-steady axisymmetric cavities. *Tsagi* **1907**, 1-17.

PARISHEV E.V. –1978– Theoretical study of the stability and pulsations of axisymmetric cavities. *Tsagi* **1907**, 17-40.

ROWE A. –1979– Evaluation of a three-speed hydrofoil with wetted upper side. *J. Ship Res.* **23**(1), 55-65.

SAVCHENKO Y.N. –2001– Experimental investigation of supercavitating motion of bodies. *VKI/RTO Special Course on Supercavitation.* Von Karman Institute for Fluid Dynamics, Brussels (Belgium).

SEMENENKO V.N. –1998– Instability and oscillation of gas-filled supercavities. *Proc. 3rd Int. Symp. on Cavitation*, vol. 2, Grenoble (France), 25-30.

SEMENENKO V.N. –2001– Artificial supercavitation. Physics and calculation. *VKI/RTO Special Course on Supercavitation.* Von Karman Institute for Fluid Dynamics, Brussels (Belgium).

SEMENOV Y.A. –1998– Exact solution of problem of unsteady cavitation flow past wedge. *Proc. 3rd Int. Symp. on Cavitation*, vol. 2, Grenoble (France), 55-59.

SEREBRYAKOV V. –1973– Asymptotic solution of the problem of slender axisymmetric cavity. *Rpt NAS of Ukrainia A* **12**, 1119-1122.

SILBERMAN E. & SONG C.S. –1961– Instability of ventilated cavities. *J. Ship Res.* **5**(1), 13-33.

SONG C.S. –1962– Pulsation of ventilated cavities. *J. Ship Res.* **5**(4), 1-20.

VASIN A.D. –2001– The principle of independency of the cavity sections expansion as the basis for investigations on cavitation flows. *VKI/RTO Special Course on Supercavitation.* Von Karman Institute for Fluid Dynamics, Brussels (Belgium).

VERRON J. –1977– Écoulements cavitants autour d'ailes d'envergure finie en présence d'une surface libre. *J. Méc.* **12**(4), 745-774.

VERRON J. & MICHEL J.M. –1984– Base-vented hydrofoils of finite span under a free surface: an experimental investigation. *J. Ship Res.* **28**(2), 90-106.

10. VORTEX CAVITATION

Coherent vortices are observed in many flow situations. Generally, rotational structures generate low pressure regions inside the liquid itself, whereas the minimum pressure occurs at the liquid boundary for irrotational flows. Such pressure drops can be very intense, so that vortex cavitation often starts for high values of the cavitation number in comparison with other types of cavitation.

Different kinds of vortices exist according to their mode of production. We can distinguish the case of well-shaped, almost steady state vortices (treated in the present chapter), from the rotational coherent structures observed in shear flows, which are deeply affected by turbulence (see chap. 11). While the former are usually attached to solid bodies which continuously supply them with circulation, the latter are free and their life time can be short due to viscous dissipation.

Examples of the first of these include apex vortices developed along the leading edge of delta wings, hub vortices trailing downstream of a propeller hub, tip vortices which occur at the tip of lifting foils and propeller blades. From the second group there are the alternate BÉNARD-KÁRMÁN vortices behind bluff bodies, the vortices formed in shear layers at the frontiers of wakes and jets and the vapor filaments shed by developed cavities.

In the first section of the present chapter, we recall basic results on vorticity dynamics which are guidelines for the understanding of vortex cavitation. Simple models of cavitating vortices are also presented. Section 10.2 is devoted to a physical description of the non-cavitating vortex created at the tip of a three-dimensional wing, with the aim of finding the location of the minimum pressure point. This case is representative of the helicoidal vortices shed in the wake of propeller blades for which cavitation inception is often a major problem since it can be an important source of noise in the marine environment. Finally, section 10.3 presents experimental results on tip vortex cavitation inception and its connection with its original non-cavitating structure. Further effects such as the influence of nuclei and confinement are also discussed.

10.1. THEORETICAL RESULTS

10.1.1. BASIC VORTICITY THEOREMS

For an inviscid, barotropic fluid (certainly the case for an incompressible liquid) submitted to potential force fields, KELVIN's theorem states that the circulation:

$$\Gamma = \oint_{\gamma(t)} \vec{V}.d\vec{\ell} \tag{10.1}$$

round any closed material curve $\gamma(t)$ is invariant with time.

If $\vec{\omega} = \text{curl}\,\vec{V}$ is the local vorticity of the fluid and $S(t)$ a material surface bounded by the closed material curve $\gamma(t)$, we have the following relation (STOKES's theorem):

$$\oint_{\gamma(t)} \vec{V}.d\vec{\ell} = \iint_{S(t)} \vec{\omega}.\vec{\nu}\, dS \tag{10.2}$$

where $\vec{\nu}$ is the unit vector normal to S. The vorticity flux is constant as is the circulation Γ. A consequence of this is the STOKES-LAGRANGE theorem according to which, if the initial vorticity is zero at any point, it will remain zero at any subsequent time for the same body of fluid.

The HELMHOLTZ theorem states that a vortex line, i.e. a line which is everywhere tangent to the vorticity vector $\vec{\omega}$, is a material line. The same can be said of a vortex tube which is the surface made of all vortex lines passing through a closed curve $\gamma(t)$. This is the basis of the concept of coherent structures.

The strength of a vortex tube, defined by equation (10.1) or (10.2), is independent of the cross-section of the tube considered. This prevents a vortex tube from ending inside the fluid. It is also constant with time when the vortex tube evolves with the flow.

For an element of a vortex tube of length $\delta\ell$ and area δs, the conservation of mass and circulation are written respectively as $\rho.\delta s.\delta\ell = \text{Constant}$ and $\omega.\delta s = \text{Constant}$, such that the ratio $\omega/\delta\ell$ remains constant for an incompressible liquid. As a consequence, if a vortex tube is stretched, its cross-section decreases and the mean vorticity across the section increases. This result can also be obtained via the conservation of the angular momentum of an axisymmetric filament.

10.1.2. THE MAIN EFFECTS OF CAVITATION ON ROTATIONAL FLOWS

The characteristics of a vortex change due to the inception of cavitation. To generate a cylindrical vapor core of diameter d_v, conservation of mass shows that the corresponding cylindrical volume of liquid, before phase change, has a diameter d_ℓ such that their ratio is:

$$\frac{d_\ell}{d_v} \approx \sqrt{\frac{\rho_v}{\rho_\ell}} \tag{10.3}$$

For water at room temperature, the ratio of liquid to vapor densities is close to 58,000. For a vapor core of about one millimeter, which is a typical value at cavitation inception, the diameter d_ℓ is of the order of 4 μm. In other words, liquid particles situated initially at 2 μm from the axis are ejected at 0.5 mm because of cavitation. Thus, the geometry of the flow in the close neighborhood of the vortex axis is strongly modified. The ratio (10.3) depends significantly upon the fluid since, for liquid hydrogen at 50°K, for example, the ratio of liquid to vapor densities is only 31.

Phase change occurs in a short time, typically smaller than 0.1 ms, so that the mean radial velocity of the liquid particles ejected from the vortex axis by the inception of cavitation is of the order of 5 m/s for a vapor core of 1 mm in diameter. Furthermore, the pressure field itself is obviously modified since the pressure in the cavitating vortex is fixed at the vapor pressure and cannot fall below it.

Hence, when cavitation develops in a liquid vortex, the geometry together with the pressure and the velocity fields are usually drastically changed, so that cavitation cannot be considered as a passive means of visualization of rotational flows. At most, the very first cavitation events can give a qualitative idea of the pre-existing rotational structures.

A second point concerns the effect of stretching on vorticity. As mentioned in the previous section, the rotation rate of a pure liquid vortex is increased when stretched. Generally speaking, this result no longer holds in cavitating conditions since the angular momentum is still constant but the inertia is changed. Indeed, the external pressure plays the role of an additional free parameter and the original link between the elongation rate of a liquid vortex tube and its rotation rate is broken by cavitation.

The evolution of a cavitating vortex depends on both the external pressure and the self-induced pressure drop due to its rotation. During its life-time, a cavitating vortex generally experiences simultaneously changes in length and in ambient pressure. If the ambient pressure is constant, stretching induces an increase in the rotation rate and hence an increase in the vapor core radius. If the length of the vortex filament is kept constant, the effect of the ambient pressure is twofold. On one hand, an increase in pressure results in a reduction of the vapor core radius, while on the other hand, the volume reduction is accompanied by an increase in the rate of rotation. This secondary, antagonistic effect can give rise to natural oscillations in the case of isolated vortices, as seen in both examples to be presented in sections 10.1.3 and 10.1.4.

A third point concerns the motion, relative to the liquid medium, of cavitating vortices which undergo volume variations. Although no theoretical result is available, experiments have clearly demonstrated that volume variations affect the translation velocity of a vortex cavity via a virtual mass effect (see chap. 11).

Finally, we can ask how a cavitating vortex returns to the non-cavitating state. Two modes of collapse are expected, an axial mode especially for vortices ending on solid walls, and a radial mode. The latter requires viscous dissipation since, in an inviscid liquid, the rotation rate of the particles at the interface would become infinite as they reach the axis.

10.1.3. AXISYMMETRIC CAVITATING VORTEX

Consider an axisymmetric cavitating vortex limited by an external circular cylinder of time-dependent radius $\mathcal{R}(t)$. The assumption of a finite external radius is necessary to avoid the singular logarithmic behavior which classically arises in two-dimensional configurations (see eq. 10.6).

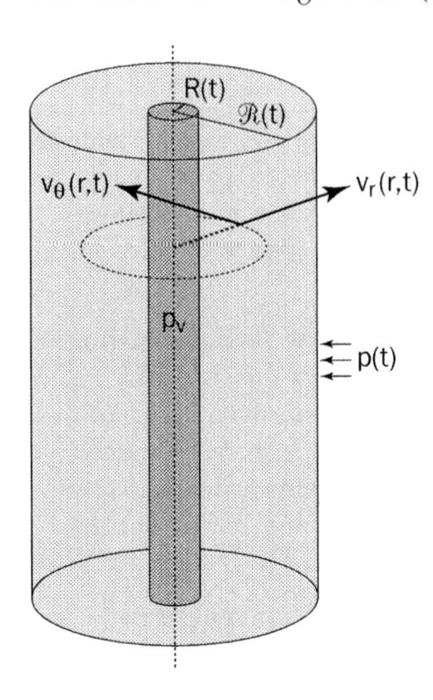

10.1
Schematic of the axisymmetric cavitating vortex

The radius of the internal vapor core is $R(t)$. The annular liquid vortex is subject to an external pressure $p(t)$ at its outer boundary $r = \mathcal{R}(t)$ and to the vapor pressure p_v at its internal boundary $r = R(t)$. The liquid is assumed inviscid and the flow irrotational. The mass of liquid in this annular vortex is assumed constant, so that the external and internal radii are such that $\mathcal{R}^2 - R^2$ is a constant at any time. This somewhat academic configuration allows us to obtain simply some basic results.

The velocity inside the liquid has two components, a radial one $v_r(r,t)$ and a tangential one $v_\theta(r,t)$. Because of the absence of vorticity in the liquid, the tangential velocity has the form:

$$v_\theta(r,t) = \frac{\Gamma}{2\pi r} \tag{10.4}$$

where Γ is a constant which measures the circulation round any curve enclosing the vortex center. A constant value for the circulation can also be deduced from KELVIN's theorem.

As for the radial velocity, the condition of an incompressible fluid gives:

$$v_r(r,t) = \frac{R(t)\,\dot{R}(t)}{r} \tag{10.5}$$

Integration of the radial EULER equation leads to the following differential equation for the evolution of the vapor core radius $R(t)$:

$$\left[R\ddot{R} + \dot{R}^2\right] \ln\frac{\mathcal{R}}{R} - \frac{1}{2}\left[\dot{R}^2 + \frac{\Gamma^2}{4\pi^2 R^2}\right]\left[1 - \frac{R^2}{\mathcal{R}^2}\right] = \frac{p_v - p(t)}{\rho} \tag{10.6}$$

Equation (10.6) plays the role of the RAYLEIGH-PLESSET equation for a cavitating vortex. As mentioned in section 10.1.2, the complete collapse of the vortex up to $R = 0$ requires an infinite pressure difference as long as the circulation Γ is non-zero.

The equilibrium condition under a given constant pressure p is given by:

$$\frac{\Gamma^2}{8\pi^2 R^2}\left[1 - \frac{R^2}{\mathcal{R}^2}\right] = \frac{p - p_v}{\rho} \tag{10.7}$$

This equation expresses the balance between the pressure difference and the centrifugal force which tends to increase the vortex size. The equilibrium is stable and any deviation is accompanied by natural oscillations of the radii. Linearizing equation (10.6) around the equilibrium leads to the following expression for the period of oscillation:

$$T = \frac{4\pi^2 R^2}{\Gamma}\sqrt{\ln\frac{\mathcal{R}}{R}} \tag{10.8}$$

For example, with $R = 1$ mm, $\mathcal{R} = 10$ mm and $v_\theta(R) = \Gamma / 2\pi R = 10$ m/s, the oscillation frequency is $f = 1/T = 1{,}049$ Hz, while equation (10.7) gives a pressure difference close to 0.5 bar in the case of water.

10.1.4. TOROIDAL CAVITATING VORTEX

Toroidal vortices are encountered at the periphery of submerged round liquid jets. Usually, they are produced at a regular frequency f corresponding to a STROUHAL number $S = fd/V$ close to 0.3 (V and d stand respectively for the velocity and the diameter of the jet). For various purposes and especially for erosion enhancement, it may be of interest to reinforce the strength of the vortices by exciting the jet at the same frequency. Then, depending chiefly on the ambient pressure and the jet velocity, cavitation can appear in the core of these structures. The following section covers the basic theoretical results on the dynamics of an isolated toroidal cavitating vortex [CHAHINE & GENOUX 1983, GENOUX & CHAHINE 1983].

10.2
Schematic of the cavitating toroidal vortex

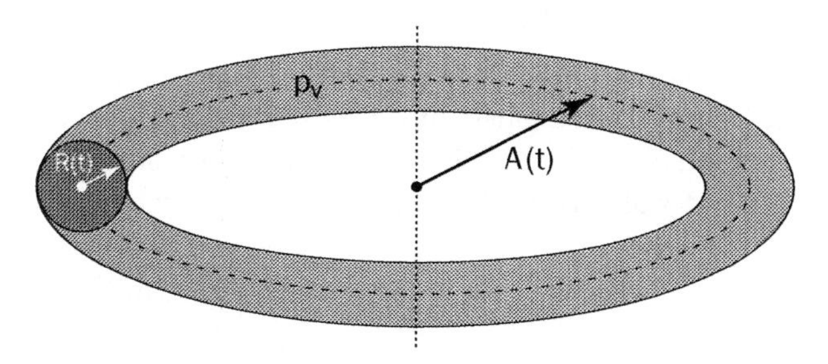

The ring bubble (fig. 10.2) is supposed initially at equilibrium in an infinite medium where the ambient pressure at infinity is p_∞. It has a vortical motion of circulation Γ. The fluid is assumed inviscid and incompressible, so that Γ remains constant in time. The bubble contains a mix of non-condensable gas with partial pressure p_{g0} and vapor with pressure p_v. The total internal pressure balances the external pressure on the bubble surface and surface tension effects are characterized by the constant S.

The cross-sectional radius R_0 is assumed to be small with respect to the overall ring radius A_0. In other words, the parameter

$$\varepsilon = \frac{R_0}{A_0} \tag{10.9}$$

is assumed much smaller than unity so that the bubble has a circular cross-section at equilibrium. Because of the two length scales, a matched asymptotic approach can be used. In the outer region, the torus is modelled by a distribution of singularities on a moving circle of radius $A(t)$. In the inner region, it is reduced to a cylinder of variable section radius $R(t)$.

A first result concerns the self-induced velocity of translation along the axis of symmetry in the static case which is given by:

$$V = \frac{\Gamma}{4\pi R_0}\left\{\varepsilon\left[\ln\frac{8}{\varepsilon} - \frac{1}{2} + \frac{S/R_0}{\rho(\Gamma/2\pi R_0)^2}\right] + O(\varepsilon)\right\} \tag{10.10}$$

This translation velocity is of order $\varepsilon \ln(8/\varepsilon)$. It can be shown that it is negligible with respect to the collapse velocity, so that the ring can be assumed motionless to first approximation.

In the dynamic case, the evolution of the radius $R(t)$ at the same order $\varepsilon \ln(8/\varepsilon)$ is given by the following differential equation which is analogous to the RAYLEIGH-PLESSET equation:

$$\rho\left[R\ddot{R} + \dot{R}^2\right]\ln\frac{8A_0}{R} - \frac{1}{2}\rho\dot{R}^2 = p_v - p_\infty(t) + \frac{1}{2}\rho\left[\frac{\Gamma}{2\pi R}\right]^2 + p_{g0}\left(\frac{R_0}{R}\right)^{2\kappa} - \frac{S}{R} \tag{10.11}$$

In this equation, κ stands for the polytropic exponent which characterizes the thermodynamic evolution of the gas.

The time evolution of the ring bubble diameter after a sudden pressure increase Δp at infinity is presented in figure 10.3 for various vortex strengths. As expected, circulation tends to resist the collapse, in a similar way as non-condensable gas. The order of magnitude of the collapse time is:

$$\tau \cong R_0\sqrt{\frac{\rho}{\Delta p}}\sqrt{\ln\frac{8}{\varepsilon}} \tag{10.12}$$

This expression is similar to the usual RAYLEIGH time for a spherical bubble except for the logarithmic factor which causes the collapse to be longer for a bubble ring than for a spherical bubble of the same initial radius R_0.

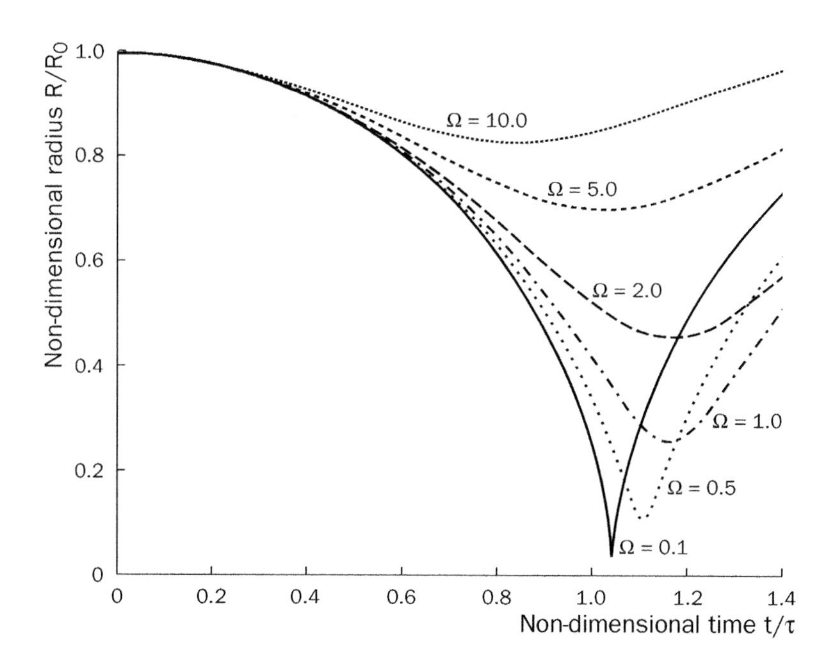

10.3 - Influence of vortex strength on the collapse of a ring bubble in the case of a sudden increase in pressure at infinity ($\varepsilon = 0.05$, $\kappa = 1.4$, $p_{g0} = 0.01\ \Delta p$)

The Weber number defined by $We = \dfrac{\Delta p}{S / R_0}$ *is 200. The parameter* $\Omega = \dfrac{\rho(\Gamma / 2\pi R_0)^2}{\Delta p}$ *measures the non-dimensional pressure drop due to the vortical motion [from CHAHINE & GENOUX, 1983].*

A last result concerns the stability of a ring bubble evolving quasi-statically due to a slow pressure change at infinity. The equilibrium equation resulting from the dynamic equation (10.11) in which the gas evolution is assumed isotherm ($\kappa = 1$) is:

$$p_\infty - p_v = \frac{1}{2}\rho\left[\frac{\Gamma}{2\pi R}\right]^2 + p_{g0}\left(\frac{R_0}{R}\right)^2 - \frac{S}{R} \tag{10.13}$$

For the initial conditions, we have that:

$$p_{\infty 0} - p_v = \frac{1}{2}\rho\left[\frac{\Gamma}{2\pi R_0}\right]^2 + p_{g0} - \frac{S}{R_0} \tag{10.14}$$

and equation (10.13) becomes:

$$p_\infty - p_v = \left[p_{\infty 0} - p_v + \frac{S}{R_0}\right]\left[\frac{R_0}{R}\right]^2 - \frac{S}{R} \tag{10.15}$$

This equation can be written in non-dimensional form:

$$\frac{p_\infty - p_v}{p_{\infty 0} - p_v} = \left[1 + \frac{1}{We}\right]\left[\frac{R_0}{R}\right]^2 - \frac{1}{We}\frac{R_0}{R} \tag{10.16}$$

in which We is a WEBER number defined by:

$$We = \frac{R_0 (p_{\infty 0} - p_v)}{S} \tag{10.17}$$

Figure 10.4 illustrates equation (10.16). On the whole, the static evolution of the cavitating vortex ring is similar to the case of the spherical nuclei described in chapter 2. In particular, critical conditions for stability correspond to the minima of the curves and are given by:

$$\begin{cases} \dfrac{R_c}{R_0} = 2(1 + We) \\ \dfrac{p_c - p_v}{p_{\infty 0} - p_v} = -\dfrac{1}{4We(1 + We)} \end{cases} \tag{10.18}$$

Small values of We give better stability. Coming back to the dimensional form of the equilibrium equation, it turns out that, as expected, large values of surface tension and small values of the circulation Γ will promote stability.

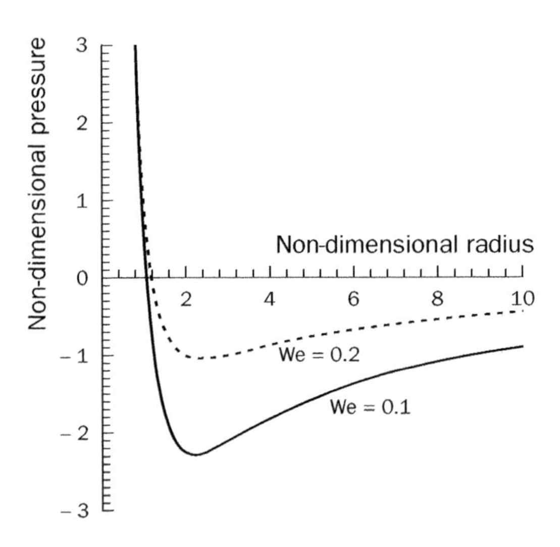

10.4 - Equilibrium curves for the vortex bubble ring
[from CHAHINE & GENOUX, 1983]

10.2. THE NON-CAVITATING TIP VORTEX

10.2.1. TIP VORTEX FORMATION

Consider the three-dimensional flow around an elliptic planform as shown schematically on figure 10.5. For a positive angle of attack, the pressure on the lower side is higher than that on the upper suction side and a circulation is generated round the foil. At the tip, this pressure difference must vanish and thus the circulation also vanishes. Hence, the circulation varies along the span from a maximum value Γ_0 at midspan to zero at the tip.

If the aspect ratio[1] is high enough, the flow can be interpreted using the lifting line theory of PRANDTL (fig. 10.5). In order to satisfy KELVIN's theorem, PRANDTL modelled the 3D flow by a set of U-shaped line vortices of infinite length, as indicated in figure 10.5. The parts of these vortices which are attached to the foil account for the spanwise variation of circulation whereas the free parts form a vortex sheet in the wake.

The vortex sheet is unstable and tends to roll-up because of its self-induced velocity. This mechanism leads to the formation of the tip vortices as explained by WESTWATER (1936). The circulation of the tip vortex progressively increases downstream as the roll-up process advances. Far enough from the foil, when roll-up is completed, the whole circulation Γ_0 is found around each tip vortex.

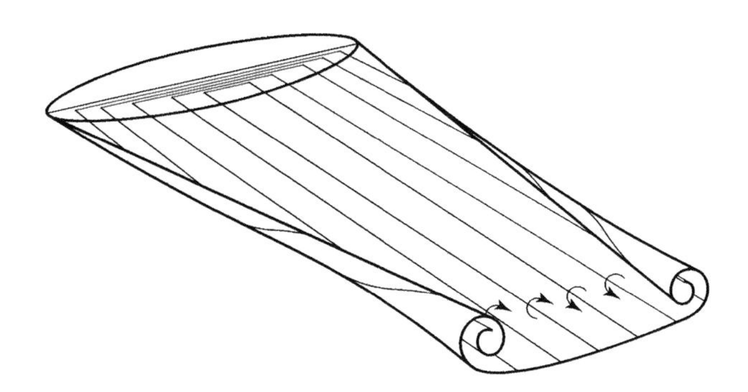

10.5 - **Schematic of tip vortex formation**

1. The aspect ratio of a foil is defined by $AR = \dfrac{(\text{span length})^2}{\text{planform area}}$. In the case of an elliptic planform of maximum chord length c, the aspect ratio is $AR = \dfrac{4b}{\pi c}$.

Such a simple model does not take into account viscous effects. It is more suited to a description of the tip vortex in the far wake. However, experiments show that cavitation begins at a short distance downstream of the wing (within one chord length at most), and that the roll-up process is already significantly initiated at such a relatively short distance.

The evolution of the strength and size of a non-cavitating tip vortex along its path is a complex phenomenon, governed by both viscous diffusion and the capture of vortex lines. Its modeling and the prediction of the minimum pressure coefficient are difficult, especially in the vicinity of the wing where the tip vortex is far from being axisymmetric.

10.2.2. *VORTEX MODELS IN VISCOUS FLUIDS*

Whatever the origin of a vortex, it is characterized by two zones:

- its core where vorticity is effectively constant and viscous effects are dominant,
- an outer region where the motion is mainly irrotational and the fluid can be considered as inviscid.

The simplest model of a vortex is that proposed by RANKINE. The core $r \leq a$ is assumed to be in solid body rotation so that the tangential velocity v_θ at a distance r from the vortex axis is given by:

$$v_\theta(r) = \frac{\Gamma}{2\pi a^2} r \tag{10.19}$$

The rotation rate is $\omega = \Gamma / 2\pi a^2$.

In the outer potential region $r > a$, the tangential velocity is:

$$v_\theta(r) = \frac{\Gamma}{2\pi r} \tag{10.20}$$

Γ is the vortex strength or the circulation of the velocity. The core radius a is the distance to the vortex axis where the tangential velocity is maximum.

BURGERS model is derived from OSEEN's solution [LAMB 1932] of the NAVIER-STOKES equations which accounts for the diffusion of vorticity in the case of a two-dimensional single vortex. The tangential velocity is given by:

$$v_\theta(r) = \frac{\Gamma}{2\pi r}\left(1 - e^{-1.256\, r^2/a^2}\right) \tag{10.21}$$

The approximate value of 1.256 ensures that the velocity is maximum for $r = a$. The corresponding maximum is equal to $0.715 \frac{\Gamma}{2\pi a}$ and so, is 0.715 times the maximum velocity of an equivalent RANKINE vortex (fig. 10.6).

Such a velocity distribution corresponds to a solid rotation near the axis where $r \ll a$, and to the potential flow given by expression (10.20) for large values of r/a.

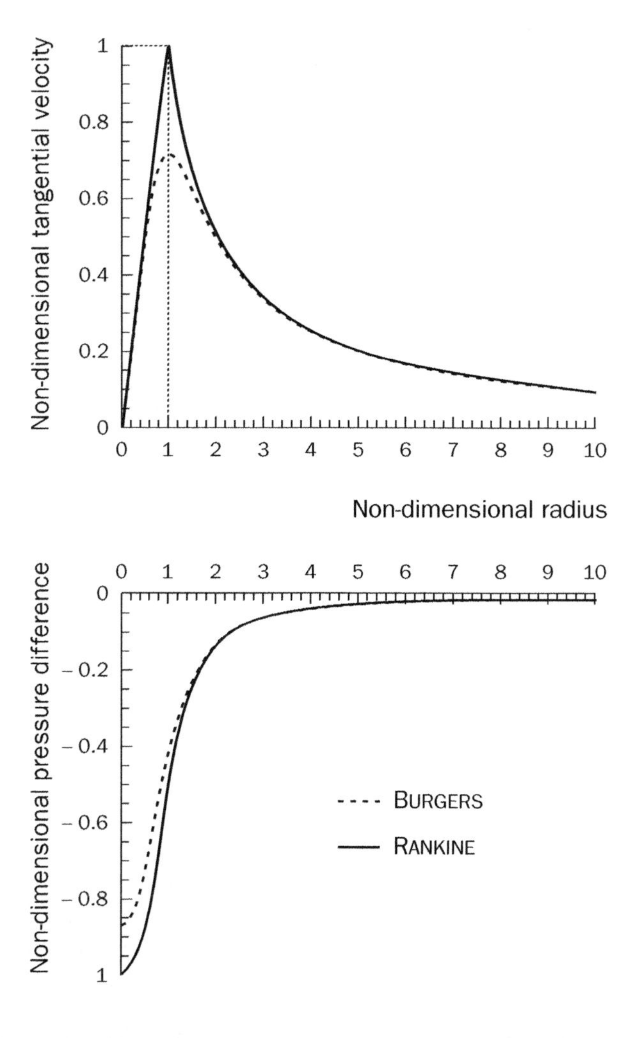

10.6 - Velocity and pressure distributions in RANKINE and BURGERS vortices

The velocity is non-dimensionalized by $\Gamma/2\pi a$, the pressure difference $p_{min} - p_{\infty}$ by $\rho(\Gamma/2\pi a)^2$ and the radius by a.

The radial equilibrium equation:

$$\frac{\partial p}{\partial r} = \rho \frac{v_\theta^2}{r} \tag{10.22}$$

allows us to compute the radial pressure distribution and more especially the minimum pressure p_{min} at the vortex center from the pressure at infinity p_∞.

For a RANKINE vortex, the minimum pressure is given by:

$$\frac{p_{min} - p_\infty}{\rho} = -\left[\frac{\Gamma}{2\pi a}\right]^2 \tag{10.23}$$

whereas, for a BURGERS vortex, it is given by:

$$\frac{p_{min} - p_\infty}{\rho} = -0.871\left[\frac{\Gamma}{2\pi a}\right]^2 \tag{10.24}$$

Cavitation occurs on the vortex axis when the minimum pressure falls below the vapor pressure p_v. The velocity and pressure distributions for both models are compared in figure 10.6.

10.2.3. *TIP VORTEX STRUCTURE*

Tangential velocity

In the past, experimentation has been the only means of obtaining information on the vortex structure. STINEBRING *et al.* (1991) were the first to measure the velocity field at a small distance from the tip, in the case of a trapezoidal lifting surface. They showed that the vortex is fully three-dimensional in the close wake of the wing.

FRUMAN *et al.* (1991, 1992a, 1992b, 1993) and PAUCHET *et al.* (1993) conducted systematic measurements of the axial and tangential components of the velocity at various stations within a short distance downstream of the tip. Figure 10.7 presents the evolution of the tangential velocity profiles along the tip vortex for an elliptical foil. Let us recall that the tip vortex flow is not axisymmetric in the vicinity of the tip and that figure 10.7 gives only a partial idea of the vortex structure.

A central zone with solid body rotation is clearly visible. The rotation rate is very high, larger than 1,000 revolutions per second for the present operating conditions. The maximum velocity first increases rapidly, reaches a maximum at a distance of about 0.125 c_{max} before decreasing slowly.

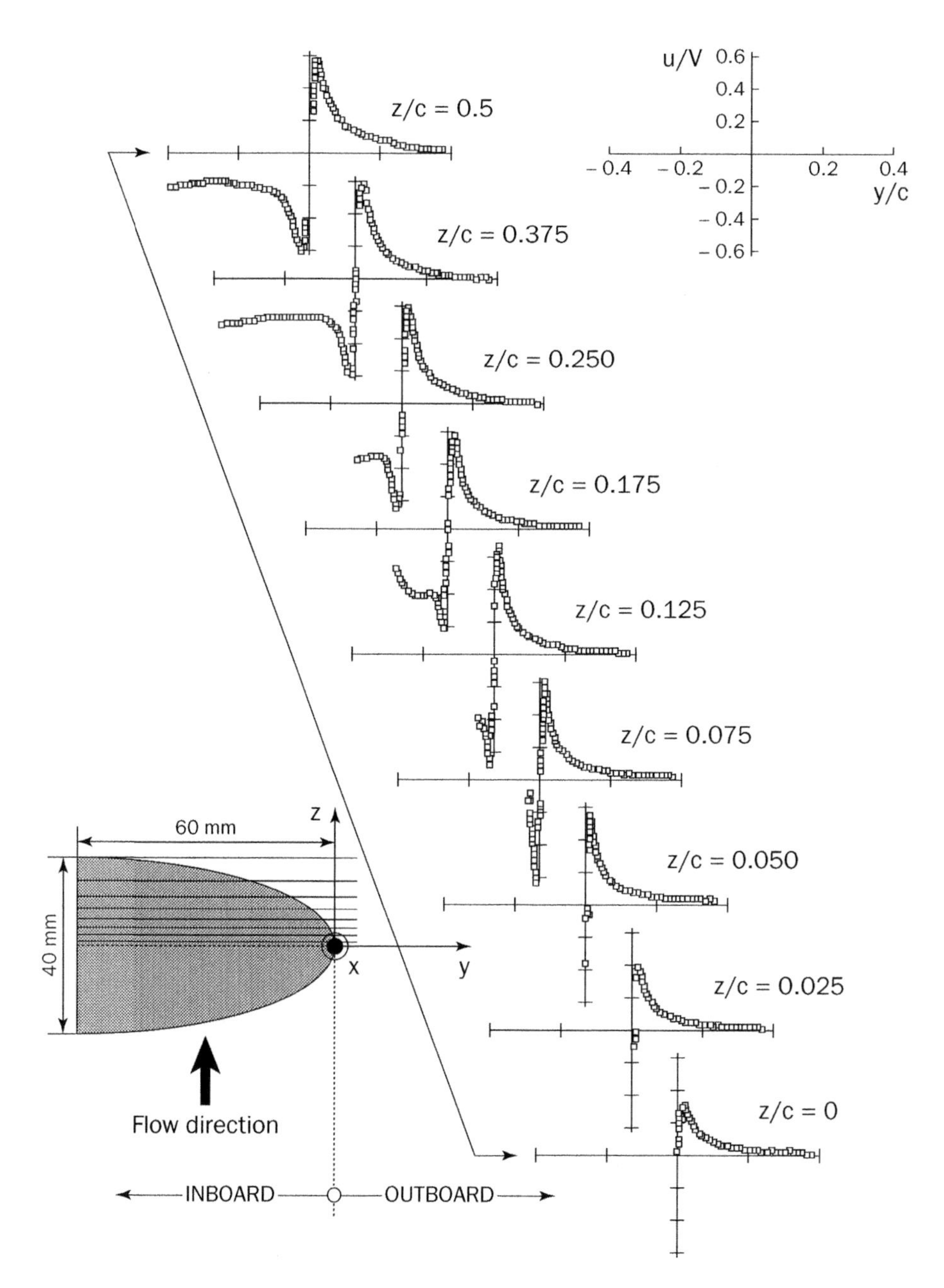

10.7 - Tangential velocity profiles at different stations z along the vortex path

(NACA 16020, maximum chord length at root $c = 40$ mm, $AR = 3.8$, incidence $= 10.6°$ and $V = 9$ m/s)

The component presented here is that along the x-axis. It is non-dimensionalized by the incoming velocity V and plotted as a function of the distance y from the vortex center [from FRUMAN et al., 1992b].

Angular momentum

The angular momentum, i.e. the product of the tangential velocity, u, and the distance, y, to the vortex axis, is plotted as a function of the distance to the vortex axis on figure 10.8. FRUMAN *et al.* (1992b) showed that a nearly constant value is achieved outside the core ($y > 0.03\,c$) on the outboard side $y > 0$. This value allows us to determine the vortex strength Γ at each measuring station. On the inboard side and for $y < -0.05\,c$, the angular momentum increases roughly linearly with the distance, which indicates an almost constant induced velocity along the span, as predicted by PRANDTL's lifting line theory.

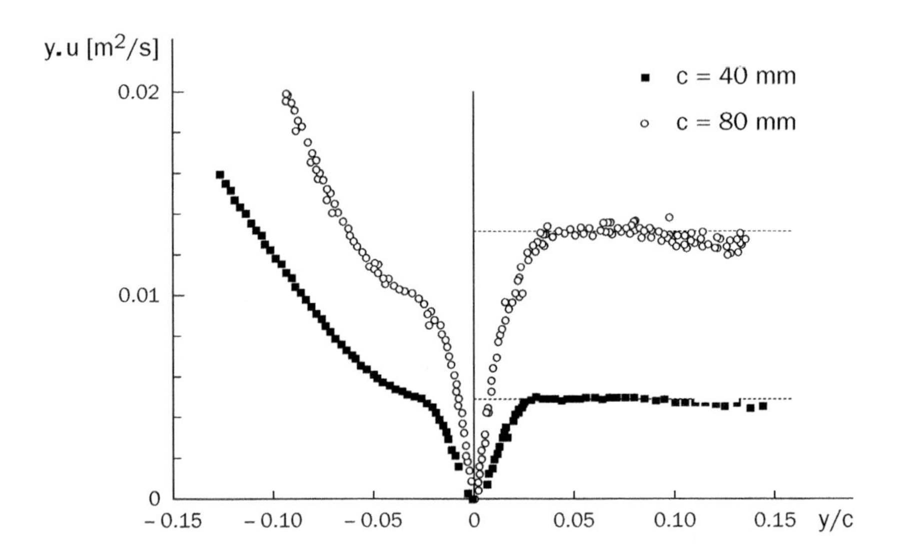

10.8 - Angular momentum (y.u) of the tangential velocity component u measured along the x-axis for z/c = 0.25

Two series of measurements for the same elliptic foil at two different length scales in two cavitation tunnels are presented [from FRUMAN et al., 1992b].

Tip vortex strength and core radius

For each velocity profile, once the vortex strength Γ has been determined as explained above, the viscous core radius, a, is adjusted in order to fit the experimental velocity profile using a BURGERS vortex model. The adjustment is conducted on the outboard side which is less affected by the wake of the hydrofoil.

Figure 10.9 shows the experimental data and the fitted curves for three of the velocity profiles. The adjustment is almost perfect on the outboard side while there is some divergence on the inboard side.

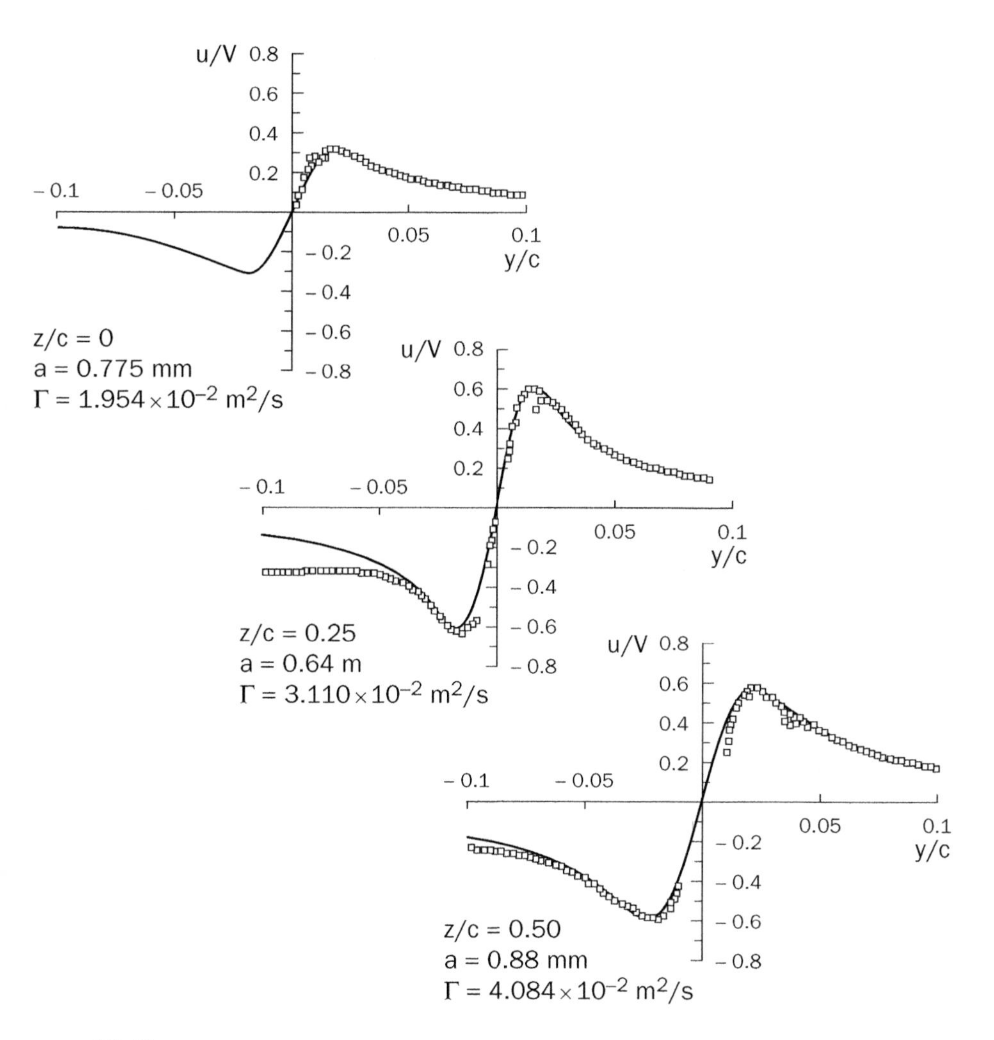

10.9 - **Example of fitting of the experimental data, shown in figure 10.7, with a BURGERS velocity profile as given by equation (10.21)**
[from FRUMAN et al., 1992b]

The local vortex strength Γ is non-dimensionalized by the mid-span bound circulation Γ_0 computed from the lift coefficient C_L as follows:

$$\Gamma_0 = \frac{1}{2} C_L V c \tag{10.25}$$

As for the vortex core radius a, it is non-dimensionalized by the boundary layer thickness at mid-span, δ, calculated as for a fully turbulent boundary layer over a flat plate of length the maximum chord c [SCHLICHTING 1987]:

$$\frac{\delta}{c} = 0.37 \left[\frac{Vc}{\nu}\right]^{-0.2} = \frac{0.37}{Re^{0.2}} \tag{10.26}$$

Such a correlation between the vortex core radius and the boundary layer thickness is supported by visualization which shows that, for this kind of elliptic planform, the tip vortex is the continuation of the laminar separation bubble attached to the foil leading edge. In the vicinity of the wing tip, the vortex structure is controlled mainly by viscous effects, as suggested by MCCORMICK (1962) who was the first to correlate the core radius of the tip vortex to the boundary layer thickness.

Variations of vortex strength and vortex core radius with the downstream distance from the tip are given on figure 10.10. In spite of the variety of flow conditions in terms of chord length and REYNOLDS number, these results appear in good quantitative agreement with each other.

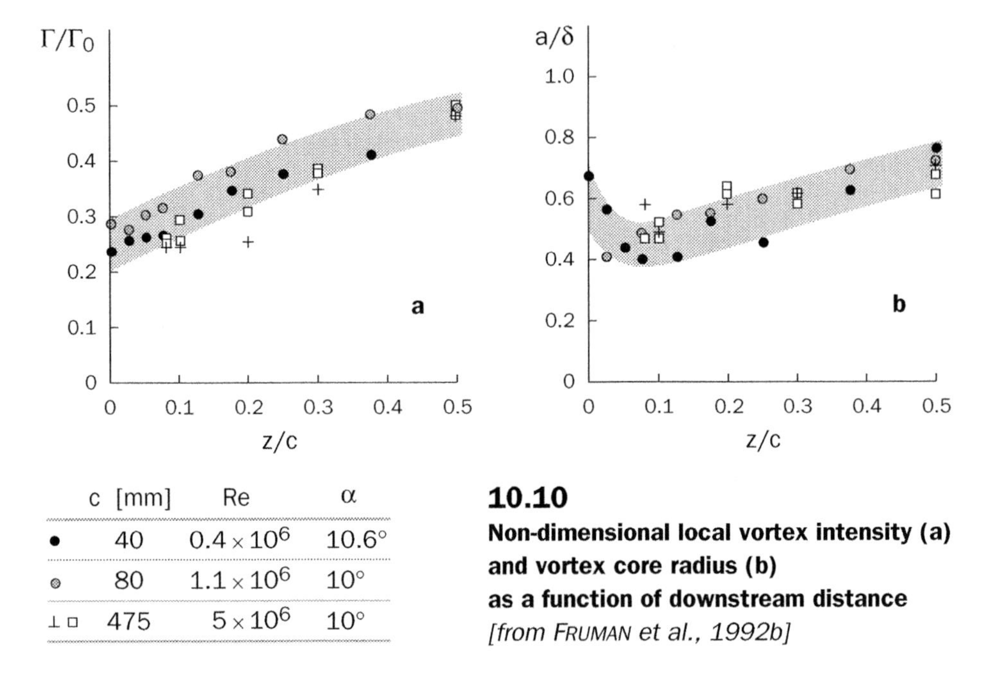

10.10
Non-dimensional local vortex intensity (a) and vortex core radius (b) as a function of downstream distance
[from FRUMAN et al., 1992b]

They show that the roll-up process is significantly initiated at the foil tip since the local vortex strength is already equal to 25% of the bound circulation Γ_0. Half a chord length downstream, the tip vortex strength has increased and is almost doubled.

The vortex core radius is minimum in the vicinity of the tip, at about 10% of the chord length. It increases downstream and reaches roughly 70% of the boundary layer thickness at half a chord length downstream. Such orders of magnitude confirm that the turbulent boundary layer thickness is actually a relevant length scale for the core radius.

The minimum value of the core radius at about 10% of the chord length indicates that the pressure on the vortex axis is also minimum at the same location along the vortex path. This conclusion is corroborated by the observations of cavitation inception by MAINES and ARNDT (1993), who found that cavitation nuclei are actually activated around this point.

10.3. CAVITATION IN A TIP VORTEX

10.3.1. SCALING LAWS FOR CAVITATION INCEPTION

Whatever the considered model of the vortex, the minimum pressure in the vortex center can be written, in non-dimensional form, as (see eq. 10.23 or 10.24):

$$Cp_{min} = \frac{p_{min} - p_{\infty}}{\frac{1}{2}\rho V^2} = -k\left(\frac{\Gamma}{Va}\right)^2 \tag{10.27}$$

where the parameter k is a constant.

The previous results have shown that the relevant scales for the local vortex intensity Γ and the core radius a are respectively the mid-span bound circulation Γ_0 and the boundary layer thickness δ, so that relation (10.27) can be rewritten:

$$Cp = -k\left[\frac{\Gamma/\Gamma_0}{a/\delta}\right]^2 \left(\frac{\Gamma_0}{V\delta}\right)^2 \tag{10.28}$$

Using relations (10.25) and (10.26), the pressure coefficient can be expressed as:

$$Cp = -k\left[\frac{\Gamma/\Gamma_0}{0.74\,(a/\delta)}\right]^2 C_L{}^2\, Re^{0.4} \tag{10.29}$$

The previous expression is valid at any station along the vortex path. In particular, the minimum pressure coefficient along the whole vortex path has a similar expression from which the critical value of the cavitation number at inception can be deduced:

$$\sigma_{vi} = k\left[\frac{(\Gamma/\Gamma_0)_{min}}{0.74\,(a/\delta)_{min}}\right]^2 C_L{}^2\, Re^{0.4} \tag{10.30}$$

$(\Gamma/\Gamma_0)_{min}$ and $(a/\delta)_{min}$ are the non-dimensionalized values of the vortex strength and of the core radius respectively at the point along the tip vortex where the pressure is minimum. If it is assumed that these values are independent of the incidence for a given foil, cavitation inception data should follow the scaling rule:

$$\sigma_{vi} = KC_L{}^2\, Re^{0.4} \tag{10.31}$$

where K is a constant.

This scaling law was proposed by several authors (in particular BILLET & HOLL 1979 and FRUMAN *et al.* 1992b). It takes into account both viscous and potential effects. The early semi-empirical work of MCCORMICK (1962) suggested an exponent of about 0.35 for the influence of the REYNOLDS number on the desinent cavitation number for rectangular planforms (fig. 10.11).

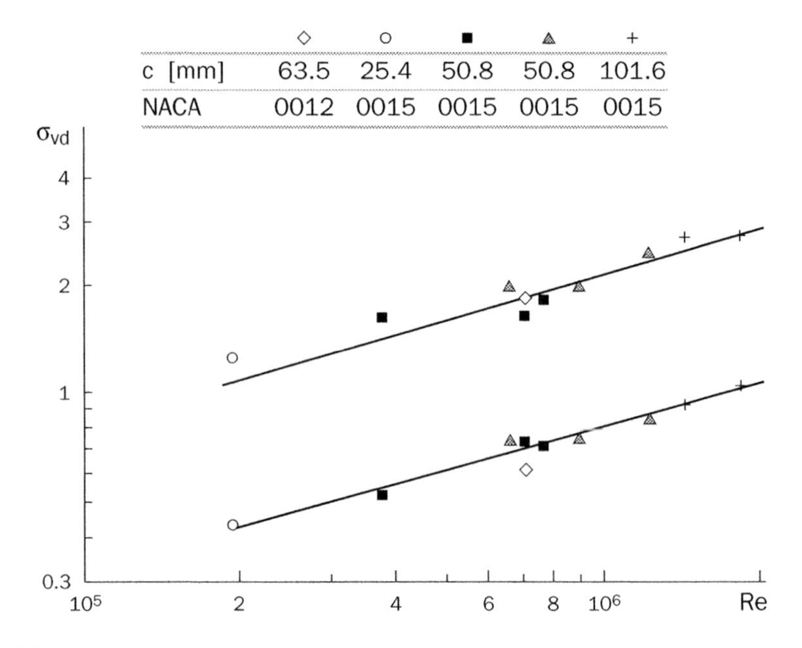

10.11 - Critical desinence cavitation number as a function of REYNOLDS number for rectangular wings at 4° and 8° incidences *[from MCCORMICK, 1962]*

10.3.2. CORRELATION OF CAVITATION DATA WITH THE LIFT COEFFICIENT

Measurements of critical cavitation numbers for tip vortex cavitation together with lift coefficients allow us to estimate the validity of equation (10.31). Usually, desinent cavitation numbers σ_d are preferred to incipient cavitation numbers because of a better reproducibility. The desinent cavitation number is defined as that value of σ_v above which cavitation can no longer be sustained.

The correlation between σ and $C_L{}^2\,Re^{0.4}$ is linear for only a limited range of the lift coefficient, typically less than 0.6, as shown on figure 10.12. Above this value a significant deviation from linearity is observed. FRUMAN *et al.* (1992b) even report that, depending upon the foil shape, a correlation with $C_L{}^2$ does not necessarily provide for a better linearity than a correlation with C_L.

Moreover, when the relationship (10.31) is actually linear, the slope may be different for different foils as shown on the enlargement of figure 10.12. The dependency of the constant of proportionality K on the foil cross-section was confirmed by MAINES and ARNDT (1997) who showed that the slope K is 0.073, 0.068 and 0.059 for elliptical hydrofoils whose cross-sections are respectively a NACA 4215, a NACA 66_2-415 and a NACA 16020 hydrofoil (see fig. 10.13).

On the whole, it appears that correlation (10.31) is not a universal one since, when it is valid, the coefficient of proportionality K appears to depend upon the foil section.

The hydrofoil planform has also a significant influence on tip vortex roll-up and cavitation as proved by FRUMAN *et al.* (1995). These authors conducted tests with three wings of the same aspect ratio and the same distribution of chord length along the span, but with a tip which is moved from a forward to a rear position. These experiments have shown that the effect of planform is far from negligible.

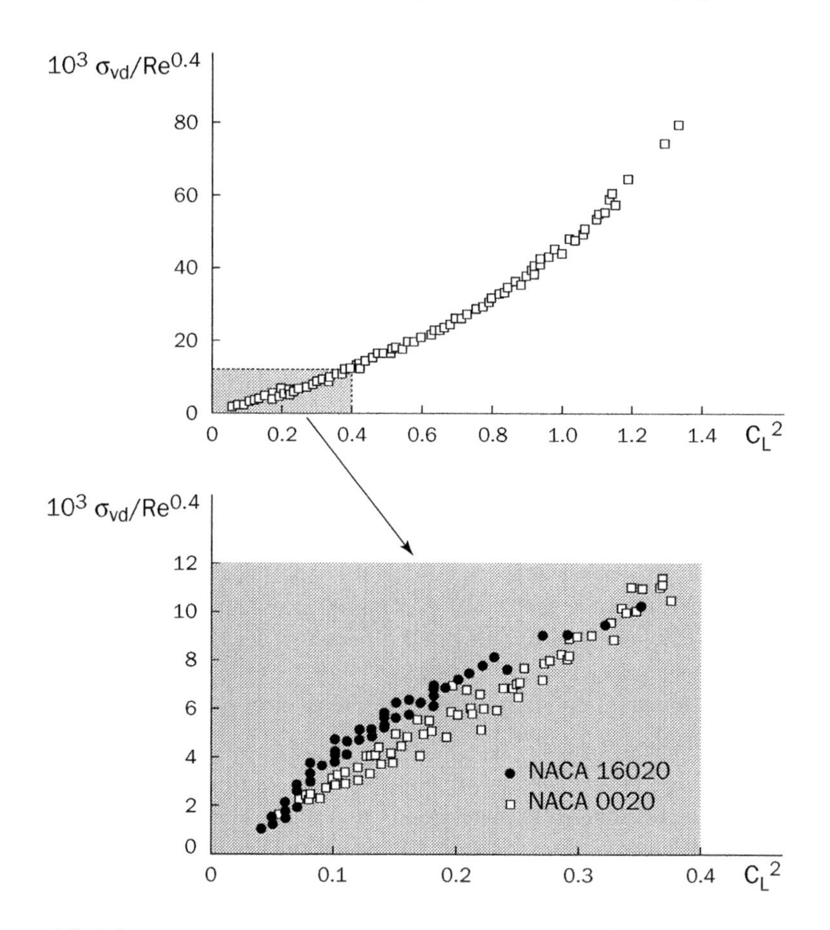

10.12 - Desinent cavitation number divided by $Re^{0.4}$ as a function of the square of the lift coefficient for NACA 16020 and 0020 elliptic foils
[measurements by PAUCHET et al., 1994]

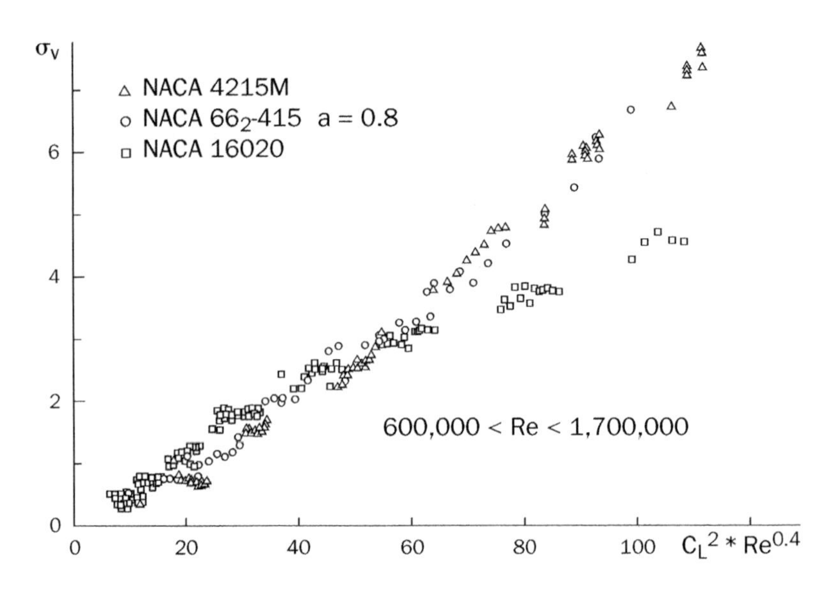

10.13 - Correlation of cavitation data with measured lift for three different foil sections *[from MAINES & ARNDT, 1997]*

10.3.3. EFFECT OF NUCLEI CONTENT

Usually, the very first cavitation events are elongated bubbles which appear intermittently in the vortex core, at some distance downstream of the tip [MAINES & ARNDT 1993]. This observation confirms that the minimum pressure does not occur at the very tip of the foil but somewhat downstream (see § 10.2.3). The stability of initially spherical bubbles in the pressure field of a vortex was studied by LIGNEUL (1989) and LIGNEUL & LATORRE (1989). Criteria for tip vortex cavitation inception based on the very first cavitation events depend significantly upon the nuclei content [BOULON *et al.* 1997].

A second criterion, based on the attachment of a continuous vapor tube to the wing tip, can be considered. It corresponds to a slightly more developed stage of cavitation and the order of magnitude of the difference in cavitation parameters between both criteria usually lies around unity. Contrary to the first, this second criterion is effectively insensitive to nuclei content.

The characteristic time of capture of a nucleus by a tip vortex is a fundamental parameter in the inception of tip vortex cavitation, especially under unsteady conditions. It depends on the nucleus size, the mode of feeding (radial capture or axial feeding) and the nuclei density. In the case of axial feeding, the characteristic time for a nucleus to enter the vortex core from upstream through a circular cross-section whose radius is equal to the vortex core radius a is of the order of:

$$\tau_a \approx \frac{1}{n.\pi a^2.V} \tag{10.32}$$

where V is the freestream axial velocity and n the nuclei density. This characteristic time depends strongly on n. For example, with $a = 0.9$ mm, $V = 8$ m/s, τ_a can vary between about 1 second for deaerated water with a nuclei density n of the order of 0.05 nuclei/cm^3 to about 5 milliseconds in the case of strong nuclei seeding at a concentration of 10 nuclei/cm^3.

Similar estimates can be obtained for radial capture, on the basis of the HSIEH equation (see § 4.4 and LIGNEUL & LATORRE 1989) and the RANKINE vortex model. The nucleus radius R is generally small enough so that it can be assumed that the pressure gradient balances the viscous STOKES drag. An estimate of the time τ_r required for a nucleus situated initially at a distance r_0 from the vortex axis to reach the axis is:

$$\tau_r = \frac{3}{4}\frac{\nu}{\omega^2 R^2}\left(\frac{r_0}{a}\right)^4 \tag{10.33}$$

In this equation, a and ω are connected to the circulation Γ by the relation $\Gamma = 2\pi\omega a^2$ and can be obtained from LDV measurements. Typical values are $a = 0.9$ mm and $\omega = 10{,}000$ rd/s. The initial position r_0 depends on the nuclei density n and on the typical length of the portion of the vortex path along which the capture of a nucleus can actually trigger cavitation. For a rough estimate, we suppose that inception can occur over a length of the order of the maximum chord length c. Then, on average, a cylinder of radius:

$$r^* = \frac{1}{\sqrt{\pi n c}}$$

contains one nucleus whose mean distance from the vortex axis is $r_0 \approx \frac{2}{3} r^*$. Taking the same values as previously for the nuclei densities, i.e. 0.05 and 10 nuclei/cm^3, and considering nuclei of radius $R = 5\,\mu$m and a chord length of 60 mm, one obtains $\tau_r = 0.98$ s and $\tau_r = 26\,\mu$s respectively [BOULON *et al.* 1997].

Thus, for both modes, the capture times are around one second in the case of deaerated water, and they are considerably reduced if the nuclei content is increased (e.g. using nuclei seeding). The frequency of cavitation events for vortex cavitation strongly depends on the nuclei density, which explains that the incipent cavitation parameter σ_{vi}, often defined on the basis of a threshold rate of events, also depends considerably on water quality.

The effect of nuclei content is particularly important in the case of unsteady tip vortex cavitation, such as on an oscillating foil [BOULON *et al.* 1997]. Tip vortex cavitation is then expected to occur only if nuclei have enough time to reach the vortex core during the period of oscillation. It requires that the previous capture times be much smaller than the characteristic period of oscillation, i.e. that the nuclei concentration is high. Otherwise, a significant delay in cavitation inception is observed in comparison with the quasi-steady case and the effect of water quality becomes important.

10.3.4. *EFFECT OF CONFINEMENT*

The effect of confinement is an important topic in rotating machinery. For example, axial pump impellers operate in the close vicinity of the casing and similarly, ship propellers can be placed inside a duct in order to improve the quality of the incoming flow.

In axial flow pumps, two cavitation patterns can occur, tip clearance cavitation and tip vortex cavitation. The former is caused by the separation of the flow as it passes between the blade tip and the casing wall. It can be generally prevented by rounding the clearance edge on the pressure side of the blade, as shown by LABORDE *et al.* (1995). On the other hand, tip vortex cavitation develops in the low pressure region of the vortices attached to the blade tips.

GEARHART & ROSS (1991) and FARELL & BILLET (1994) indicated an optimum tip clearance for cavitation performance. Below this optimum value, a reduction in tip clearance leads to an increase in the inception cavitation number σ_{vi}. Above it, σ_{vi} increases asymptotically with tip clearance, up to a value which is characteristic of a free, unshrouded impeller.

A different trend was pointed out by BOULON *et al.* (1999) in the case of an elliptical wing confined by a perpendicular flat plate in a cavitation tunnel. They observed that the confinement induces a large advance in tip vortex cavitation inception as shown on figure 10.14.

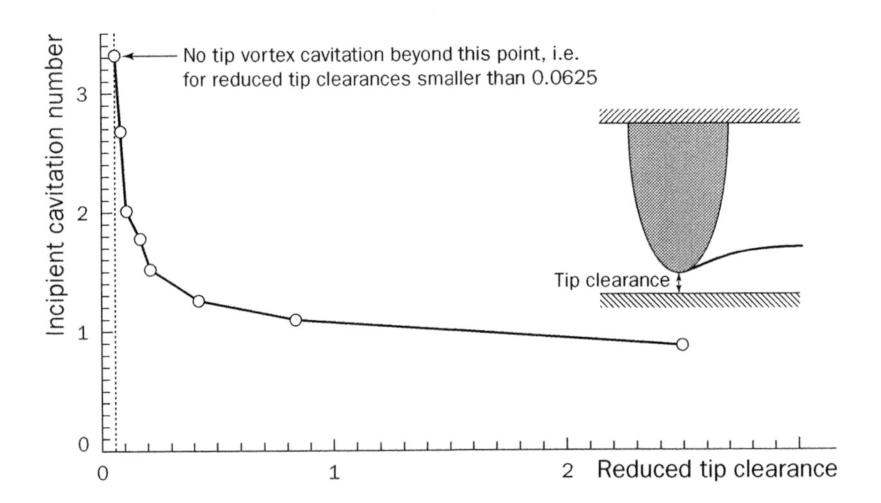

10.14 - Effect of tip clearance on critical cavitation number for a 6° angle of attack and a flow velocity of 7 m/s

The foil has a NACA 16020 cross-section and an elliptical planform. Its maximum chordlength is 0.12 m and its span length is 0.18 m. The tip clearance is non-dimensionalized by the maximum foil thickness 0.024 m [from BOULON *et al., 1999].*

This is due to a purely potential effect, which can be explained by considering the image vortex, symmetrical to the actual tip vortex of the wing with respect to the solid plate. The image vortex induces an increase in the angle of attack and in the lift coefficient, and consequently an increase in the tip vortex strength. This explains the premature development of cavitation in confined situations. No significant interaction between the tip vortex and the boundary layer was observed by BOULON *et al.* since the boundary layer was so thin that the vortex escaped very rapidly following contraction.

For very small values of tip clearance (see fig. 10.14), the large increase in angle of attack induced by the strong confinement can force the flow to stall at the tip. In this case, tip vortex cavitation was no longer observed by the authors and inception occurred in the form of unsteady small vapor structures which are characteristic of separated flows.

Finally, when the cavitation number σ_v is lowered below the critical value for inception, a partial attached cavity often develops on the wing in addition to the cavitating tip vortex. Typical views of interactions between tip vortex cavitation and a partial cavity can be found in BOULON *et al.* (1997, 1999).

REFERENCES

ARNDT R.E.A. & DUGUÉ C. –1992– Recent advances in tip vortex cavitation research. *Proc. Int. Symp. on Propulsors and Cavitation,* Hamburg (Germany), June, 22-25.

BATCHELOR G.K. –1967– An introduction to fluid dynamics. *Cambridge University Press.*

BOULON O., FRANC J.P. & MICHEL J.M. –1997– Tip vortex cavitation on an oscillating hydrofoil. *J. Fluids Eng.* **119**, 752-758.

BOULON O., CALLENAERE M., FRANC J.P. & MICHEL J.M. –1999– An experimental insight into the effect of confinement on tip vortex cavitation of an elliptical hydrofoil. *J. Fluid Mech.* **390**, 1-23.

BRIANÇON-MARJOLLET L. & MICHEL J.M. –1990– The hydrodynamic tunnel of I.M.G.: former and recent equipment. *J. Fluids Eng.* **112**, 338-342.

CHAHINE G.L. & GENOUX P.F. –1983– Collapse of a cavitating vortex ring. *J. Fluids Eng.* **105**, 400-405.

FARRELL K.J. & BILLET M.L. –1994– A correlation of leakage vortex cavitation in axial-flow pumps. *J. Fluids Eng.* **116**, 551-557.

FRANC J.P. –1982– Étude de cavitation. Sillage cavitant d'obstacles épais. *PhD Thesis,* Institut National Polytechnique de Grenoble (France), 66-121.

FRUMAN D.H., CERRUTTI P., PICHON T & DUPONT P. –1995– Effect of hydrofoil planform on tip vortex roll-up and cavitation. *J. Fluids Eng.* **117**, 162-169.

FRUMAN D.H., DUGUÉ C. & CERRUTTI P. –1991– Tip vortex roll-up and cavitation. *ASME Cavitation and Multiphase Flow Forum,* FED **109**, 43-48.

FRUMAN D.H., DUGUÉ C. & CERRUTTI P. –1992a– Enroulement et cavitation de tourbillon marginal. *Revue Scientifique et Technique de la Défense,* 133-141.

FRUMAN D.H., DUGUÉ C., PAUCHET A., CERRUTTI P. & BRIANÇON-MARJOLLET L. –1992b– Tip vortex roll-up and cavitation. *Proc. 19th Int. Symp. on Naval Hydrodynamics,* Seoul (Korea), August, 24-28.

GENOUX P. & CHAHINE G.L. –1983– Équilibre statique et dynamique d'un tore de vapeur tourbillonnaire. *J. Méc. Théor. Appl.* **2**, 829-857.

GEARHART W.S. & ROSS J.R. –1991– Tip leakage effects. *J. Fluids Eng.* **109**, 159-164.

LABORDE R., CHANTREL P., RETAILLEAU A., MORY M. & BOULON O. –1995– Tip clearance cavitation in an axial flow pump. *Proc. Int. Symp. on Cavitation,* Deauville (France), 173-180.

LAMB H. –1932– Hydrodynamics. *Cambridge University Press,* London and New York.

LIGNEUL P. –1989– Theoretical tip vortex cavitation inception threshold. *Eur. J. Mech. B/Fluids* **8**, 495-521.

LIGNEUL P. & LATORRE R. –1989– Study on the capture and noise of spherical nuclei in the presence of the tip vortex of hydrofoils and propellers. *Acoustica* **68**, 1-14.

MCCORMICK B.W. –1962– On vortex produced by a vortex trailing from a lifting surface. *J. Basic Eng.,* September, 369-379.

MILNE-THOMSON L.M. –1968– Theoretical hydrodynamic, 5th ed. *Macmillan,* 377-380.

MAINES B.H. & ARNDT R.E.A. –1993– Bubble dynamics of cavitation inception in a wing tip vortex. *ASME,* FED **153**, 93-97.

MAINES B.H. & ARNDT R.E.A. –1997– Tip vortex formation and cavitation. *J. Fluids Eng.* **119**, 413-419.

PAUCHET A., BRIANÇON-MARJOLLET L. & FRUMAN D.H. –1993– Recent results on the effect of cross-section on hydrofoil tip vortex cavitation occurrence at high Reynolds numbers. *ASME,* FED **153**, 81-86.

PAUCHET A., BRIANÇON-MARJOLLET L., GOWING S., CERRUTTI P. & PICHON T. –1994– Effects of foil size and shape on tip vortex cavitation occurrence. *Proc. 2nd Int. Symp. on Cavitation*, Tokyo (Japan), April, 133-139.

STINEBRING D.R., FARRELL K.J. & BILLET M.L. –1991– The structure of a three-dimensional tip vortex at high Reynolds numbers. *J. Fluids Eng.* **113**, 496-503.

WESTWATER F.L. –1936– *Aero. Res. Coun.,* Rpt and Mem. n° 1692.

11. *SHEAR CAVITATION*

The present chapter is devoted to cavitation in turbulent shear flows and more especially in wakes and submerged jets at high REYNOLDS number. Other flow situations such as the separated regions which develop on foils at large angles of attack are also relevant of this kind of flow.

The inception and development of cavitation in shear flows are mainly controlled by their non-cavitating structure. Such flows are limited by regions of high shear where vorticity is produced. As a result, coherent rotational structures are formed and the pressure level drops in the core of the vortices which become potential sites of cavitation.

From a theoretical viewpoint, the pressure field can be computed from the velocity field using POISSON equation:[1]

$$\frac{\Delta p}{\rho} = \frac{1}{2}\Omega^2 - e_{ij}\, e_{ij}$$

where $\vec{\Omega} = \mathrm{curl}\,\vec{V}$ is the vorticity and e_{ij} the deformation rates. Numerical analysis shows the prominent role of vorticity in this equation compared to deformation rate. In other words, the regions of minimum pressure coincide markedly with the coherent structures in which vorticity concentrates. As a result, the minimum pressure can be very different from the mean pressure. This is shown by the probability density function of pressure fluctuations, which is far from being symmetrical and expands much farther on the negative side than on the positive one (see e.g. DOUADY *et al.* 1991, MÉTAIS & LESIEUR 1992).

Such a deterministic approach based on coherent structures is complementary to the classical statistical approach of turbulence based on the determination of root mean square values of velocity fluctuations $\overline{u'^2}$. In the statistical approach, pressure fluctuations are considered as proportional to $\rho\,\overline{u'^2}$, the constant of proportionality often being determined semi-empirically. The consideration of coherent structures in turbulent flows is more recent. They were first revealed by experimentation (e.g. BROWN & ROSHKO 1974, WINANT & BROWAND 1974) and are now simulated numerically with increasing reliability (e.g. LESIEUR 1993).

From an erosion view point, cavitating vortices shed by attached cavities are often considered as particularly aggressive [SELIM & HUTTON 1983] and can cause significant damages when collapsing in regions of adverse pressure gradient (see e.g. SOYAMA *et al.* 1992).

---

1. This equation is obtained by applying the divergence operator to the NAVIER-STOKES equation.

11.1. *JET CAVITATION*

Jet cavitation appears at the periphery of submerged jets e.g. in boat propulsion or in discharge control valves and, more generally, at the frontier of any separated flow. It is an important topic, for instance in the search for noiseless energy dissipators.

Inception of cavitation in jets depends primarily upon the structure of the non-cavitating flow. Axisymmetric jets, for instance, are rather complicated, since several kinds of vortices are formed such as toroidal, linear streamwise and helix vortices.

Indeed, little is known on this topic, and experimental results on cavitation inception are extremely scattered. The values of σ_{vi} can vary between 0.1 and 2 typically. Such large differences are due to various causes including differences in nozzle geometries, in the inception criteria considered experimentally and mainly in the test procedures, in particular in the control of water quality which strongly affects jet cavitation.

Considering the absence of consensus in this field, we limit the scope of our review here to the presentation of a few experimental results which illustrate the complexity of the phenomenon and to basic elements of theoretical analysis.

11.1.1. *SOME EXPERIMENTAL RESULTS*

Influence of water quality

OOI (1985) demonstrated the important effect of water quality on jet cavitation. In his experiments, he measured the dissolved air content together with the nuclei content by means of holography. Table 11.1 shows that the nuclei concentration is clearly correlated to the dissolved air content and that the critical cavitation parameter decreases when the air content is decreased. For smaller jets, the cavitation inception number changes considerably with the air content.

PAUCHET *et al.* (1992) conducted experiments in which the water quality was controlled by means of a nuclei seeding device. The mode of nuclei seeding could be changed by altering the geometry of the injector through which engassed water was discharged upstream of the nozzle together with the pressure at which water was engassed. PAUCHET *et al.* measured the frequency of cavitation events detected by an accelerometer and found, as expected, that this frequency increases when the cavitation number decreases.

For degassed water without nuclei injection, this increase is very rapid, so that the cavitation number σ_{vi} at inception is well defined. Conversely, if nuclei seeding is turned on, the increase in the frequency of cavitation events is much more gradual when σ_v decreases, and the domain of σ_{vi}-values is greatly enlarged.

If the cavitation inception number σ_{vi} is defined using a threshold value of the non-dimensional frequency of cavitation events, it appears from figure 11.1 that this critical value depends considerably upon the water quality.

Dissolved air content (ppm)	σ_{vi} Jet diameter (mm)			Nuclei concentration (per cm³) Nuclei size	
	3.17	4.76	6.35	$5\,\mu m < R < 10\,\mu m$	$10\,\mu m < R < 20\,\mu m$
14.1	0.16	0.15	0.27	150	25
10.9	0.08	0.11	0.24	74	15
7.7	0.06	0.07	0.23	24	8.2
4.2	0.03	0.06	0.22	5.3	2.1

Table 11.1 - Measured σ_{vi}-values and measured nuclei concentrations on the jet axis, at an abscissa equal to two diameters *[from OOI, 1985]*

Cavitation inception was detected acoustically, using a critical threshold of five cavitation events per second.

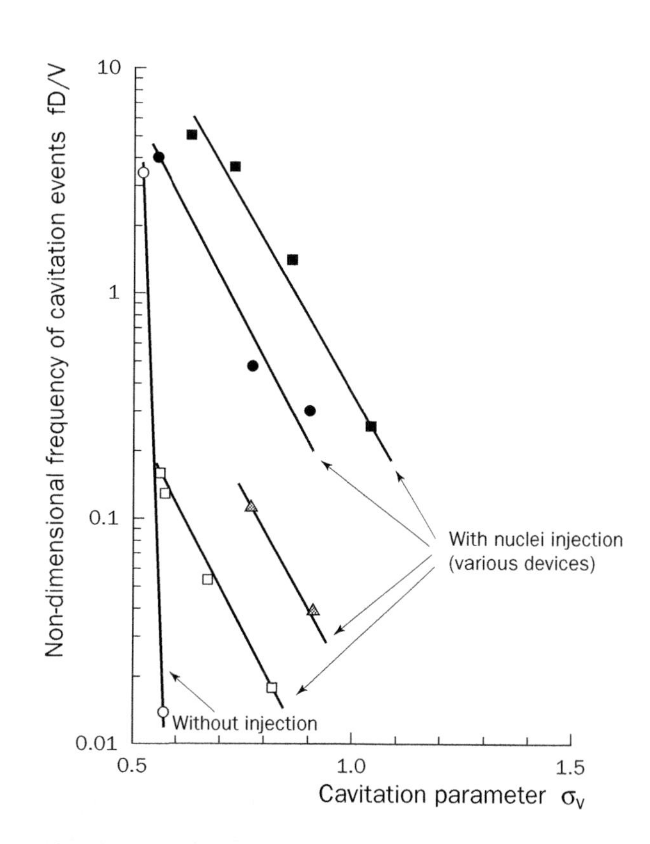

11.1 Variation of the frequency of cavitation events with the cavitation number, for several water qualities obtained with different devices of nuclei injection

The frequency is non-dimensionalized using the jet velocity $V = 20\ m/s$ and diameter $D = 40\ mm$. The decrease in σ_v is obtained by lowering the reference pressure, the velocity V being kept constant. The cavitation number is defined by the ratio $2(p - p_v)/\rho V^2$, where p is the pressure at the nozzle oulet [from PAUCHET et al., 1992].

Although each curve in figure 11.1 corresponds to identical technical procedures of nuclei injection, the nuclei spectrum cannot be the same since this depends upon the pressure at the outlet of the engassed water injector, i.e. upon the σ_V-parameter. In practice, it is almost impossible to generate a given and invariable nuclei spectrum whatever the operating conditions of the cavitation tunnel.

Influence of jet velocity and nozzle diameter

According to OOI's results, for a given nozzle geometry, the cavitation inception number is almost independent of the REYNOLDS number based on the jet diameter and the exit velocity.

This conclusion was not confirmed by PAUCHET (1991) who found that the flow velocity has an important influence on the variation of the non-dimensional frequency with cavitation number. Surprisingly, effects are found opposite for the larger tested nozzle (D=40 mm) and the smaller one (D=20 mm). In the latter case, at a given value of σ_V, the non-dimensional frequency increases with the velocity, whereas for the large diameter the trend is opposite.

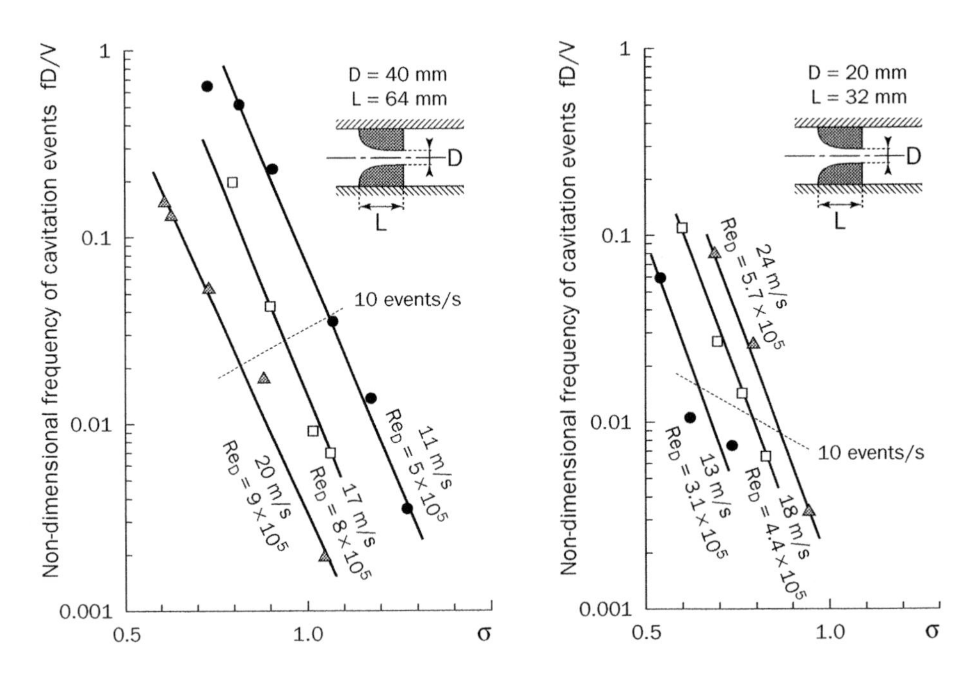

11.2 - Influence of jet velocity on the non-dimensionalized frequency of cavitation events for the 6 hole injector at an engassing pressure of 16 bars

These seeding conditions ensure that big enough nuclei are injected, with a critical pressure close to the vapor pressure, so that no significant delay in the nuclei growth is expected [from PAUCHET et al., 1992].

OOI's results (see tab. 11.1) show that, for the same air content, the σ_{vi}-values are approximately the same for the two smallest jets, while the biggest one gives higher values which do not depend greatly on the air content. Moreover, OOI observes that, for small jets, cavitation appears only downstream of the potential cone, while for the biggest one, cavitation may also appear in the shear layer close to the nozzle exit.

Other influences were also studied. The rate of turbulence of the inlet flow seems to have little importance when varied from 3% to 15% [PAUCHET 1991]. As for the nozzle length, this might also influence jet cavitation since it controls the boundary layer thickness at the nozzle exit.

In conclusion, it is difficult to obtain a comprehensive view and to separate the effects of the various parameters including the water quality, so that the empirical approach remains often the preferential one in jet cavitation.

11.1.2. *SOME ELEMENTS OF ANALYSIS OF JET CAVITATION*

The classical analysis of cavitation inception in jets is based on a statistical modeling of the turbulent flow. The liquid pressure is considered as the superposition of its mean value $\overline{p}$ and its fluctuation p', so that the local condition for the destabilization of a nucleus of critical pressure p_c is:

$$\overline{p} - p' < p_c \tag{11.1}$$

Here, we write $-p'$ (with $p'>0$) since we are interested in the negative values of the pressure fluctuations. If we use the pressure coefficient:

$$C_{\overline{p}} = \frac{\overline{p} - p_{ref}}{\frac{1}{2}\rho V^2} \tag{11.2}$$

equation (11.1) takes the following non-dimensional form:

$$\sigma_v < -C_{\overline{p}} - \frac{p_v - p_c}{\frac{1}{2}\rho V^2} + \frac{p'}{\frac{1}{2}\rho V^2} \tag{11.3}$$

The first term on the right hand side takes into account the mean pressure field of the turbulent flow.

The second expresses the static delay to cavitation inception due to the difference between the vapor pressure and the actual critical pressure of the nuclei. Here, the critical pressure of the biggest nuclei, i.e. the liquid susceptibility pressure (see § 2.4), should be considered. The static delay decreases when the velocity increases, and is usually negligible in industrial situations, unless the liquid is strongly deaerated.

The third term expresses the advance to cavitation inception due to the pressure fluctuations. The modeling of this term is the biggest difficulty in the prediction of jet cavitation inception. The simplest way of overcoming it is to assume that the RMS value of p' is correlated to the turbulent kinetic energy k:

$$k = \frac{1}{2}\left[\overline{u'^2} + \overline{v'^2} + \overline{w'^2}\right] \tag{11.4}$$

via a linear relationship:

$$\sqrt{\overline{p'^2}} = C.\frac{2}{3}\rho k \tag{11.5}$$

In case of isotropic turbulence, the three RMS values of the velocity fluctuations are equal and the previous relation becomes:

$$\sqrt{\overline{p'^2}} = C.\rho\, u'^2 \tag{11.6}$$

where the constant C is approximately 0.7 [HINZE 1959]. For turbulent shear flows, the coefficient C is larger. For example, PAUCHET (1991, 1993) reported C-values of the order of 4 to 6 in order to give a correct account of experimental data. Unfortunately, the use of a unique value of C does not apply to all experimental results. Further insight into the coherent structures which form in submerged jets appears to be necessary for a better understanding and prediction of jet cavitation, as this is attempted below for wake cavitation.

11.2. *WAKE CAVITATION*

As distinct from slender bodies, the separation of the boundary layer on bluff bodies give rise to wakes of large size in which various kinds of vortices form, develop and interact with each other (see e.g. MORKOVIN 1964, BERGER & WILLE 1972). The cores of these vortices are the location of pressure drops which will cause the inception of cavitation.

In this section, we will mainly consider obstacles with sharp edges, such as discs or wedges, for which the separation point of the boundary layer is fixed. The situation may be somewhat more complicated in the case of a circular cylinder or a sphere for instance, because of the unsteady nature of the separation point.

11.2.1. *CAVITATION INCEPTION IN THE WAKE OF CIRCULAR DISCS*

The inception of cavitation in the wake of a circular disk was investigated by KERMEEN and PARKIN (1957). Figure 11.3 shows the dependence of the inception cavitation number, σ_{vi}, versus the REYNOLDS number for several sharp-edged disks of various diameters. The REYNOLDS number varies from 0.8×10^5 up to 6×10^6.

All points lie above the value $-C_{pb} = 0.44$ for the pressure coefficient measured at the base of the discs, which confirms that smaller pressure levels are found in the wake due to coherent vortices.

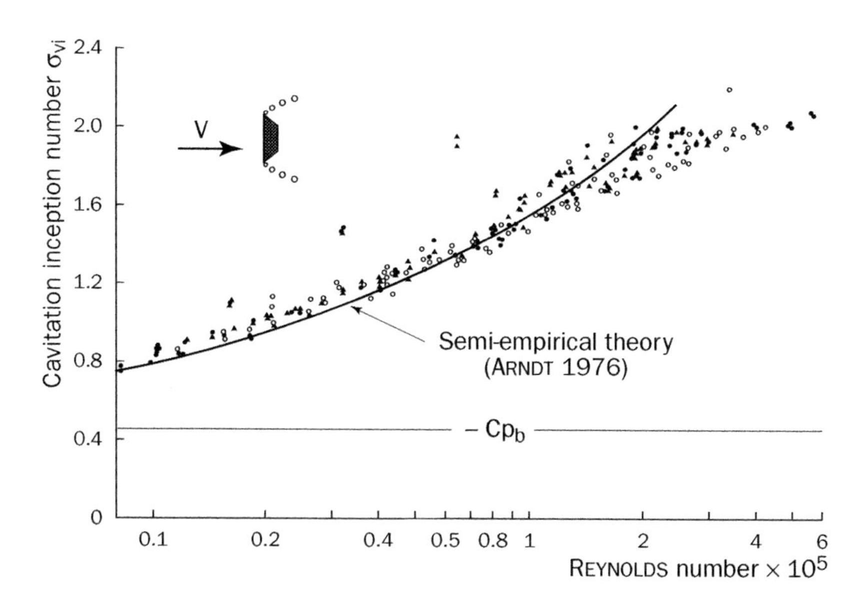

11.3 - Variation of the cavitation inception number σ_{vi} with the REYNOLDS number for sharp-edged circular disks *[from KERMEEN & PARKIN, 1957]* **and correlation with theory** *[from ARNDT, 1976]*

The cavitation inception parameter σ_{vi} increases monotonically with the REYNOLDS number, up to a REYNOLDS number of about 2×10^6 after which a plateau value is approached.

From visualization by dye injection, KERMEEN and PARKIN showed that the shear layer is made of a succession of ring vortices which are shed at a frequency F which is independent of the REYNOLDS number. Pairing of successive vortices was observed, and KERMEEN and PARKIN found that the non-dimensional frequency is inversely proportional to the distance x from the separation point:

$$S = \frac{Fd}{V_\infty} \cong \frac{1.35}{x/d} \tag{11.7}$$

In this relation, d stands for the disk diameter and V_∞ the free stream velocity.

These authors observed that cavitation inception occurs in the shear layer in the form of elongated filaments. It begins with the explosive growth of relatively large nuclei which become distorted cavities.

11.2.2. *MODELING OF WAKE CAVITATION INCEPTION*

Here we give a detailed account of ARNDT's analysis (1976) by which the experimental results of KERMEEN and PARKIN were reinterpreted in order to clarify the σ_{vi} dependence on REYNOLDS number shown on figure 11.3. ARNDT's model can be applied to the analysis of cavitation inception in the wake of any bluff body. It was used successfully by BELAHADJI *et al.* (1995) to interpret cavitation inception data behind a two-dimensional wedge (see § 11.2.3).

The analysis is focused on the near-wake and is mainly concerned with the ring vortices which are issued from the disk edge. It is assumed that cavitation occurs when the pressure in the core of these vortices (the sum of the static pressure at the base of the disc and the additional pressure drop due to vortex rotation), is equal to the vapor pressure. A RANKINE model is used for the vortex, so that the condition of cavitation inception is (see § 10.2.2):

$$\sigma_{vi} = -C_{pb} + 2\left(\frac{\Gamma}{2\pi a V_\infty}\right)^2 \tag{11.8}$$

In this relation, a is the vortex core radius and Γ the circulation. These are the main unknowns of the problem. The base pressure coefficient is supposed known from direct measurements.

The boundary layer is the location of vorticity production. At the separation point, the vorticity produced all along the wall is shed into the wake in the form of discrete vortices (fig. 11.4). ARNDT suggests that both a and Γ be calculated from the equations of conservation for mass and vorticity.

At the separation point, the volume flowrate of fluid issued from the boundary layer is, per unit span length:

$$\int_0^\delta u.dy = V(\delta - \delta_1) \tag{11.9}$$

where V is the external velocity at separation, $u(y)$ the velocity profile inside the boundary layer and δ and δ_1 the thickness and displacement thickness of the boundary layer respectively. The external velocity at separation V can be estimated by:

$$V = V_\infty \sqrt{1 - C_{pb}} \tag{11.10}$$

As for the flowrate of vorticity introduced into the shear layer, it is given by, per unit span length:

$$\int_0^\delta \frac{\partial u}{\partial y}.u.dy = \frac{1}{2}V^2 \tag{11.11}$$

since $\frac{\partial u}{\partial y}$ is the dominant term in the vorticity expression $\vec{\omega} = \text{curl}\,\vec{V}$.

As vorticity is contained in discrete vortices of core radius a and circulation Γ which are shed at a regular frequency F, the conservation of mass and vorticity is written (using eq. 11.9 and 11.11):

$$\pi a^2 F = V(\delta - \delta_1) \tag{11.12}$$

$$F\Gamma = \frac{1}{2}V^2 \tag{11.13}$$

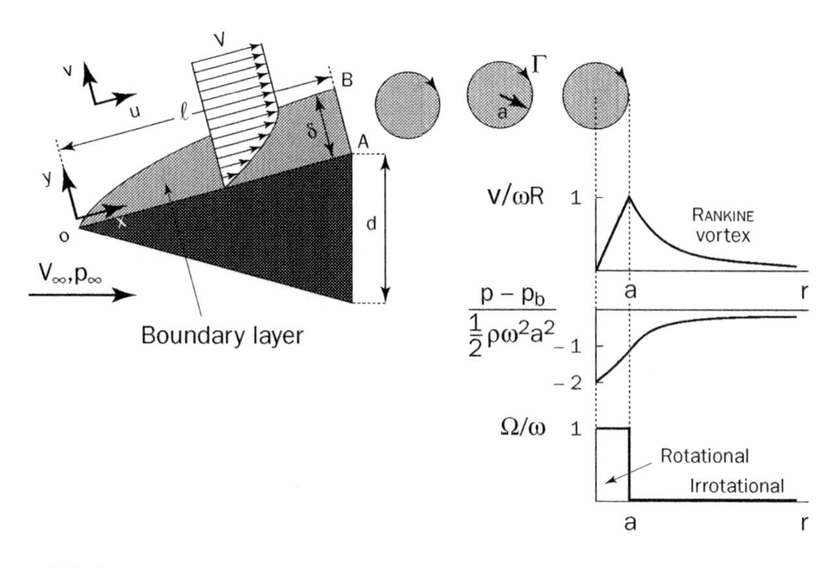

11.4 - **Theoretical analysis of cavitation in the wake of a bluff body**
[from ARNDT, 1976]

Assuming that the shedding frequency is known, equation (11.12) allows us to estimate the core radius a and equation (11.13) the circulation Γ. Replacing the previous expressions for a and Γ in equation (11.8) and taking into account equation (11.10), we finally obtain:

$$\sigma_{vi} = -C_{pb} + \frac{(1 - C_{pb})^{3/2}}{8\pi S} \frac{d}{\delta - \delta_1} \tag{11.14}$$

where S is the STROUHAL number defined by equation (11.7).

Using classical estimates of the boundary layer thicknesses in laminar axisymmetric flow:

$$\begin{cases} \delta = \dfrac{2.51\,d}{\sqrt{Re}} \quad \text{with } Re = \dfrac{V_\infty d}{\nu} \\ \dfrac{\delta_1}{\delta} = 0.27 \end{cases} \tag{11.15}$$

equation (11.14) finally gives, for discs:

$$\sigma_{vi} = -C_{pb} + 0.0217 \frac{(1 - C_{pb})^{3/2} Re^{1/2}}{S} \tag{11.16}$$

This relationship shows the influence of the REYNOLDS number on the cavitation inception parameter.

Using the measured value of the base pressure coefficient $C_{pb} = -0.44$, ARNDT assumed the value 10.4 for the STROUHAL number in order to get the best fit of the theoretical curve to the experimental data of KERMEEN and PARKIN. The theoretical curve, shown on figure 11.3, agrees fairly well with experimental results, for REYNOLDS numbers up to about 2.10^5.

ARNDT notes that, from equation (11.7), the value 10.6 of the STROUHAL number gives x/d equal to about $1/8$, which actually corresponds to the point at which cavitation appears in the experimental situation.

11.2.3. *CAVITATION IN THE WAKE OF A TWO-DIMENSIONAL WEDGE*

Overview of the structure of the cavitating wake

The sketch in figure 11.5 shows the three kinds of cavitating vortices observed by BELAHADJI *et al.* (1995) in the wake of a two-dimensional wedge.

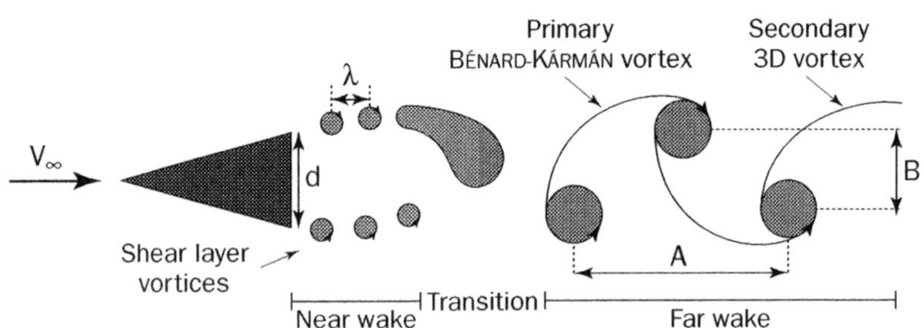

11.5 - Typical structure of a cavitating wake (in the sketch, the scales are distorted for clarity)

In the near wake, whose length is about 0.7 d (d denotes the wedge base), small-scale vortices are periodically shed in the two shear layers which originate in the wedge trailing edges. These vortices are comparable to the ring vortices observed by KERMEEN and PARKIN in the wakes of discs (see § 11.2.1). They result from a KELVIN-HELMHOLTZ instability.

The far wake is made up of the classical 2D BÉNARD-KÁRMÁN vortices. They are connected together by streamwise 3D vortex filaments which were described, under non-cavitating conditions, by TOWNSEND (1979), MUMFORD (1983), LASHERAS *et al.* (1986) and LASHERAS & CHOI (1988), among many others.

Between the near wake and the far wake, a transition region is observed which, in most cases, is made up of a two-phase mixture. In this region, the 2D small scale vortices of the near wake give rise to the large scale 2D BÉNARD-KÁRMÁN vortices.

The far wake begins at a distance from the trailing edges of the wedge which varies between 0.9 d, when cavitation is moderately developed, up to 4.4 d for small values of the cavitation parameter. Such estimates were also found by RAMAMURTHY and BALACHANDAR (1990) in their study of the near wakes of cavitating bluff bodies.

There is abundant literature devoted either to the BÉNARD-KÁRMÁN primary vortices or to the three-dimensional streamwise vortices in wakes or jets at moderate and high REYNOLDS numbers. Therefore, we will essentially focus here on the changes brought about by the development of cavitation.

BENARD-KARMAN vortices

The shedding frequency of the BÉNARD-KÁRMÁN vortices is strongly affected by the development of cavitation. The STROUHAL number can exceed its value in the non-cavitating regime by 30% (fig. 11.6). Thus, cavitation has an important influence upon the dynamics of BÉNARD-KÁRMÁN vortices.

The existence of the maximum in the curve $S(\sigma_v)$ is well established and was reported by YOUNG & HOLL (1966), FRANC (1982) and RAMAMURTHY & BALACHANDAR (1990). For high values of σ_v, the STROUHAL number approaches a constant value which characterizes the non-cavitating flow at the considered REYNOLDS number.

The geometry of the vortex street is also drastically changed by the development of cavitation. For low values of the cavitation number, the distance between the two rows decreases by about 80% in comparison with the non-cavitating case (fig. 11.7), so that the counter-rotating BÉNARD-KÁRMÁN vortices appear almost lined up when cavitation is sufficiently developed. This vortex street narrowing is connected to a transformation of the structure of the cavitating cores, which change from a two-phase mixture to a more vaporous core as cavitation develops.

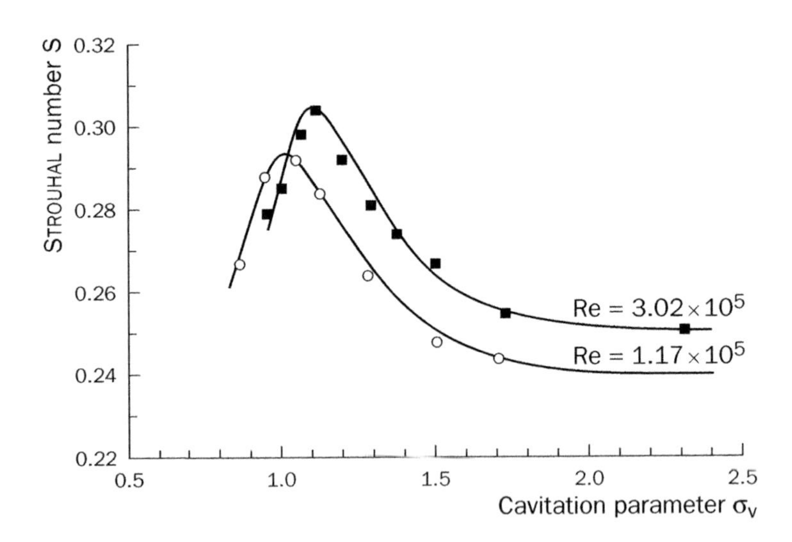

11.6 - Shedding frequency of the cavitating BÉNARD-KÁRMÁN vortices as a function of the cavitation parameter for a wedge of base $d = 35$ mm at two different REYNOLDS numbers

The measurements were obtained under stroboscopic lighting. The uncertainty on the STROUHAL number S defined by $S = fd/V$ does not exceed 2.4% [from FRANC, 1982].

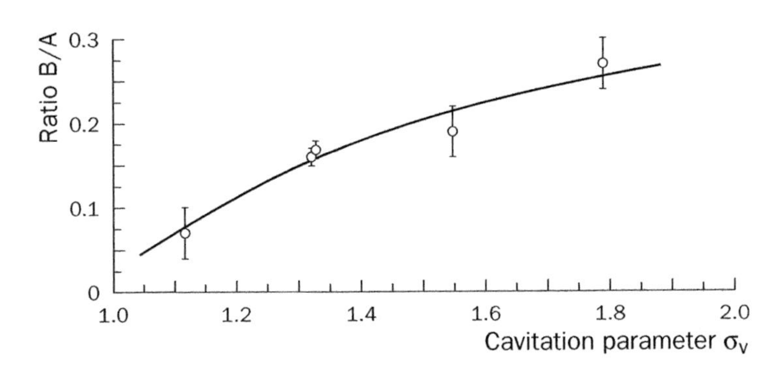

11.7 - Influence of the cavitation number on the ratio B/A (B: distance between the two vortex rows, A: wavelength of the BÉNARD-KÁRMÁN vortices) in the cavitating wake of a wedge (see figure 11.5)

The measurements were made at the beginning of the far wake [from BELAHADJI et al., 1995].

From high speed movies, BELAHADJI *et al.* (1995) determined the time-dependent evolution of the advection velocity of the BÉNARD-KÁRMÁN vortices together with the diameter of their cavitating core. Figure 11.8 shows that the size of the cavitating core fluctuates at a frequency twice the shedding frequency with an amplitude which decreases downstream.

The maximum diameter corresponds to the instant of shedding of those vortices from the region of transition. Hence, the shedding of the BÉNARD-KÁRMÁN vortices induces pressure fluctuations which naturally influence the size of their cavitating core.

Figure 11.8 also shows the time-dependent evolution of the advection velocity of the vortices. Their entrainment by the external flow is clearly visible since the velocity, initially affected by the wake deficit profile, approaches the free stream velocity.

Similarly to the diameter, the instantaneous advection velocity of the BÉNARD-KÁRMÁN vortices is affected by the pressure fluctuations, but in an opposite phase. High velocities coincide with minimum diameters. This is probably the result of an added mass effect (see § 10.1.2).

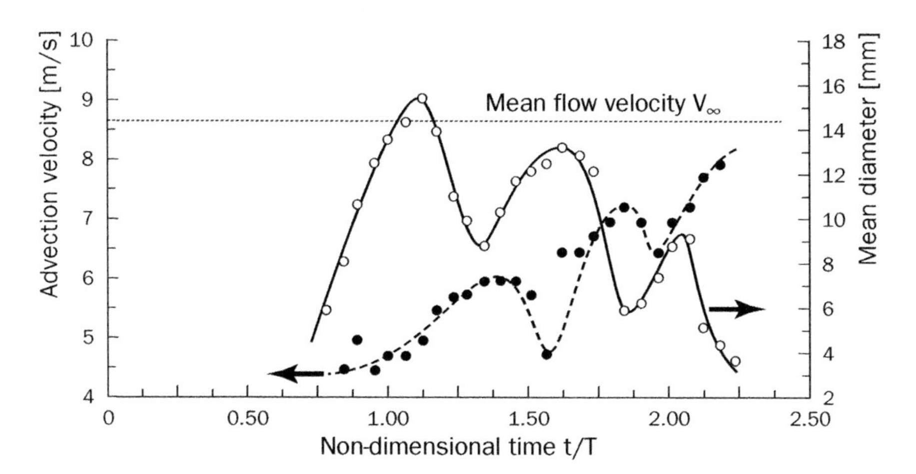

11.8 - Time variation of the diameter (○) and advection velocity of BÉNARD-KÁRMÁN cavitating vortices (●) in the far wake of a wedge

($Re = 3.1 \times 10^5$; $\sigma_v = 1.33$; $S = 0.293$; $f = 72$ Hz)

The time t is non-dimensionalized using the shedding period T. The origin of time is arbitrary [from BELAHADJI et al., 1995].

Streamwise vortices

The streamwise vortices are contained in planes roughly perpendicular to the BÉNARD-KÁRMÁN spanwise vortices. They are stretched between two consecutive counter-rotating vortices. Because of their high vorticity, they are usually the first structures which cavitate in the wake (see fig. 11.9).

For many years, particular attention has been devoted to streamwise vortices, especially in plane, free shear layers. The computational studies by CORCOS and LIN (1984) and the experimental ones by LASHERAS, CHO and MAXWORTHY (1986) show that these streamwise structures are due to an instability of the strain regions (braids) formed when the two-dimensional, KELVIN-HELMHOLTZ instability develops. Visualizations by BELAHADJI, FRANC and MICHEL (1995) on cavitating wakes present close similarities with the observations of LASHERAS and CHOI (1988) on plane shear layers.

BELAHADJI *et al.* measured the mean spacing s between these cavitating streamwise vortices. They showed that the spacing s non-dimensionalized by the distance $A/2$ between two consecutive counter rotating BÉNARD-KÁRMÁN vortices $\xi = 2s/A$ is about 0.64. It is practically independent of the level of cavitation development. The same value is obtained by BERNAL and ROSHKO (1986) in the case of a mixing layer.

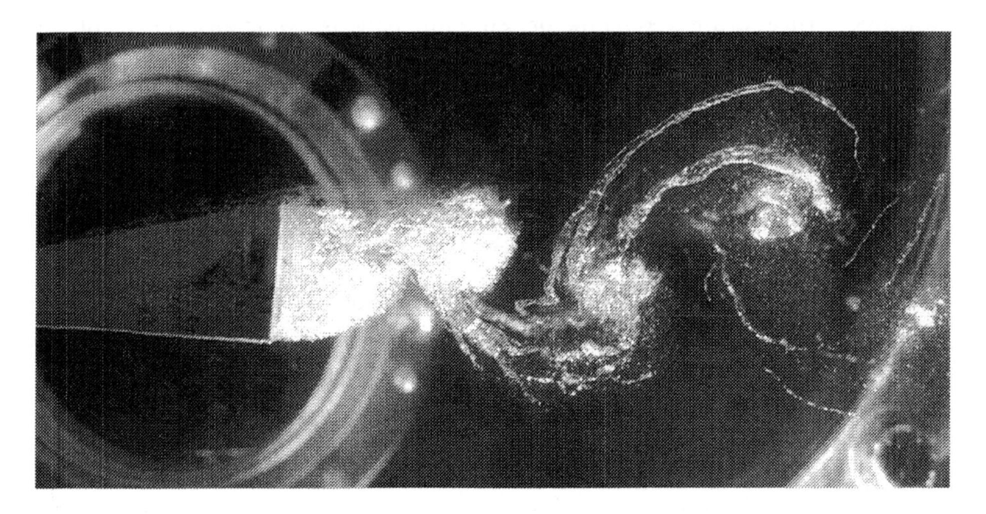

11.9 - Cavitating streamwise vortices

Shear layer vortices

The shear layer vortices are periodically produced by the two boundary layers which develop on the sides of the wedge. Their rotation rate is high, so that cavitation occurs in their core well before the primary BÉNARD-KÁRMÁN vortices.

The shear layer vortices collect in the transition region where, by successive pairing, they give rise to the BÉNARD-KÁRMÁN vortices. They are produced in an essentially periodic way at a frequency F which can be measured on high-speed movies as soon as they are made visible by cavitation.

BELAHADJI *et al.* have shown that the shedding frequency F of the shear layer vortices is between 7 and 12 times the shedding frequency f of the BÉNARD-KÁRMÁN vortices, according to the degree of development of the cavitation (fig. 11.10). This last value agrees with those obtained by KOURTA *et al.* (1987) for the non-cavitating wake of a circular cylinder at moderate REYNOLDS number.

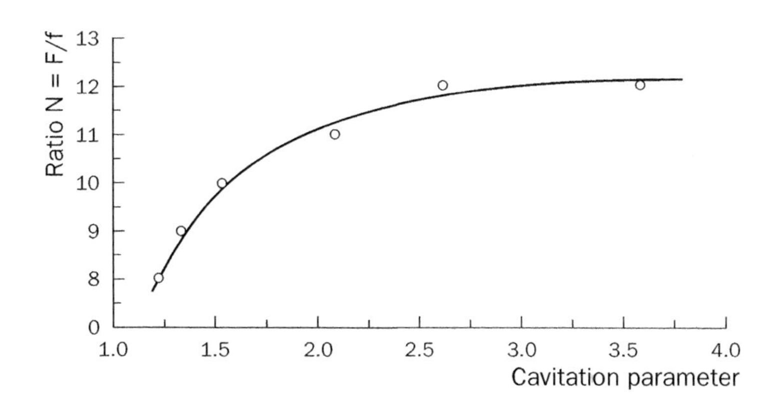

11.10 - Ratio N = F/f versus cavitation parameter in the cavitating wake of a wedge. F is the shedding frequency of the shear layer vortices and f that of the BÉNARD-KÁRMÁN vortices
[from BELAHADJI et al., 1995]

The cavitating shear layer vortices are accelerated by the external flow, so that their wavelength λ, i.e. the distance between two consecutive vortices, increases when they are advected downstream.

Their wavelength λ_0, immediately at the exit of the wedge, is classically non-dimensionalized by the boundary layer thickness δ to give the non-dimensional wavelength $\alpha = \pi\delta/\lambda_0$ used in the stability theory of free shear layers. Visualizations by BELAHADJI *et al.* showed that this non-dimensionalized wavelength α is about 0.38 (at $Re = 3.15 \times 10^5$ and $\sigma_v = 1.33$). This value is in agreement with classical results from the KELVIN-HELMHOLTZ instability theory (see e.g. DRAZIN & REID 1981), which shows that a shear layer is unstable for perturbations of α in the range 0-0.64 and that the maximal amplification rate occurs for $\alpha = 0.4$.

Prediction of cavitation inception

BELAHADJI *et al.* (1995) used ARNDT's model (see § 11.2.2) to analyze their cavitation inception data on a two-dimensional wedge. They applied equation (11.14) in which the boundary layer thicknesses were approximated by classical flat plate formulae. Using the measured values of the base pressure coefficient $C_{pb} = -1.48$ and of the shedding frequency F of the shear layer vortices, they predicted the cavitation inception number by means of equation (11.14). As shown in figure 11.11, ARNDT's model leads to a prediction of cavitation inception in good agreement with the experimental data.

To be fully predictive, ARNDT's model requires knowledge of the shedding frequency of the shear layer vortices. It can be measured experimentally as done by BELAHADJI *et al.* (1995) or it can be deduced from the stability theory in case counting is not possible.

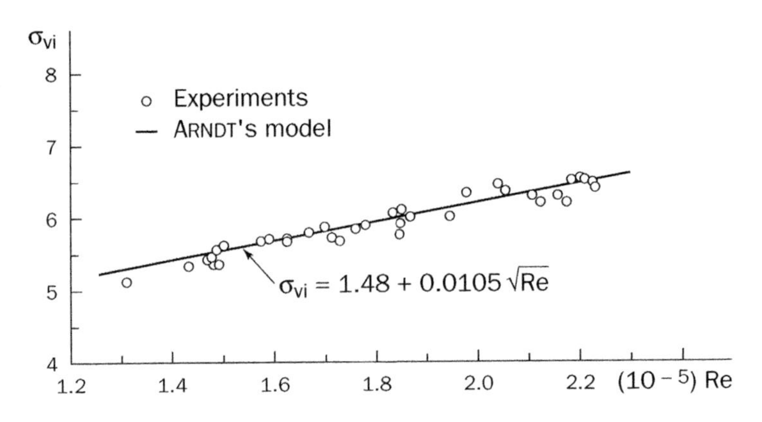

11.11 - Cavitation inception number in the wake of a wedge versus REYNOLDS number

Comparison between ARNDT's model and experimental data. [from BELAHADJI et al., 1995]

Although the model can be questioned as regards some of its details (for example, the choice of the RANKINE model or the way to estimate the boundary layer thicknesses using e.g. flat plate formulae), it takes into account the essential mechanisms involved in the production of near wake cavitating vortices.

The model can also be used to estimate the viscous core radius, a, and the rotation rate, ω, of the shear layer vortices. For the cavitating wedge at a REYNOLDS number of 1.91×10^5, we find $a = 1.5$ mm and $\omega = 4{,}800$ rd/s, i.e. 765 revolutions per second. Such a high value of the rotation rate together with a rather small size of the vortex core shows that coherent vortical structures are the location of high pressure drops in the wake. This is the reason why high ambient pressure levels are usually required to avoid cavitation in the wake of bluff bodies.

REFERENCES

ARNDT R.E.A. –1976– Semi-empirical analysis of cavitation in the wake of a sharp-edged disk. *Trans. ASME I – J. Fluids Eng.* **98**, 560-562.

BELAHADJI B., FRANC J.P. & MICHEL J.M. –1995– Cavitation in the rotational structures of a turbulent wake. *J. Fluid Mech.* **287**, 383-403.

BERGER E. & WILLE R. –1972– Periodic flow phenomena. *Ann. Rev. Fluid Mech.* **4**, 313-340.

BERNAL L.P. & ROSHKO A. –1986– Streamwise vortex structure in plane mixing layers. *J. Fluid Mech.* **170**, 499-525.

BILLET M.L. & HOLL W.J. –1979– Scale effects on various types of limited cavitation. *Proc. Int. Symp. on Cavitation Inception, ASME Winter Annual Meeting*, New York (USA), December 2-7, 11-23.

BRIANÇON-MARJOLLET L. & MICHEL J.M. –1990– The hydrodynamic tunnel of I.M.G.: former and recent equipment. *J. Fluids Eng.* **112**, 338-342.

BROWN G.L. & ROSHKO A. –1974– On density effects and large structure in turbulent mixing layers. *J. Fluid Mech.* **64**, 775-816.

CORCOS G.M. & LIN S.J. –1984– The mixing layer: deterministic models of a turbulent flow. Part 2 – the origin of the three-dimensional motion. *J. Fluid Mech.* **139**, 67-95.

DOUADY S., COUDER Y. & BRACHET M.E. –1991– Direct observation of the intermittency of intense vorticity filaments in turbulence. *Phys. Rev. Lett.* **67**, 983-986.

DRAZIN P.G. & REID W.H. –1981– Hydrodynamic stability. *Cambridge University Press.*

FRANC J.P. –1982– Étude de cavitation, tome 2: sillage cavitant d'obstacles épais. *PhD Thesis*, Institut National Polytechnique de Grenoble (France), 66-121.

FRANC J.P., MICHEL J.M. & LESIEUR M. –1982– Structures rotationnelles bi et tri-dimensionnelles dans un sillage cavitant. *CR Acad. Sci.* **295**, Paris, 773-777.

GRANT M.L. –1958– The large eddies of turbulent motion. *J. Fluid Mech.* **4**, 149-190.

HINZE J.O. –1959– Turbulence. *McGraw-Hill Book Company Ed.*

KERMEEN R.W. & PARKIN B.R. –1957– Incipient cavitation and wake flow behind sharp edged disks. *CIT Hydrodynamics Laboratory*, Rpt 85-4, August.

KNAPP R.T., DAILY J.W. & HAMMITT F.G. –1970– Cavitation. *McGraw-Hill Book Company Ed.*

KOURTA A., BOISSON H.C., CHASSAING P. & HA MINH H. –1987– Non-linear interaction and the transition to turbulence in the wake of a circular cylinder. *J. Fluid Mech.* **181**, 141-161.

LASHERAS J.C., CHO J.S. & MAXWORTHY T. –1986– On the origin and evolution of streamwise vortical structures in a plane, free shear layer. *J. Fluid Mech.* **172**, 231-258.

LASHERAS J.C. & CHOI H. –1988– Three-dimensional instability of a plane free shear layer. An experimental study of the formation and evolution of streamwise vortices. *J. Fluid Mech.* **189**, 53-86.

LESIEUR M. –1993– Turbulence in fluids, 2nd ed. *Kluwer.*

MÉTAIS O. & LESIEUR M. –1992– Spectral large-eddy simulation of isotropic and stably stratified turbulence. *J. Fluid Mech.* **239**, 157-194.

MORKOVIN M.V. –1964– Flow around circular cylinder-kaleidoscope of challenging fluid phenomena. *Proc. ASME Symp. on Fully Separated Flow.*

MUMFORD J.C. –1983– The structure of the large eddies in fully turbulent shear flows. Part 2 – the plane wake. *J. Fluid Mech.* **137**, 447.

OOI K.K. –1985– Scale effects on cavitation inception in submerged jets: a new look. *J. Fluid Mech.* **151**, 367-390.

PAUCHET J. –1991– Etude théorique et expérimentale d'un jet cavitant. *ACB-CERG*, Rpt 21-277, Grenoble (France), October.

PAUCHET J., RETAILLEAU A. & WOILLEZ J. –1992– The prediction of cavitation inception in turbulent water jets. *Proc. ASME Cavitation and Multiphase Flow Forum*, FED **135**, 149-158.

RAMAMURTHY A.S. & BALACHANDAR R. –1990– The near wake characteristics of cavitating bluff sources. *J. Fluids Eng.* **112**, 492-495.

SCHLICHTING H. –1987– Boundary layer theory. *McGraw-Hill Book Company Ed.*

SELIM S.M.A. & HUTTON S.P. –1983– Classification of cavity mechanics and erosion. *I. Mech. E. Conf.*, 41-49.

SOYAMA H., KATO H. & OBA R. –1992– Cavitation observations of severely erosive vortex streets. *NACA*, Rpt 1191.

TOWNSEND A.A. –1979– Flow patterns of large eddies in a wake and in a boundary layer. *J. Fluid Mech.* **95**, 515.

WINANT C.D. & BROWAND F.K. –1974– Vortex pairing: the mechanism of turbulent mixing-layer growth at moderate Reynolds number. *J. Fluid Mech.* **63**, 237-255.

YOUNG A.J. & HOLL W.J. –1966– Effects of cavitation on periodic wakes behind symmetric wedges. *J. Basic Eng.*, 163-176.

12. CAVITATION EROSION

It is well-known that cavitation can severely damage solid walls by removing material from the surface. The phenomenon of cavitation erosion is complex since it includes both hydrodynamic and material aspects.

From a hydrodynamic viewpoint, vapor structures are produced in the low pressure regions of a cavitating flow. They are entrained by the flow and may violently collapse when entering regions of pressure recovery, causing the erosion of the solid walls.

Erosion is due to the concentration of mechanical energy on very small areas of the walls exposed to cavitation, following the collapse of vapor structures. This energy concentration results in high stress levels which can exceed the resistance of the material, such as its yield strength, ultimate strength or fatigue limit. The response of the material to such a micro-bombardment by a myriad of collapsing vapor structures from the standpoint of continuum mechanics, solid physics and metallurgy is also a key point in cavitation erosion.

This chapter is essentially devoted to the hydrodynamic aspects of cavitation erosion with a special emphasis on the physical mechanisms involved. The numerous empirical prediction methods, although essential in practical situations, are only briefly mentioned in section 12.1. Section 12.2 is dedicated to the presentation of a few global results such as the influence of the velocity on cavitation erosion damage. The basic mechanisms of energy concentration which may cause erosion are reviewed in section 12.3. The different techniques for estimating the aggressiveness of a cavitating flow are presented in section 12.4, together with the scaling laws available to analyze the effects of flow velocity, length scale and fluid nature on aggressiveness. Section 12.5 gives an overview of the response of the material to the impact loads due to cavitation collapses.

We will focus on hydrodynamic cavitation erosion and not on vibratory cavitation erosion. This occurs without strong mean flow, such as in the cooling circuits of Diesel engines where the periodic motion exposes bubbles to periodic pressure variations, resulting in successive explosions and collapses.[1] Hydrodynamic cavitation erosion is commonly observed in fluid machinery and hydraulic structures at high flow velocities.

---

1. In situations where collapse occurs periodically and hit the wall at almost the same place, cavitation erosion damage is often confined to a rather small area and appears as a deep drilling of the wall.

12.1. *EMPIRICAL METHODS*

Empirical methods are very numerous and widely used in industry to predict cavitation erosion damage. On the whole, the empirical approach aims at developing a forward-looking method involving one or several of the following steps.

- The first step usually consists in conducting hydrodynamic tests on a model in order to identify the operating conditions for which cavitation erosion occurs and, if possible, to quantify the aggressiveness of the model cavitating flow, also called erosion potential, by pitting tests (see e.g. LECOFFRE 1985) or in terms of the stresses applied to the wall.
- The second step consists in conducting erosion tests on materials using a laboratory erosion device. The most classical one is the vibratory device (see PREECE 1979, among others, for a description of this apparatus) but other devices (such as Venturis) are also currently used. The main objective is to classify materials according to their resistance to cavitation erosion and, as far as possible, to correlate their resistance to classical mechanical properties such as, hardness, resilience, yield strength, ultimate tensile strength, strain energy... It is common practice to accelerate laboratory erosion tests in order to get long-term erosion data in a limited time.
- The last step is the transposition of the erosion data from the model to the prototype, which requires knowledge of the scaling laws governing the phenomenon.

Indeed, the various empirical methods developed over the years, together with long industrial experience on prototype machines, allow designers to choose the appropriate material for each industrial situation and to gain a global view on the influence of the main governing parameters on the cavitation erosion process.

However, there are a few serious defects:

- The classification of materials in terms of their resistance to cavitation erosion is far from being universal as it depends on the erosion device and also on the liquid.
- Correlations between the erosion resistance of materials and their mechanical properties are often very scattered and valid only within a limited range of flow parameters such as the liquid velocity and the cavitation parameter as well as for limited groups of materials.
- The scaling laws used for the transposition from model to prototype are not yet fully determined.

From the designer's viewpoint, the ideal situation would be to have at his disposal a fully predictive computational method which could dispense with model tests and be used at a preliminary design step. This is far beyond present capabilities and erosion tests will still have to be carried out for some time to come.

12.2. SOME GLOBAL RESULTS

12.2.1. INFLUENCE OF FLOW VELOCITY

Among the global results is the rapid increase of cavitation erosion damage with the liquid velocity relatively to the wall. The dependence of the mass loss rate $\dot{m}$ with the flow velocity V is usually expressed by a power law:

$$\dot{m} = k(V - V_0)^n \tag{12.1}$$

in which n is in the range 4 to 9, and V_0 is the threshold velocity of cavitation erosion for the material under consideration. For velocities smaller than this threshold, cavitation erosion is zero or entirely negligible. For water and stainless steel 316 L for instance, the threshold velocity lies around 20 m/s.

Relation (12.1) is not valid beyond a certain range of velocity variation. It is only relevant to hydrodynamic cavitation since there is no global motion of the liquid relative to the wall in the case of vibratory cavitation.

A high value of the exponent n is an indicator of the severe damage which can be suffered by hydraulic systems at high velocities. Indeed, spectacular damage by cavitation erosion has been observed in the past, for example to the concrete walls of the Tarbela Tunnel in Pakistan, which collapsed during discharge of the reservoir in the summer of 1974 [KENN & GARROD 1981].

12.2.2. TIME EVOLUTION OF MASS LOSS RATE

A second general feature of cavitation erosion is the typical evolution of mass loss rate with exposure time.

Initially, during a so-called incubation period, no mass loss is observed. Then, the mass loss rate progressively increases (an acceleration period) and finally reaches a steady state. For long duration tests, an attenuation period is often observed during which the erosion rate decreases.

In fact, such an evolution of the erosion rate with the exposure time is very schematic and many different types of curves have been obtained experimentally depending upon the erosion device and the material.

During the incubation period, the hydrodynamic impacts produce only a pitting of the material surface, i.e. plastic deformation in the form of permanent indents, without mass loss. As pits are rather small, the probability of overlapping is initially small, so that pitting tests of short duration are often used to analyze the flow aggressiveness. The material itself is then considered as a "virtual" transducer, whose threshold for pitting is usually identified with its yield strength.

KNAPP (1955) and STINEBRING (1976) found that the pitting rate obeys a power law of the type V^{α}, with an exponent α close to 6, whereas the total volume of pits follows approximately a V^5 law, for velocities up to 60 m/s.

Several experimentalists found that the duration T_i of the incubation period is correlated to the maximum erosion rate $\dot{m}_{max}$ during the steady period. The product $\dot{m}_{max} \cdot T_i$ was found to be approximately constant, the value of the constant depending upon the erosion test device.

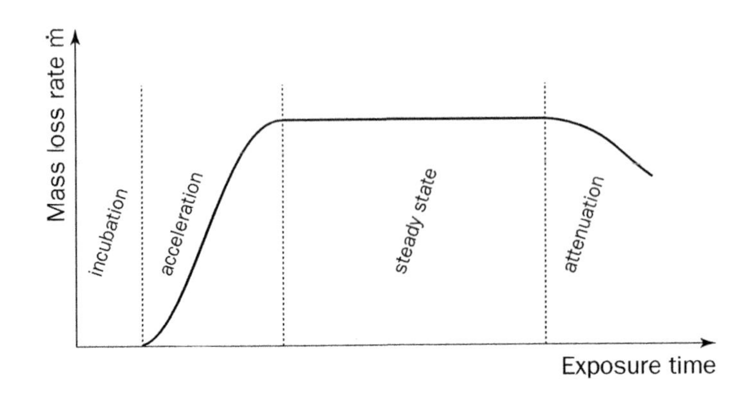

12.1 - Typical evolution of mass loss rate versus exposure time

12.2.3. *MISCELLANEOUS COMMENTS*

Generally, when the cavitation parameter is lowered for a fixed velocity, erosion reaches a maximum and then decreases when cavitation becomes well-developed. The decrease in flow aggressiveness may be a consequence of a smaller production of vapor structures or of weaker collapses due to decrease in the adverse pressure gradient.

The aggressiveness of cavitating flows may be significantly increased by flow unsteadiness, which can amplify the violence of the collapses depending upon the characteristic frequencies. The collective collapse of bubbles contained in cavitation clouds, like the ones shed by self-oscillating partial cavities, can also be, according to several authors, far more violent than individual collapses.

In practical situations, erosion may be limited and even avoided by means of flow ventilation. This is a current practice in the field of dam hydraulics by the use of spillways aerators.

In the case of corrosive liquids, the damage can be enhanced by a synergetic effect of erosion and corrosion. Cavitation has the effect of mechanically removing the passivating corrosion film and hence opening fresh highly reactive corrosion sites, so that the total wear rate is greater than the addition of the purely corrosive wear rate and the purely cavitation erosion rate.

As a final comment here, several definitions of the "erosion intensity" are found in the literature. Among these, one finds:

- the pitting rate, i.e. the pit density per unit time and unit surface area,
- the total mass loss rate $\dot{m} = \frac{dm}{dt}$ of the whole damaged specimen, or
- the corresponding volume loss rate,
- the mean depth of penetration rate, MDPR, which is the volume loss rate per unit surface area, as proposed initially by KATO.

Each of these definitions integrates hydrodynamics and material characteristics.

12.3. BASIC HYDRODYNAMIC MECHANISMS OF ENERGY CONCENTRATION

The main elementary mechanisms of energy concentration which may result in cavitation erosion are now discussed. They are all characterized by the high values of the impact load they produce, the small area of the impacted surface and the short duration of the impulse.

12.3.1. COLLAPSE AND REBOUND OF A SPHERICAL BUBBLE

Numerous studies (see for example the work of FUJIKAWA & AKAMATSU 1980, presented in section 5.4), have shown that the end of the collapse of a spherical bubble is marked by high values of the internal temperature and pressure, and is followed by the emission of a pressure wave of high intensity.

In the vicinity of the bubble center, the pressure wave duration is of the order of one microsecond and the amplitude of the wave decreases spatially approximately as r^{-1}. FUJIKAWA and AKAMATSU actually measured very high pressure levels of the order of 100 MPa at the instant of impact of the pressure wave on the sensitive surface of transducers.

12.3.2. MICROJET

A microjet is produced when a bubble collapses under non-symmetrical conditions (see § 4.3). If a solid wall is close enough to the bubble, the microjet is directed towards the wall.

Current estimates of microjet velocity V_j, normal to the wall, give values of the order of 100 m/s. The pressure rise due to the impact of such a high-speed microjet on the solid wall can be estimated using the water-hammer formula of JOUKOWSKI and ALLIEVI:

$$\Delta p = \rho c V_j \tag{12.2}$$

where c stands for the velocity of sound. We typically obtain $\Delta p = 150$ MPa for water, i.e. the same order of magnitude as the pressure wave amplitude. The duration of the pressure pulse is fixed by the jet diameter d and is of the order of d/2c. For a bubble of 1 mm initial diameter, the jet diameter is about 0.1 mm, which leads to a very small value for the duration of the pressure pulse, i.e. about 0.03 μs.

Thus, it appears that both hydrodynamic mechanisms –the shock wave and the microjet– give rise to high pressure pulses, with the same order of magnitude as the yield strength of usual metals. Attention has also to be paid to the pulse duration, since the transfer of energy from the liquid to the solid requires a minimum time to happen. From this point of view, the pressure wave and the microjet issued from the collapse of an isolated bubble are of very short duration. The value will however increase with the bubble size.

12.3.3. *COLLECTIVE COLLAPSE*

Collective effects occur with the collapse of a cloud of bubbles. In vibratory devices for example, the rapid oscillation of a small specimen in the liquid simultaneously generates many bubbles which collapse together near the specimen.

Collective collapse is also observed following the initial collapse of a single bubble close to a solid wall. The microjet which pierces the bubble leads to the formation of a vapor torus. This torus often splits into several smaller bubbles which undergo a subsequent collective collapse.

Collective collapses are typically characterized by cascades of implosions [TOMITA & SHIMA 1986, ALLONCLE *et al.* 1992, PHILIPP & LAUTERBORN 1998]. The pressure wave emitted by the collapse and rebound of a particular bubble tends to enhance the collapse velocities of the neighboring bubbles, thus increasing the amplitude of their own pressure waves.

The collapse of a cloud of bubbles, as formed by the periodic break-up of a sheet cavity, was also studied by REISMAN, WANG and BRENNEN (1998). Depending upon the characteristics of the cloud, they showed that the shock wave which propagates inward may strengthen considerably near the cloud centre because of geometric focusing and so enhance the erosive potential.

12.3.4. *CAVITATING VORTICES*

Cavitating vortices appear in shear flows (such as submerged jets and the wakes of bluff bodies) and also at the rear of partial cavities which shed cavitating vortical structures (see chap. 11 and 7 respectively). These appear to be responsible for severe erosion in fluid machinery, as described by OBA (1994).

Some test devices are based upon the repetitive axial collapse of a cavitating vortex, such as the vortex generator (LECOFFRE *et al.* 1981, AVELLAN & FARHAT 1989; see also DOMINGUEZ-CORTAZAR *et al.* 1997, FILALI *et al.* 1999a-b for another version of the apparatus). Axial collapse velocities larger than 100 m/s can be generated in such an apparatus so that the stress level, as estimated by the water hammer formula, is at least of the same order of magnitude as the one produced by shock waves and microjets.

Two main features seem to be at the origin of the high erosive potential of cavitating vortices:

- the formation of a foamy cloud at the end of the axial collapse in which cascade mechanisms can occur, and
- the rather long duration of the loading time, which is typically several tens of microseconds.

12.4. *AGGRESSIVENESS OF A CAVITATING FLOW*

The aggressiveness of a cavitating flow can be characterized in two different ways:

- ♦ By conducting pitting tests limited to the incubation period in which pits do not overlap. In such a technique, the solid wall is considered as a virtual transducer, whose threshold corresponds to its yield stress.
- ♦ By the direct measurement of the impact loads applied to the wall using suitable transducers [FRANC *et al.* 1994, OKADA & HATTORI 1994].

Besides pitting and force measurements, other techniques are used, namely acoustic and electro-chemical methods. The former is of easy use for the detection of cavitation, as explained in section 5.2.3, but the difficulty is to discriminate erosive from non-erosive cavitation. Such a non-intrusive technique is very useful for the monitoring of hydraulic machines.

As for the electro-chemical method, it is based upon the measurement of the current due to the repassivation of the eroded surface. It gives a signal proportional to mass loss rate and requires the mounting of an electrode where cavitation erosion is expected to occur [SIMONEAU 1995].

Before examining the corresponding techniques and the main results of both approaches, we present in the next section a simplified approach [LECOFFRE 1995] which is useful to estimate the aggressiveness of a collapsing isolated bubble and highlight the basic influence of the bubble size and the adverse pressure gradient.

12.4.1. *AGGRESSIVENESS OF A COLLAPSING BUBBLE*

Consider a spherical bubble fixed in space which collapses due to a linearly increasing pressure going from the vapor pressure p_v to the pressure at infinity p_∞, over a characteristic time T (fig. 12.2-a). This represents the practical situation where the bubble is supposed to move in an adverse pressure gradient, so that the parameter T can be considered as an indicator of the steepness of this gradient. The shorter T, the higher it is.

The classical RAYLEIGH model is used to compute the bubble collapse:

$$R\ddot{R} + \frac{3}{2}\dot{R}^2 = \frac{p_v - p(t)}{\rho}$$

This model gives an infinite final velocity as R tends to zero. To avoid this problem, it is possible to numerically compute the interface velocity $\dot{R}$ for a small, conventional value of R, near the end of the collapse, e.g. for $R = 0.01\, R_0$.

The variation of the interface velocity with the parameter R_0/T is shown on figure 12.2-b where R_0 is the initial bubble radius. The interface velocity $\dot{R}$ is non-dimensionalized by a reference interface velocity $\dot{R}_0$, also computed for $R = 0.01\, R_0$, but corresponding to the limiting case $T = 0$ of an instantaneous pressure increase or an infinite pressure gradient.

Two main conclusions can be drawn from figure 12.2-b:

- for decreasing T (i.e. increasing adverse pressure gradients), the interface velocity grows, and the aggressiveness of the collapse increases;
- for a given adverse pressure gradient, the aggressiveness increases with the initial size of the bubble.

Furthermore, according to this model, the lifetime of the bubble increases with its size. This is the reason why, in the case of an attached cavity shedding a large population of bubbles, the size of erosion pits generally increases downstream.

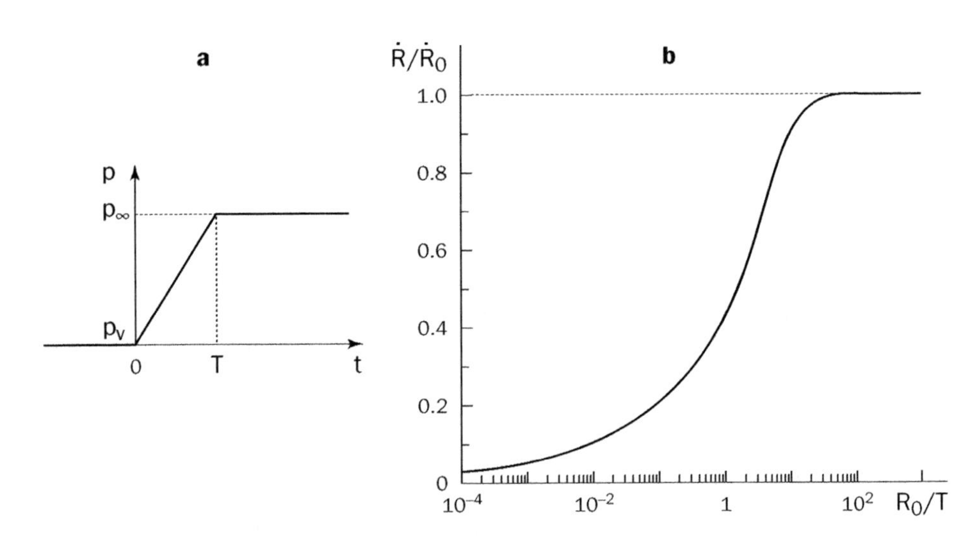

12.2 - (a) Time evolution of the applied pressure and (b) resulting interface velocity as a function of R_0/T

12.4.2. *PITTING TESTS*

Erosion pits usually have small depth, so that they look like shallow indents, and, as a consequence, their measurement and the estimation of their size is difficult.

Apart from classical roughness meters, several techniques were specially developed to analyze erosion pits. Among them are:

- Interferential techniques such as that developed by BELAHADJI *et al.* (1991) using a metallographic microscope with interferential MIRAU type objectives. This technique gives an image of the pits in the form of roughly concentric interference fringes, which represent lines of equal depth. The accuracy in depth is directly connected to the wavelength of the monochromatic light.
- A second technique involves focusing a laser beam on the surface to be measured. By displacing the tested target under the beam, a complete mapping of the surface is achieved. The accuracy on the depth measurement is 0.06 μm.

Erosion pits on metals are mostly circular with a diameter ranging typically from a few micrometers to one millimeter. The ratio of their maximum depth to their diameter is of the order of 1% in most cases, particularly for water. However, it can exceptionally reach 10% for example in the case of cavitating mercury.

The shape of the cross-section can usually be well approximated by a gaussian curve. Estimates of pit diameter are often based upon a conventional definition, wherein the diameter is taken to be that value corresponding to a local depth equal to 10% of the maximum depth.

Figure 12.3 shows typical cumulative histograms of pit diameter. The vertical axis values represent the number of pits per unit time and unit surface area whose diameter is larger than the corresponding value on the horizontal axis.

Cumulative histograms have usually no inflexion point, so that the distribution functions (which are indeed the derivatives of the cumulative histograms, see § 2.4) do not show any maximum. In other words, the distribution is not centered around a mean or typical size. The smaller the pits, the higher the pitting rate.

From figure 12.3 (whose axes are both logarithmic), it is clear that all histograms have the same shape and can be more or less superposed by appropriate translations along the vertical axis. Hence, the flow velocity affects the absolute value of the pitting rate but does not significantly change the distribution of pits in each class of size, at least within the present range of flow velocities.

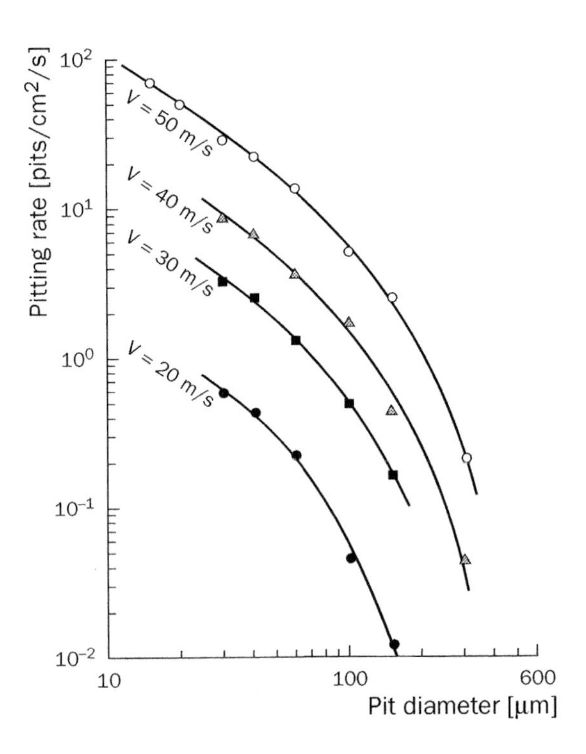

12.3 - Typical cumulative histograms of pit diameter on 316 L stainless steel specimens eroded at different velocities by a cavitating water flow through a Venturi of diameter 40 mm

The plotted data correspond to the point of maximum erosion. [from BELAHADJI et al., 1991]

It is also possible to build histograms of pit area or pit volume, if pit volumes are actually measured or at least estimated from the maximum depth and an assumed shape of the pit cross-section. Generally, such histograms present a maximum for an intermediate size, contrary to histograms of pit diameter.

As an example, figure 12.4 presents the contribution of each class of diameter to the total eroded area, for the same conditions as figure 12.3. The curves reveal a characteristic size (around 70 μm in the present case) for which the contribution to the eroded surface is maximum. Smaller pits, although very numerous, have a minor contribution. The larger pits nevertheless make only a minor contribution to erosion damage because of their small density.

From examination of histograms, it can be inferred that the pitting rate grows approximately as a power law of the flow velocity, with an exponent between 4 and 6 (see § 12.2.2).

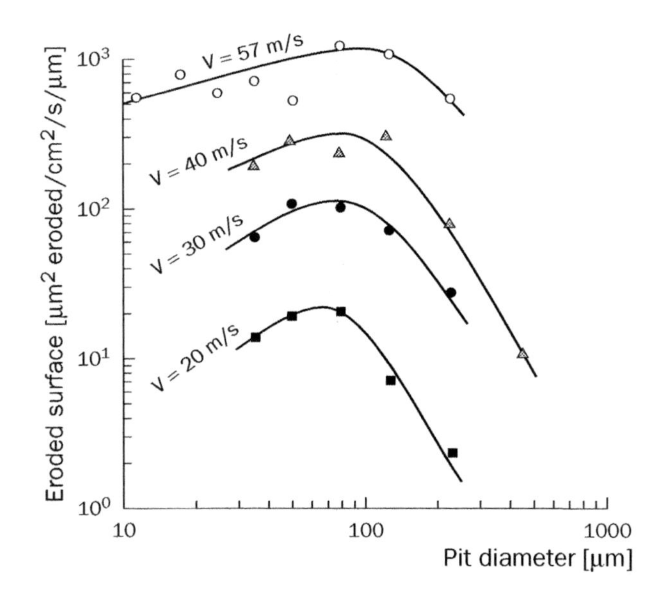

12.4 - Contribution of each class of pit size to the total eroded area on 316 L stainless steel specimens eroded at different velocities by a cavitating water flow through a 40 mm diameter Venturi

The plotted data correspond to the point of maximum erosion (unpublished results from the authors).

12.4.3. *FORCE MEASUREMENTS*

Pitting tests cannot give a value of the normal stresses applied to the solid wall but only compare it to the material yield strength. For a more precise quantification of flow aggressiveness, it is essential to directly measure the applied stresses by appropriate pressure transducers, flush mounted on the wall.

Such measurements require the use of high performance transducers. As their sensitive part is directly exposed to the cavitation attack, they must be resistant enough not to be damaged. In addition, the very short rise time of the pressure pulses due to bubble collapse demands a high natural frequency, greater than at least one Megahertz.

A major difficulty in the interpretation of measurements is due to the difference in size between the sensitive membrane of the pressure transducer (typically of the order of one millimeter) and the pressure impact (from a few micrometers to several hundred micrometers). Thus, the pressure is far from being uniform on the whole surface of the transducer and the physical quantity which is actually measured is the total force F and not the pressure. If the signal output is interpreted in terms of pressure, the computation of the ratio of the measured force F to the area of the sensitive membrane δS can induce an error of the order of the ratio $\delta S/\delta s$, where δs

is the area of the impacted surface. This ratio can be quite large and is particularly difficult to quantify. Thus, it is better to interpret the output signal as a measurement of the normal force which is applied on the sensitive membrane.

Special transducers have been build in order to meet those requirements. MOMMA (1991) and FILALI *et al.* (1999) developed transducers made of piezo-electric films simply glued on the solid wall. Their small thickness (typically between 25 μm and 150 μm) ensures a high natural frequency of the order of 10 MHz. They have to be protected by an appropriate coating which changes their theoretical response and makes a dynamic calibration necessary.

The piezo-electric material may be mounted between two metallic rods, as originally proposed by EDWARDS and JONES (1960) (fig. 12.5). One of them, the detection rod, is exposed to the cavitating flow and transmits the impact load to a piezo-electric ceramic, while the second rod, by a correct matching of the acoustic impedances, limits the distortion of the signal by reflected waves. The major advantage of this technique is the use of a detection rod which simultaneously allows the measurement of the forces and the observation of pits or mass loss on the detection rod itself [OKADA *et al.* 1995]. By associating the larger pit to the higher measured impact load, it is possible to estimate the actual pressure. Values of the order of a few GPa are currently obtained using this procedure.

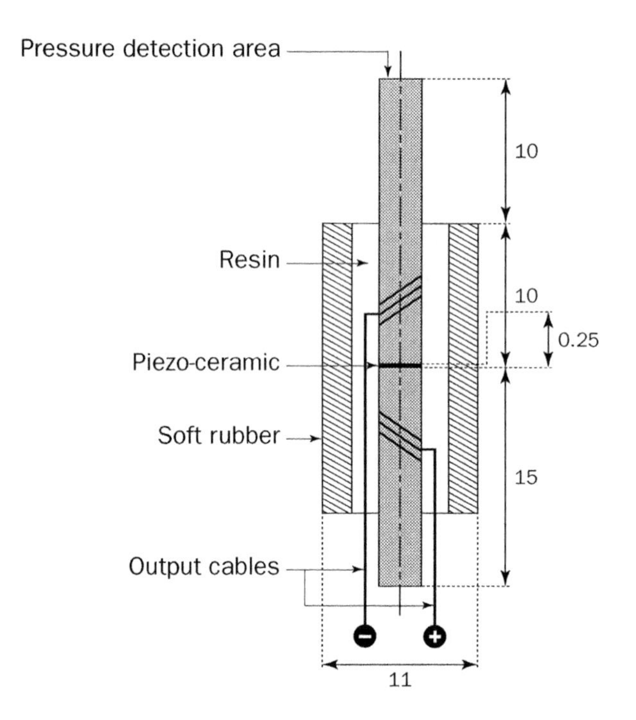

12.5 - Schematic design of a pressure transducer used for flow aggressiveness measurements *[from OKADA et al., 1995]*

Other physical processes have been used for the measurement of forces such as the development of dislocation lines in magnesium monocrystals [HATTORI *et al.* 1986, OKADA *et al.* 1994]. As the crystal response depends on the direction of the force, either normal or tangential to the surface, this technique helps confirm that cavitation forces responsible for the material attack are chiefly normal to the wall [FILALI *et al.* 1999b].

The dynamic calibration of transducers requires that the amplitude and rise time of the forces applied for calibration be of the same order of magnitude as the ones encountered in cavitation erosion. Two main methods are commonly used:

- The fall and rebound of a steel ball, for which the force can be evaluated on the basis of the equation of momentum, when the duration of the shock and the rebound height of the ball are measured.
- The breaking of a pencil lead under progressive loading, up to its rupture. In this case, the rapid discharge of the transducer is used for calibration [SOYAMA *et al.* 1998].

Figure 12.6 presents a typical example of a cumulative histogram of forces, measured by piezo-electric films. The result in terms of the pressure pulse rate $\dot{n}$ per unit surface area versus the force F is almost independent of the size of the pressure transducer. Note that the forces lie in the range 10 N to 100 N and that the pressure pulse rate (vertical axis) covers almost three orders of magnitude. The linearity of the curve in logarithmic scales suggests a power law of the type $\dot{n} = F^{-m}$. The exponent m is here close to 3.3. A similar value $m = 3.5$ was obtained by KARIMI in another test device, the vortex generator.

12.6
Typical example of cumulative histograms of forces measured by three transducers of different diameter (5, 3 and 1 mm)

The results collapse onto a single curve. They were obtained on a Venturi device placed in a mercury loop at a velocity of 7 m/s [from FRANC et al., 1994].

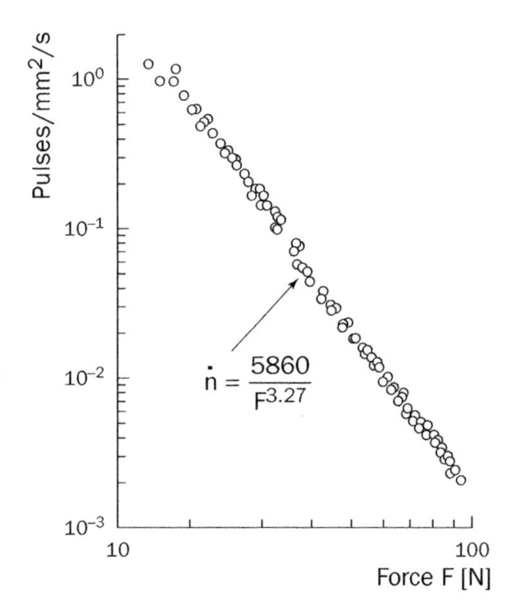

12.4.4. SCALING LAWS FOR FLOW AGGRESSIVENESS

Basic scaling laws can be derived from simple considerations. The hydrodynamic aggressiveness of a cavitating flow is supposed to be characterized, at a given point, by the density $\dot{n}$ (per unit time and unit surface area) of pressure pulses of amplitude greater than a given value Δp. In the present approach, the size of the impacted area is disregarded, although this parameter should be added on to the amplitude for a better description of the flow aggressiveness.

Among all the physical parameters which may affect the pressure pulse height spectra, we shall consider, to first approximation, only the following:

- the flow velocity V;
- a geometric length scale L. In fact, only geometrically similar flows are considered, which implies that the cavitation number is the same in both flows for similar extents of cavitation;
- the acoustic impedance, ρc, of the liquid, which is assumed to be the main liquid property.

It was shown in section 7.4.5 that the classic rules of dimensional analysis lead to the following scaling law for pressure pulse height spectra:

$$\frac{\dot{n}L^3}{V} = \text{function}\left(\frac{\Delta p}{\rho c V}\right) \tag{12.3}$$

This scaling law allows us to predict the effects of changes in flow velocity, length scale and fluid nature on the aggressiveness of geometrically similar flows.

It applies to pressure pulse height spectra, and can be extended to pitting rate on a given material by assuming that the material is characterized by a unique threshold parameter Δp^*. This threshold, often considered as its yield strength, is such that a pit is formed only if $\Delta p \geq \Delta p^*$. If this is the case, the pitting rate coincides with the pulse rate obtained in replacing Δp by the material threshold Δp^* in equation (12.3). Scaling laws for pitting rate can then be easily deduced.

The main trends to be extracted from equation (12.3) are:

- The impact pressure due to the collapse of a cavitating structure is scaled by a water hammer type formula $\rho c V$.
- If $\Delta p / \rho c V$ is kept constant (i.e. pressure pulse amplitudes in both model and prototype follow the water hammer scaling law), the frequency f of pressure pulses on two geometrically similar surface areas (which is proportional to $\dot{n}L^2$) is then scaled according to a classic STROUHAL scaling law:

$$\frac{fL}{V} = \text{Constant} \tag{12.4}$$

This relation applies to the frequency of the small scale vapor structures responsible for cavitation erosion. In the case of cloud cavitation, the large scale vapor structures which are periodically shed by a sheet cavity also obey a STROUHAL scaling law (see chap. 7). In the framework of the present assumptions, this basic scaling law can then be extended to smaller scales.

- For a given material, a given fluid and a given velocity, the pitting rate is inversely proportional to the third power of the length scale.

Figure 12.7 summarizes the three scaling laws which can be deduced from equation (12.3) when an initial flow characterized by the parameters V, ρc and L, is transposed into a similar one characterized by the parameters αV, $\gamma \rho c$ and λL. In logarithmic scales, they are represented by three translations which must be successively applied to the initial histogram. These scaling laws were proposed by LECOFFRE *et al.* (1985), following the work of KATO (1975).

Figure 7.12 (see chap. 7) brings a partial validation to this scaling when only velocity effects are considered.

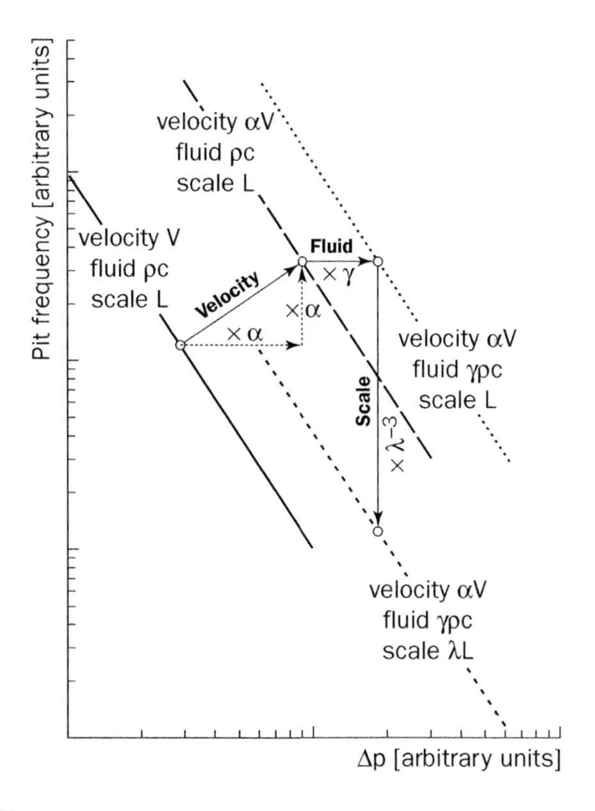

12.7 - Principle of the transposition of cumulative histograms of pressure pulse amplitudes *[from LECOFFRE et al., 1985]*

Concerning the change in length scale, figure 12.8 shows histograms of pit size for two similar Venturi test sections of diameters 40 and 120 mm, i.e. with a length scale ratio $\lambda = 3$. If we suppose that pit size (and bubble size) are proportional to the length scale (see e.g. FRANC & MICHEL 1997), figure 12.8 shows that pitting rate can be scaled as λ^n with n lying between −2 and −3. Hence, the scaling law given by equation (12.3) which predicts a transposition with power −3 appears acceptable, especially considering the difficulties in measuring pit size due to their shallowness, particularly for small pits.

It must be kept in mind that the scaling laws represented by equation (12.3) are based upon a first order model, taking into account only inertia and liquid compressibility, and neglecting all other physical phenomena such as viscosity, surface tension, dissolved gas content and metallurgical properties of the solid wall.

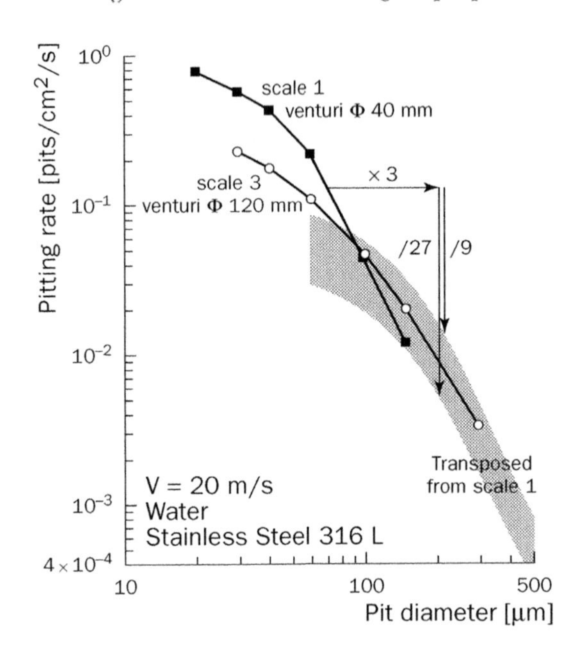

12.8 - Effect of length scale on histograms of pit size *[from FRANC & MICHEL, 1997]*

12.4.5. *ASYMPTOTIC BEHAVIOR OF PITTING RATE AT HIGH VELOCITIES*

The behavior of the pitting rate at high flow velocities deserves special consideration. Let $\dot{n}_V(\Delta p)$ be the aggressiveness of a cavitating flow with velocity V. If the material threshold is Δp^*, the pitting rate is $\dot{n}_V(\Delta p^*)$.

From the previous analysis, the pressure pulse height Δp increases proportionally to the flow velocity V, so that, for high enough velocities, it can be considered that all vapor structures give a pressure pulse greater than the material threshold Δp^*, and so actually produce a pit. Hence, the pitting rate is close to $\dot{n}_V(0)$ since $\Delta p >> \Delta p^*$.

For a velocity αV, we have, from equation (12.3):

$$\dot{n}_{\alpha V}(\alpha\,\Delta p) = \alpha\,\dot{n}_V(\Delta p) \qquad (12.5)$$

and for the limiting case of high velocities:

$$\dot{n}_{\alpha V}(0) = \alpha\,\dot{n}_V(0) \qquad (12.6)$$

Thus, the pitting rate should be proportional to the flow velocity V for high enough velocities, i.e. the pitting rate should asymptotically follow a power law V^n with an exponent n equal to unity. This behavior has been confirmed by experiments in a Venturi using mercury as the cavitating liquid. As shown in figure 12.14, at a velocity of approximately 6 m/s, a change in the exponent n is observed and the pitting rate becomes proportional to the flow velocity.

If equation (12.3) is used to transpose this characteristic velocity from mercury to water, and considering the ratio of about 13 between the acoustic impedances of the two liquids, it can be expected that this change of trend with the flow velocity should be observed at around 78 m/s in the case of water and stainless steel. This has not yet been confirmed since erosion tests above such a velocity in water are unusual.

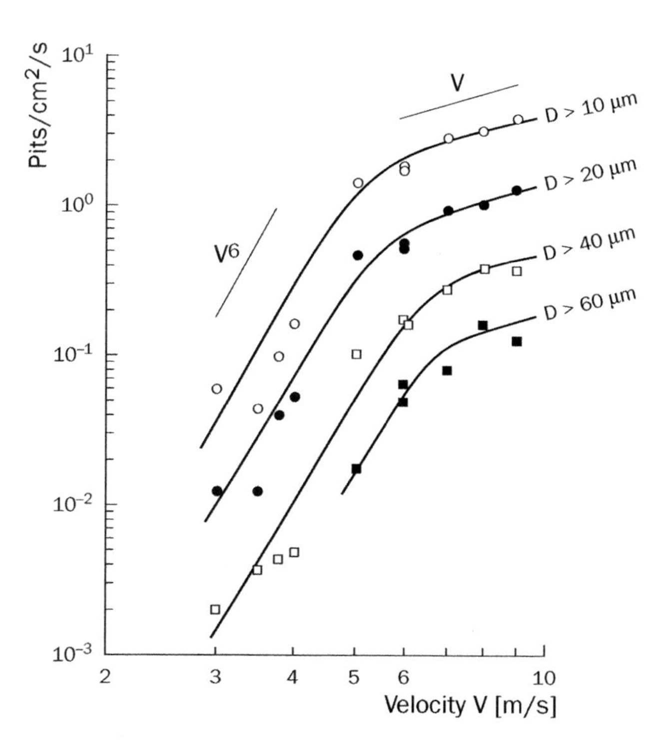

12.9 - Influence of flow velocity on pitting rate on 316 L stainless steel, for a cavitating 40 mm Venturi in mercury
[from BELAHADJI et al., 1991]

12.5. INSIGHT INTO THE MATERIAL RESPONSE

12.5.1. INTERACTION BETWEEN THE LIQUID FLOW AND A SOLID WALL

When a shock wave, emitted in the liquid, impacts on a solid wall with an elastic type behavior, the wall recoils and the shock wave intensity is weakened. At the same time, part of the incident wave is reflected and part is transmitted into the solid.

The phenomenon can be approached simply by considering a liquid jet (diameter d, velocity V_ℓ, density ρ_ℓ, sound velocity c_ℓ) impacting normally on a solid wall initially at rest (density ρ_S, sound velocity c_S). The conservation of mass and momentum in both the liquid and the solid allows us to compute the interface velocity V:

$$\frac{V}{V_\ell} = \frac{1}{1 + \dfrac{(\rho c)_s}{(\rho c)_\ell}} \tag{12.7}$$

and the actual impact pressure:

$$\Delta p = \frac{(\rho c)_\ell V_\ell}{1 + \dfrac{(\rho c)_\ell}{(\rho c)_s}} \tag{12.8}$$

The numerator $(\rho c)_\ell V_\ell$ in (12.8) is the impact pressure on a rigid wall (i.e. the denominator is equal to 1), as given by the classic water hammer formula. For an elastic material, this impact pressure Δp is then dampened depending on the ratio of the liquid and the solid acoustic impedances ρc.

In the case of water impacting upon usual metals such as aluminium or stainless steel, the incident pressure wave is not significantly attenuated by the elastic behavior of the wall since the acoustic impedance of water is usually much smaller than that of the wall (tab. 12.1). On the contrary, if the cavitating liquid is mercury, the shock wave loses one third of its initial amplitude upon impact against a stainless steel wall. The liquid/solid interaction is also predominant in the case of an elastic coating of small acoustic impedance which will considerably dampen the impact pressure and so may be used to limit cavitation erosion damage.

From the above value of the interfacial velocity V, we can estimate the total recoil distance of the wall, given the duration of the impact load. For example, for a water jet of diameter $d = 5\ \text{mm}$ and velocity $V_\ell = 100\ \text{m/s}$ impacting against a stainless steel wall, we find a recoil velocity of $V = 3.6\ \text{m/s}$ and an impact duration $\Delta t = d/2c_\ell = 1.7\ \mu\text{s}$. The total recoil of the wall is then less than 6 µm. Of course, these values are rough estimates since the elastic behavior of the solid material is a simplified modeling of its actual behavior.

	Water	Mercury	Aluminium	Stainless steel
ρ (kg / m^3)	1,000	13,550	2,700	7,800
c (m / s)	1,500	1,450	5,000	5,200
ρc (kg / m^2 s)	1.5×10^6	19.6×10^6	13.5×10^6	40.6×10^6

Table 12.1 - Density, sound velocity and acoustic impedance of some liquids and metals

12.5.2. *CAVITATION EROSION AND STRAIN RATE*

A fundamental characteristic of the impact load due to cavitation is the high value of the strain rate, connected to the short duration.

In the case of tensile tests, the strain rate (with units s^{-1}) is classically defined as:

$$\dot{\varepsilon} = \frac{d\varepsilon}{dt} = \frac{1}{\ell_0}\frac{d\ell}{dt} = \frac{v}{\ell_0} \tag{12.9}$$

where ℓ_0 is the initial length of the test bar and v the relative velocity of its extremities.

For a rough estimate of the strain rate in the case of cavitation impact, we can choose the velocity of the liquid vapor interface for v, which is typically of the order of 100 m/s.

As for the length scale ℓ_0, it should be said that successive impact loads due to bubble collapse lead, for most metal alloys, to an increase in the superficial hardness via a work-hardening process. This phenomenon is shown on figure 12.10 which presents a measured profile of hardness on the cross-section of a stainless steel specimen eroded by cavitation. Such a profile was obtained by a classical microhardness measurement technique. To estimate the strain rate, the thickness of the hardened layer is chosen as the characteristic length scale ℓ_0. Considering a typical value of 200 µm for this parameter, we find, with $v = 100$ m/s, a strain rate of the order of $5.10^5\ s^{-1}$.

As a reference, it can be noted that strain rates for tensile tests usually lie in the range 10^{-2} to $10^2\ s^{-1}$. Hence, classical tensile tests cannot be considered as fully representative of the cavitation erosion process.

The high value of the strain rate in cavitation erosion makes it rather comparable to explosions or projectile impacts. However, some crucial differences exist mostly concerning the volume of deformation, which is very limited in cavitation erosion, and the repetitive nature of the impact loads, which means that fatigue mechanisms have often to be expected. Those features indicate that cavitation erosion behaves as a specific damage mechanism and that correlations with data taken from classical tests have to be considered with care.

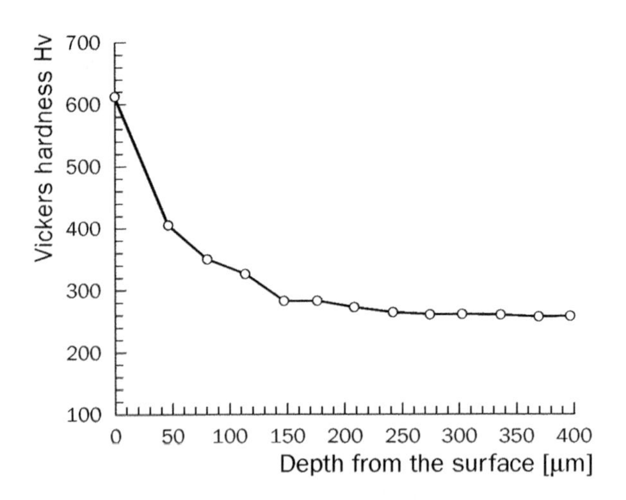

12.10 - Profile of hardness in a cross-section of an eroded 316 L stainless steel specimen measured using microhardness techniques *[from BERCHICHE et al., 2002]*

12.5.3. CORRELATION OF VOLUME LOSS WITH IMPACT ENERGY

In section 12.4.3, the technique of characterization of the aggressiveness of a cavitating flow by force measurements was presented. The signal given by pressure transducers is made up of a succession of pulses corresponding to impact loads of various maximum amplitudes F_i.

OKADA *et al.* (1995) suggested computing the quantity $\sum_i F_i^2$. For each impact, F_i^2 is a measure of the impact energy due to a collapsing vapor structure, and the summation $\sum_i F_i^2$ is a measure of the cumulative impact energy of all the vapor structures which have collapsed during the measurement time. The results of OKADA *et al.* (1995) show a dependency of this parameter on the material of the detection rod used in their pressure transducers (see fig. 12.5), although it should normally be considered as a purely hydrodynamic parameter.

From experiments on both a cavitating Venturi and a vibratory device at various operating conditions, OKADA *et al.* (1995) showed that, for aluminium and copper, the time evolution of the impact energy is similar to the time evolution of mass loss. Furthermore, they found a linear correlation between the volume loss and the impact energy, as shown on figure 12.11.

It is remarkable to observe that this correlation depends only upon the material. It does not depend upon the type of apparatus, Venturi or vibrating device, nor on the test conditions (flow velocity in the case of Venturi cavitation; distance between the vibrating disk and the specimen in the case of vibratory cavitation).

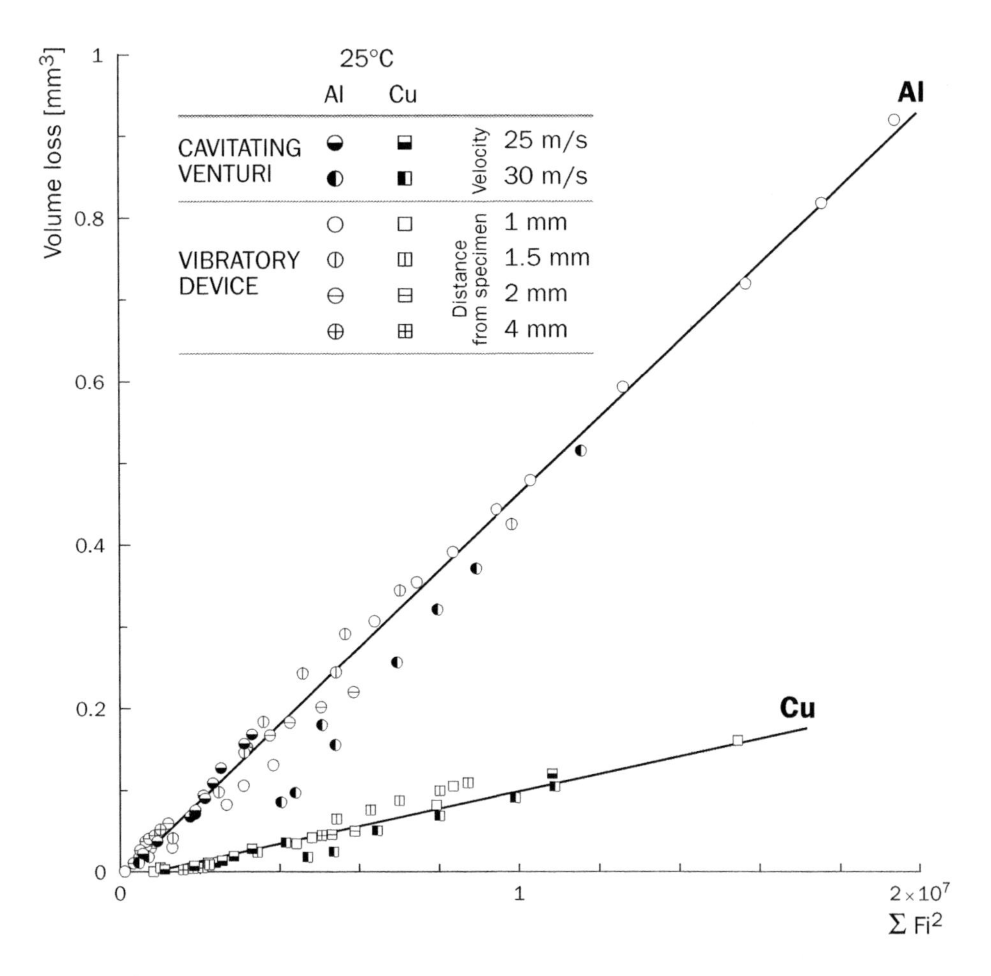

12.11 - Relation between impact energy and volume loss for Venturi and vibratory erosion tests under various operating conditions

The materials used for the detection rod of the pressure transducers are copper and aluminium – Reprinted from OKADA et al. (1995) with permission from Elsevier.

12.5.4. PHENOMENOLOGICAL MODEL FOR MASS LOSS PREDICTION

By way of example, we present here the analytical model developed by KARIMI and LEO (1987) for the prediction of cavitation erosion of ductile materials. In principle, it allows us to compute the mass loss when the flow aggressiveness in terms of impact loads, and the main mechanical and metallurgical properties of the material are known.

Three different classes of impact loads have to be considered according to their amplitude.

- Pressure pulses whose amplitude is lower than the yield strength are supposed to cause no damage and the material is supposed to return to its original state after unloading. This means that fatigue mechanisms due to a large number of repetitive impacts with an amplitude smaller than the yield strength are not taken into account in the model. As a consequence, this phenomenological model is only appropriate to cavitating flows of high aggressiveness.
- Pressure pulses whose amplitude is between the yield strength σ_Y and the ultimate tensile strength σ_U will contribute to the hardening of the material surface.
- Pressure pulses with an amplitude greater than the ultimate tensile strength are supposed to be the major cause of mass loss. These impacts are characterized by:
  - a mean value $\overline{\sigma} > \sigma_U$ of their amplitude,
  - a mean value $\overline{S}$ of the surface area of each individual impact,
  - and their rate $\dot{N}$ per unit time and unit surface area.

  In the present model, those three parameters are supposed to characterize the aggressiveness of the cavitating flow.

At the beginning of cavitation attack, the virgin material is very ductile and able to absorb a large part of the impact energy by plastic deformation. As mentioned in section 12.5.2, the accumulation of deformations induces a progressive hardening of the superficial layers of the material. It progressively becomes more fragile, its deformation ability diminishes and the probability of crack appearance and micro-ruptures increases. During this initial stage of cavitation erosion, the surface strain increases from zero, for the virgin material, up to the ultimate strain ε_U corresponding to the ultimate strength σ_U (fig. 12.12).

From that time, steady state erosion begins and mass loss remains constant. The strain profile inside the eroded material also remains constant. It is given by the following power-law:

$$\varepsilon = \varepsilon_U \left[1 - \frac{x}{L}\right]^{\theta} \tag{12.10}$$

where L is the depth of the hardened layer, θ a metallurgical parameter which measures the steepness of the hardening gradient and x the distance from the material surface. The metallurgical parameters L and θ can be estimated from micro-hardness measurements on a cross-section of the eroded specimen (see fig. 12.10). The θ-values lie typically between 2 and 5 for ordinary metals and reach 7 to 8 for industrial alloys. The thickness of the hardened layer is usually of the order of a few hundred micrometers for metal alloys.

The time necessary for a complete covering of the material surface by the impact loads of mean amplitude $\overline{\sigma}$ and surface area $\overline{S}$ at the rate $\dot{N}$ is clearly:

$$\Delta t = \frac{1}{\dot{N}\,\overline{S}} \tag{12.11}$$

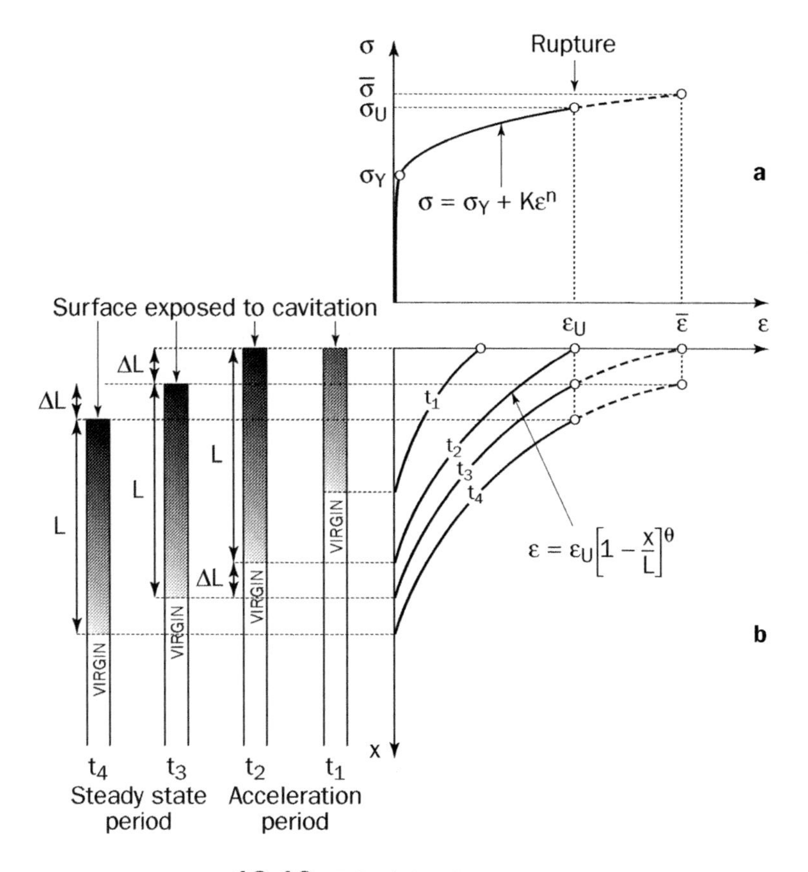

12.12 - Principle of the model

(a) stress-strain relationship - (b) evolution of the strain profile inside the material and of the hardening as a function of the exposure time [from KARIMI & LEO, 1987].

After this time, the surface strain is supposed to have uniformly reached the value $\bar{\varepsilon}$ corresponding to the stress $\bar{\sigma}$. Assuming a LUDWIK type consolidation relationship of the form:

$$\sigma = \sigma_Y + K\varepsilon^n \tag{12.12}$$

we obtain:

$$\bar{\varepsilon} = \left[\frac{\bar{\sigma} - \sigma_Y}{K}\right]^{1/n} = \varepsilon_U \left[\frac{\bar{\sigma} - \sigma_Y}{\sigma_U - \sigma_Y}\right]^{1/n} \tag{12.13}$$

The stress/strain relationship can be obtained by classical tensile tests, although such tests are made at a strain rate usually much smaller than that encountered in cavitation erosion (see § 12.5.2). Let us observe that this relationship is extrapolated here beyond the ultimate tensile strength.

After one complete covering of the material surface by cavitation impacts (between time t_2 and t_3 in figure 12.12, t_3 and t_4 and so on), the thickness of the hardened layer is incremented by ΔL and the strain profile inside the material is supposed to retain the same shape (see eq. 12.10):

$$\varepsilon = \bar{\varepsilon}\left[1 - \frac{x}{L + \Delta L}\right]^{\theta} \tag{12.14}$$

In the layer of thickness ΔL just below the material surface, the strain is greater than the ultimate strain ε_U. As the material cannot sustain a strain greater than its ultimate strain, the layer of depth ΔL is removed by cavitation. The previous equation gives:

$$\varepsilon_U = \bar{\varepsilon}\left[1 - \frac{\Delta L}{L + \Delta L}\right]^{\theta} \tag{12.15}$$

so that the thickness of the removed layer is:

$$\Delta L = L\left[\left(\frac{\bar{\varepsilon}}{\varepsilon_U}\right)^{1/\theta} - 1\right] \tag{12.16}$$

or, taking into account equation (12.13):

$$\Delta L = L\left[\left(\frac{\bar{\sigma} - \sigma_Y}{\sigma_U - \sigma_Y}\right)^{1/n\theta} - 1\right] \tag{12.17}$$

Finally, the volume loss rate per unit surface area or mean depth of penetration rate is given by:

$$\mathrm{MDPR} = \frac{\Delta L}{\Delta t} = \dot{N}\bar{S}L\left[\left(\frac{\bar{\sigma} - \sigma_Y}{\sigma_U - \sigma_Y}\right)^{1/n\theta} - 1\right] \tag{12.18}$$

The previous equation is valid only for the steady state regime of cavitation erosion. The computation of the acceleration period including the incubation time can be found in the original paper of KARIMI and LEO (1987). A somewhat improved version of the model taking into account pressure pulses of different amplitudes and different sizes was developed by BERCHICHE *et al.* (2002).

Two different kinds of parameter appear in equation (12.18):

- the mean amplitude $\bar{\sigma}$ and size $\bar{S}$ of the impact loads, which are purely hydrodynamic parameters, and
- the yield strength σ_Y, the ultimate strength σ_U, the exponent n of the stress-strain relationship, the thickness of the hardened layer L and the shape factor θ of the strain profile, which are mechanical or metallurgical parameters characterizing the material response.

Several attempts have been made to compare the prediction given by equation (12.18) to experiments [KARIMI & LEO 1987, FRANC *et al.* 1994, BERCHICHE *et al.* 2002]. The agreement is far from being perfect but it is reasonably good considering the significant difficulties in this field. In particular, it should be observed that the order of magnitude of the erosion rate in the steady state period is consistent with tests, in so far as no adjustable parameter is involved in the present model. Significant deviations are however unavoidable because of a few drastic oversimplifications such as neglecting the influence of strain rate or fatigue mechanisms.

As suggested by KATO *et al.* (1996), it would be interesting to couple this kind of model of material behavior with a theoretical model of prediction of hydrodynamic aggressiveness to allow the development of a fully predictive method of computation of cavitation damage.

REFERENCES

ALLONCLE A.P., DUFRESNE D. & TESTUD P. –1992– Étude expérimentale de bulles de vapeur générées par laser. *La Houille Blanche* **7/8**, 539-544.

AVELLAN F. & FARHAT M. –1989– Shock pressure generated by cavitation vortex collapse. *Proc. Int. Symp. on Cavitation Noise and Erosion in Fluid Systems,* FED **88**, San Francisco (USA), December 10-15, 119-125.

BELAHADJI B., FRANC J.P. & MICHEL J.M. –1991– A statistical analysis of cavitation erosion pits. *J. Fluids Eng.* **113**, 700-706.

BERCHICHE N., FRANC J.P. & MICHEL J.M. –2002– A cavitation erosion model for ductile materials. *J. Fluids Eng.* **124**, 601-606.

DOMINGUEZ-CORTAZAR M.A., MICHEL J.M. & FRANC J.P. –1997– The erosive axial collapse of a cavitating vortex: an experimental study. *J. Fluids Eng.* **119**, 686-691.

FILALI E.G. & MICHEL J.M. –1999a– The cavermod device: hydrodynamic aspects and erosion tests. *J. Fluids Eng.* **121**, 305-311.

FILALI E.G., MICHEL J.M., HATTORI S. & FUJIKAWA S. –1999b– The cavermod device: force measurements. *J. Fluids Eng.* **121**, 312-317.

FRANC J.P., MICHEL J.M., NGUYEN TRONG H. & KARIMI A. –1994– From pressure pulses measurements to mass loss prediction: the analysis of a method. *Proc. 2nd Int. Symp. on Cavitation*, Tokyo (Japan), April 5-7, 231-236.

FRANC J.P. & MICHEL J.M. –1997– Cavitation erosion in France: the state of the art. *J. Mar. Sci. Technol.* **2**, 233-244.

FUJIKAWA S. & AKAMATSU T. –1980– Effects of non-equilibrium condensation of vapor on the pressure wave produced by the collapse of a bubble in a liquid. *J. Fluid Mech.* **97**, part 3, 481-512.

HATTORI S., MIYOSHI K., BUCKEY D.H. & OKADA T. –1986– Plastic deformation of magnesium oxide {001} surface produced by cavitation. *J. Soc. Tribol. Lubr. Eng.* **44**(1), 53-60.

KARIMI A. & AVELLAN F. –1986– Comparison of erosion mechanisms in different types of cavitation. *Wear* **113**, 305-322.

KARIMI A. & LEO W.R. –1987– Phenomenological model for cavitation rate computation. *Mat. Sci. Eng.* **95**, 1-14.

KARIMI A. –1988– Modèle mathématique pour la prédiction de la vitesse d'érosion. *La Houille Blanche* **7/8**, 571-576.

KATO H. –1975– A consideration on scaling laws of cavitation erosion. *Intern. Shipbuilding Progress* **22**(253), 305-327.

KATO H., KONNO A., MAEDA M. & YAMAGUCHI H. –1996– Possibility of quantitative prediction of cavitation erosion without model test. *J. Fluids Eng.* **118**(3), 582-588.

KENN M.J. & GARROD A.D. –1981– Cavitation damage and the Tarbela Tunnel collapse of 1974. *Proc. Inst. Civ. Eng.* **70**, part 1, 65-89.

KNAPP R.T. –1955– Recent investigations of cavitation and cavitation damage. *Trans. ASME* **77**, 1045-1054.

LECOFFRE Y., MARCOZ J.& VALIBOUSE B. –1981– Generator of cavitation vortex. *ASME Fluids Eng. Conf.*, Boulder (USA).

LECOFFRE Y., MARCOZ J., FRANC J.P. & MICHEL J.M. –1985– Tentative procedure for scaling cavitation damage. *Proc. Int. Symp. on Cavitation in Hydraulic Structures and Turbomachinery*, Albuquerque (USA), June 24-26.

LECOFFRE Y. –1995– Cavitation erosion, hydrodynamics scaling laws, practical method of long term damage prediction. *Proc. Int. Symp. on Cavitation*, Deauville (France), May 2-5, 249-256.

MOMMA T. –1991– Cavitation loading and erosion produced by a cavitating jet. *PhD Thesis*, Nottingham University (England).

NGUYEN TRONG H. –1993– Développement et validation d'une méthode analytique de prévision de l'érosion de cavitation. *PhD Thesis*, Institut National Polytechnique de Grenoble (France).

OBA R. –1994– The severe cavitation erosion. *Proc. 2nd Int. Symp. on Cavitation*, Tokyo (Japan), April 5-7, 1-8.

OKADA T., HATTORI S. & SHIMIZU M. –1994– A fundamental study of cavitation erosion using a magnesium oxide single crystal (dislocation and surface roughness). *Proc. 2nd Int. Symp. on Cavitation*, Tokyo (Japan), April 5-7, 185-190.

Okada T., Iwai Y., Hattori S. & Tanimura N. –1995– Relation between impact load and the damage produced by cavitation bubble collapse. *Wear* **184**, 231-239.

Philipp A. & Lauterborn W. –1998– Cavitation erosion by single laser-produced bubbles. *J. Fluid Mech.* **361**, 75-116.

Preece C.M. –1979– Cavitation erosion. Treatise on Materials Science and Technology. Erosion. *Academic Press* **16**, 249-308.

Reisman G.E., Wang Y.-C. & Brennen C.E. –1998– Observations of shock waves in cloud cavitation. *J. Fluid Mech.* **355**, 255-283

Sato K. & Kondo S. –1996– Collapsing behaviour of vortex cavitation bubble near solid wall: spanwise-view study. *ASME Fluids Eng. Conf.* **1**, 485-490.

Simoneau R. –1995– Cavitation pit counting and steady state erosion rate. *Proc. Int. Symp. on Cavitation*, Deauville (France), May 2-5, 265-276.

Soyama H., Lichtarowicz A., Momma T. & Williams E. –1998– A new calibration method for dynamically loaded transducers and its application to cavitation impact measurement. *J. Fluids Eng.* **120**, 712-718.

Stinebring D.R., Arndt R.E.A. & Holl J.W. –1976– Scaling laws of cavitation damage. *J. Hydronautics* **11**, 1977.

Thiruvengadam A. –1963– The concept of erosion strength. *J. Basic Eng.* **85**, *serie D*, 365-376.

Tomita Y. & Shima A. –1986– Mechanisms of impulsive pressure generation and damage pit formation by bubble collapse. *J. Fluid Mech.* **169**, 535-564.

INDEX

A

B

C

D

E

F

G

H

I

J

K

L

M

N

O

P

Q

R

S

T

V

W

Mechanics

FLUID MECHANICS AND ITS APPLICATIONS

Series Editor: R. Moreau

Aims and Scope of the Series

The purpose of this series is to focus on subjects in which fluid mechanics plays a fundamental role. As well as the more traditional applications of aeronautics, hydraulics, heat and mass transfer etc., books will be published dealing with topics which are currently in a state of rapid development, such as turbulence, suspensions and multiphase fluids, super and hypersonic flows and numerical modelling techniques. It is a widely held view that it is the interdisciplinary subjects that will receive intense scientific attention, bringing them to the forefront of technological advancement. Fluids have the ability to transport matter and its properties as well as transmit force, therefore fluid mechanics is a subject that is particularly open to cross fertilisation with other sciences and disciplines of engineering. The subject of fluid mechanics will be highly relevant in domains such as chemical, metallurgical, biological and ecological engineering. This series is particularly open to such new multidisciplinary domains.

1. M. Lesieur: *Turbulence in Fluids.* 2nd rev. ed., 1990 ISBN 0-7923-0645-7
2. O. Métais and M. Lesieur (eds.): *Turbulence and Coherent Structures.* 1991 ISBN 0-7923-0646-5
3. R. Moreau: *Magnetohydrodynamics.* 1990 ISBN 0-7923-0937-5
4. E. Coustols (ed.): *Turbulence Control by Passive Means.* 1990 ISBN 0-7923-1020-9
5. A.A. Borissov (ed.): *Dynamic Structure of Detonation in Gaseous and Dispersed Media.* 1991 ISBN 0-7923-1340-2
6. K.-S. Choi (ed.): *Recent Developments in Turbulence Management.* 1991 ISBN 0-7923-1477-8
7. E.P. Evans and B. Coulbeck (eds.): *Pipeline Systems.* 1992 ISBN 0-7923-1668-1
8. B. Nau (ed.): *Fluid Sealing.* 1992 ISBN 0-7923-1669-X
9. T.K.S. Murthy (ed.): *Computational Methods in Hypersonic Aerodynamics.* 1992 ISBN 0-7923-1673-8
10. R. King (ed.): *Fluid Mechanics of Mixing.* Modelling, Operations and Experimental Techniques. 1992 ISBN 0-7923-1720-3
11. Z. Han and X. Yin: *Shock Dynamics.* 1993 ISBN 0-7923-1746-7
12. L. Svarovsky and M.T. Thew (eds.): *Hydroclones.* Analysis and Applications. 1992 ISBN 0-7923-1876-5
13. A. Lichtarowicz (ed.): *Jet Cutting Technology.* 1992 ISBN 0-7923-1979-6
14. F.T.M. Nieuwstadt (ed.): *Flow Visualization and Image Analysis.* 1993 ISBN 0-7923-1994-X
15. A.J. Saul (ed.): *Floods and Flood Management.* 1992 ISBN 0-7923-2078-6
16. D.E. Ashpis, T.B. Gatski and R. Hirsh (eds.): *Instabilities and Turbulence in Engineering Flows.* 1993 ISBN 0-7923-2161-8
17. R.S. Azad: *The Atmospheric Boundary Layer for Engineers.* 1993 ISBN 0-7923-2187-1
18. F.T.M. Nieuwstadt (ed.): *Advances in Turbulence IV.* 1993 ISBN 0-7923-2282-7
19. K.K. Prasad (ed.): *Further Developments in Turbulence Management.* 1993 ISBN 0-7923-2291-6
20. Y.A. Tatarchenko: *Shaped Crystal Growth.* 1993 ISBN 0-7923-2419-6
21. J.P. Bonnet and M.N. Glauser (eds.): *Eddy Structure Identification in Free Turbulent Shear Flows.* 1993 ISBN 0-7923-2449-8
22. R.S. Srivastava: *Interaction of Shock Waves.* 1994 ISBN 0-7923-2920-1
23. J.R. Blake, J.M. Boulton-Stone and N.H. Thomas (eds.): *Bubble Dynamics and Interface Phenomena.* 1994 ISBN 0-7923-3008-0

Mechanics

FLUID MECHANICS AND ITS APPLICATIONS

Series Editor: R. Moreau

24. R. Benzi (ed.): *Advances in Turbulence V.* 1995 ISBN 0-7923-3032-3
25. B.I. Rabinovich, V.G. Lebedev and A.I. Mytarev: *Vortex Processes and Solid Body Dynamics.* The Dynamic Problems of Spacecrafts and Magnetic Levitation Systems. 1994 ISBN 0-7923-3092-7
26. P.R. Voke, L. Kleiser and J.-P. Chollet (eds.): *Direct and Large-Eddy Simulation I.* Selected papers from the First ERCOFTAC Workshop on Direct and Large-Eddy Simulation. 1994 ISBN 0-7923-3106-0
27. J.A. Sparenberg: *Hydrodynamic Propulsion and its Optimization.* Analytic Theory. 1995 ISBN 0-7923-3201-6
28. J.F. Dijksman and G.D.C. Kuiken (eds.): *IUTAM Symposium on Numerical Simulation of Non-Isothermal Flow of Viscoelastic Liquids.* Proceedings of an IUTAM Symposium held in Kerkrade, The Netherlands. 1995 ISBN 0-7923-3262-8
29. B.M. Boubnov and G.S. Golitsyn: *Convection in Rotating Fluids.* 1995 ISBN 0-7923-3371-3
30. S.I. Green (ed.): *Fluid Vortices.* 1995 ISBN 0-7923-3376-4
31. S. Morioka and L. van Wijngaarden (eds.): *IUTAM Symposium on Waves in Liquid/Gas and Liquid/Vapour Two-Phase Systems.* 1995 ISBN 0-7923-3424-8
32. A. Gyr and H.-W. Bewersdorff: *Drag Reduction of Turbulent Flows by Additives.* 1995 ISBN 0-7923-3485-X
33. Y.P. Golovachov: *Numerical Simulation of Viscous Shock Layer Flows.* 1995 ISBN 0-7923-3626-7
34. J. Grue, B. Gjevik and J.E. Weber (eds.): *Waves and Nonlinear Processes in Hydrodynamics.* 1996 ISBN 0-7923-4031-0
35. P.W. Duck and P. Hall (eds.): *IUTAM Symposium on Nonlinear Instability and Transition in Three-Dimensional Boundary Layers.* 1996 ISBN 0-7923-4079-5
36. S. Gavrilakis, L. Machiels and P.A. Monkewitz (eds.): *Advances in Turbulence VI.* Proceedings of the 6th European Turbulence Conference. 1996 ISBN 0-7923-4132-5
37. K. Gersten (ed.): *IUTAM Symposium on Asymptotic Methods for Turbulent Shear Flows at High Reynolds Numbers.* Proceedings of the IUTAM Symposium held in Bochum, Germany. 1996 ISBN 0-7923-4138-4
38. J. Verhás: *Thermodynamics and Rheology.* 1997 ISBN 0-7923-4251-8
39. M. Champion and B. Deshaies (eds.): *IUTAM Symposium on Combustion in Supersonic Flows.* Proceedings of the IUTAM Symposium held in Poitiers, France. 1997 ISBN 0-7923-4313-1
40. M. Lesieur: *Turbulence in Fluids.* Third Revised and Enlarged Edition. 1997 ISBN 0-7923-4415-4; Pb: 0-7923-4416-2
41. L. Fulachier, J.L. Lumley and F. Anselmet (eds.): *IUTAM Symposium on Variable Density Low-Speed Turbulent Flows.* Proceedings of the IUTAM Symposium held in Marseille, France. 1997 ISBN 0-7923-4602-5
42. B.K. Shivamoggi: *Nonlinear Dynamics and Chaotic Phenomena.* An Introduction. 1997 ISBN 0-7923-4772-2
43. H. Ramkissoon, *IUTAM Symposium on Lubricated Transport of Viscous Materials.* Proceedings of the IUTAM Symposium held in Tobago, West Indies. 1998 ISBN 0-7923-4897-4
44. E. Krause and K. Gersten, *IUTAM Symposium on Dynamics of Slender Vortices.* Proceedings of the IUTAM Symposium held in Aachen, Germany. 1998 ISBN 0-7923-5041-3
45. A. Biesheuvel and G.J.F. van Heyst (eds.): *In Fascination of Fluid Dynamics.* A Symposium in honour of Leen van Wijngaarden. 1998 ISBN 0-7923-5078-2

Mechanics

FLUID MECHANICS AND ITS APPLICATIONS

Series Editor: R. Moreau

46. U. Frisch (ed.): *Advances in Turbulence VII.* Proceedings of the Seventh European Turbulence Conference, held in Saint-Jean Cap Ferrat, 30 June–3 July 1998. 1998 ISBN 0-7923-5115-0
47. E.F. Toro and J.F. Clarke: *Numerical Methods for Wave Propagation.* Selected Contributions from the Workshop held in Manchester, UK. 1998 ISBN 0-7923-5125-8
48. A. Yoshizawa: *Hydrodynamic and Magnetohydrodynamic Turbulent Flows.* Modelling and Statistical Theory. 1998 ISBN 0-7923-5225-4
49. T.L. Geers (ed.): *IUTAM Symposium on Computational Methods for Unbounded Domains.* 1998 ISBN 0-7923-5266-1
50. Z. Zapryanov and S. Tabakova: *Dynamics of Bubbles, Drops and Rigid Particles.* 1999 ISBN 0-7923-5347-1
51. A. Alemany, Ph. Marty and J.P. Thibault (eds.): *Transfer Phenomena in Magnetohydrodynamic and Electroconducting Flows.* 1999 ISBN 0-7923-5532-6
52. J.N. Sørensen, E.J. Hopfinger and N. Aubry (eds.): *IUTAM Symposium on Simulation and Identification of Organized Structures in Flows.* 1999 ISBN 0-7923-5603-9
53. G.E.A. Meier and P.R. Viswanath (eds.): *IUTAM Symposium on Mechanics of Passive and Active Flow Control.* 1999 ISBN 0-7923-5928-3
54. D. Knight and L. Sakell (eds.): *Recent Advances in DNS and LES.* 1999 ISBN 0-7923-6004-4
55. P. Orlandi: *Fluid Flow Phenomena.* A Numerical Toolkit. 2000 ISBN 0-7923-6095-8
56. M. Stanislas, J. Kompenhans and J. Westerveel (eds.): *Particle Image Velocimetry.* Progress towards Industrial Application. 2000 ISBN 0-7923-6160-1
57. H.-C. Chang (ed.): *IUTAM Symposium on Nonlinear Waves in Multi-Phase Flow.* 2000 ISBN 0-7923-6454-6
58. R.M. Kerr and Y. Kimura (eds.): *IUTAM Symposium on Developments in Geophysical Turbulence* held at the National Center for Atmospheric Research, (Boulder, CO, June 16–19, 1998) 2000 ISBN 0-7923-6673-5
59. T. Kambe, T. Nakano and T. Miyauchi (eds.): *IUTAM Symposium on Geometry and Statistics of Turbulence.* Proceedings of the IUTAM Symposium held at the Shonan International Village Center, Hayama (Kanagawa-ken, Japan November 2–5, 1999). 2001 ISBN 0-7923-6711-1
60. V.V. Aristov: *Direct Methods for Solving the Boltzmann Equation and Study of Nonequilibrium Flows.* 2001 ISBN 0-7923-6831-2
61. P.F. Hodnett (ed.): *IUTAM Symposium on Advances in Mathematical Modelling of Atmosphere and Ocean Dynamics.* Proceedings of the IUTAM Symposium held in Limerick, Ireland, 2–7 July 2000. 2001 ISBN 0-7923-7075-9
62. A.C. King and Y.D. Shikhmurzaev (eds.): *IUTAM Symposium on Free Surface Flows.* Proceedings of the IUTAM Symposium held in Birmingham, United Kingdom, 10–14 July 2000. 2001 ISBN 0-7923-7085-6
63. A. Tsinober: *An Informal Introduction to Turbulence.* 2001 ISBN 1-4020-0110-X; Pb: 1-4020-0166-5
64. R.Kh. Zeytounian: *Asymptotic Modelling of Fluid Flow Phenomena.* 2002 ISBN 1-4020-0432-X
65. R. Friedrich and W. Rodi (eds.): *Advances in LES of Complex Flows.* Prodeedings of the EUROMECH Colloquium 412, held in Munich, Germany, 4-6 October 2000. 2002 ISBN 1-4020-0486-9
66. D. Drikakis and B.J. Geurts (eds.): *Turbulent Flow Computation.* 2002 ISBN 1-4020-0523-7
67. B.O. Enflo and C.M. Hedberg: *Theory of Nonlinear Acoustics in Fluids.* 2002 ISBN 1-4020-0572-5

Mechanics

FLUID MECHANICS AND ITS APPLICATIONS

Series Editor: R. Moreau

68. I.D. Abrahams, P.A. Martin and M.J. Simon (eds.): *IUTAM Symposium on Diffraction and Scattering in Fluid Mechanics and Elasticity*. Proceedings of the IUTAM Symposium held in Manchester, (UK, 16-20 July 2000). 2002 ISBN 1-4020-0590-3
69. P. Chassaing, R.A. Antonia, F. Anselmet, L. Joly and S. Sarkar: *Variable Density Fluid Turbulence*. 2002 ISBN 1-4020-0671-3
70. A. Pollard and S. Candel (eds.): *IUTAM Symposium on Turbulent Mixing and Combustion*. Proceedings of the IUTAM Symposium held in Kingston, Ontario, Canada, June 3-6, 2001. 2002 ISBN 1-4020-0747-7
71. K. Bajer and H.K. Moffatt (eds.): *Tubes, Sheets and Singularities in Fluid Dynamics*. 2002 ISBN 1-4020-0980-1
72. P.W. Carpenter and T.J. Pedley (eds.): *Flow Past Highly Compliant Boundaries and in Collapsible Tubes*. IUTAM Symposium held at the Univerity of Warwick, Coventry, United Kingdom, 26-30 March 2001. 2003 ISBN 1-4020-1161-X
73. H. Sobieczky (ed.): *IUTAM Symposium Transsonicum IV*. Proceedings of the IUTAM Symposium held in Göttingen, Germany, 2-6 September 2002. 2003 ISBN 1-4020-1608-5
74. A.J. Smits (ed.): *IUTAM Symposium on Reynolds Number Scaling in Turbulent Flow*. Proceedings of the IUTAM Symposium held in Princeton, NJ, U.S.A., September 11-13, 2002. 2003 ISBN 1-4020-1775-8
75. H. Benaroya and T. Wei (eds.): *IUTAM Symposium on Integrated Modeling of Fully Coupled Fluid Structure Interactions Using Analysis, Computations and Experiments*. Proceedings of the IUTAM Symposium held in New Jersey, U.S.A., 2-6 June 2003. 2003 ISBN 1-4020-1806-1
76. J.-P. Franc and J.-M. Michel: *Fundamentals of Cavitation*. 2004 ISBN 1-4020-2232-8

Kluwer Academic Publishers – Dordrecht / Boston / London

Mechanics

SOLID MECHANICS AND ITS APPLICATIONS

Series Editor: G.M.L. Gladwell

90. Y. Ivanov, V. Cheshkov and M. Natova: *Polymer Composite Materials – Interface Phenomena & Processes*. 2001 ISBN 0-7923-7008-2
91. R.C. McPhedran, L.C. Botten and N.A. Nicorovici (eds.): *IUTAM Symposium on Mechanical and Electromagnetic Waves in Structured Media*. Proceedings of the IUTAM Symposium held in Sydney, NSW, Australia, 18-22 Januari 1999. 2001 ISBN 0-7923-7038-4
92. D.A. Sotiropoulos (ed.): *IUTAM Symposium on Mechanical Waves for Composite Structures Characterization*. Proceedings of the IUTAM Symposium held in Chania, Crete, Greece, June 14-17, 2000. 2001 ISBN 0-7923-7164-X
93. V.M. Alexandrov and D.A. Pozharskii: *Three-Dimensional Contact Problems*. 2001 ISBN 0-7923-7165-8
94. J.P. Dempsey and H.H. Shen (eds.): *IUTAM Symposium on Scaling Laws in Ice Mechanics and Ice Dynamics*. Proceedings of the IUTAM Symposium held in Fairbanks, Alaska, U.S.A., 13-16 June 2000. 2001 ISBN 1-4020-0171-1
95. U. Kirsch: *Design-Oriented Analysis of Structures*. A Unified Approach. 2002 ISBN 1-4020-0443-5
96. A. Preumont: *Vibration Control of Active Structures*. An Introduction (2nd Edition). 2002 ISBN 1-4020-0496-6
97. B.L. Karihaloo (ed.): *IUTAM Symposium on Analytical and Computational Fracture Mechanics of Non-Homogeneous Materials*. Proceedings of the IUTAM Symposium held in Cardiff, U.K., 18-22 June 2001. 2002 ISBN 1-4020-0510-5
98. S.M. Han and H. Benaroya: *Nonlinear and Stochastic Dynamics of Compliant Offshore Structures*. 2002 ISBN 1-4020-0573-3
99. A.M. Linkov: *Boundary Integral Equations in Elasticity Theory*. 2002 ISBN 1-4020-0574-1
100. L.P. Lebedev, I.I. Vorovich and G.M.L. Gladwell: *Functional Analysis*. Applications in Mechanics and Inverse Problems (2nd Edition). 2002 ISBN 1-4020-0667-5; Pb: 1-4020-0756-6
101. Q.P. Sun (ed.): *IUTAM Symposium on Mechanics of Martensitic Phase Transformation in Solids*. Proceedings of the IUTAM Symposium held in Hong Kong, China, 11-15 June 2001. 2002 ISBN 1-4020-0741-8
102. M.L. Munjal (ed.): *IUTAM Symposium on Designing for Quietness*. Proceedings of the IUTAM Symposium held in Bangkok, India, 12-14 December 2000. 2002 ISBN 1-4020-0765-5
103. J.A.C. Martins and M.D.P. Monteiro Marques (eds.): *Contact Mechanics*. Proceedings of the 3rd Contact Mechanics International Symposium, Praia da Consolação, Peniche, Portugal, 17-21 June 2001. 2002 ISBN 1-4020-0811-2
104. H.R. Drew and S. Pellegrino (eds.): *New Approaches to Structural Mechanics, Shells and Biological Structures*. 2002 ISBN 1-4020-0862-7
105. J.R. Vinson and R.L. Sierakowski: *The Behavior of Structures Composed of Composite Materials*. Second Edition. 2002 ISBN 1-4020-0904-6
106. Not yet published.
107. J.R. Barber: *Elasticity*. Second Edition. 2002 ISBN Hb 1-4020-0964-X; Pb 1-4020-0966-6
108. C. Miehe (ed.): *IUTAM Symposium on Computational Mechanics of Solid Materials at Large Strains*. Proceedings of the IUTAM Symposium held in Stuttgart, Germany, 20-24 August 2001. 2003 ISBN 1-4020-1170-9

Mechanics

SOLID MECHANICS AND ITS APPLICATIONS

Series Editor: G.M.L. Gladwell

109. P. Ståhle and K.G. Sundin (eds.): *IUTAM Symposium on Field Analyses for Determination of Material Parameters – Experimental and Numerical Aspects.* Proceedings of the IUTAM Symposium held in Abisko National Park, Kiruna, Sweden, July 31 – August 4, 2000. 2003 ISBN 1-4020-1283-7
110. N. Sri Namachchivaya and Y.K. Lin (eds.): *IUTAM Symposium on Nonlnear Stochastic Dynamics.* Proceedings of the IUTAM Symposium held in Monticello, IL, USA, 26 – 30 August, 2000. 2003 ISBN 1-4020-1471-6
111. H. Sobieckzky (ed.): *IUTAM Symposium Transsonicum IV.* Proceedings of the IUTAM Symposium held in Göttingen, Germany, 2–6 September 2002, 2003 ISBN 1-4020-1608-5
112. J.-C. Samin and P. Fisette: *Symbolic Modeling of Multibody Systems.* 2003 ISBN 1-4020-1629-8
113. A.B. Movchan (ed.): *IUTAM Symposium on Asymptotics, Singularities and Homogenisation in Problems of Mechanics.* Proceedings of the IUTAM Symposium held in Liverpool, United Kingdom, 8-11 July 2002. 2003 ISBN 1-4020-1780-4
114. S. Ahzi, M. Cherkaoui, M.A. Khaleel, H.M. Zbib, M.A. Zikry and B. LaMatina (eds.): *IUTAM Symposium on Multiscale Modeling and Characterization of Elastic-Inelastic Behavior of Engineering Materials.* Proceedings of the IUTAM Symposium held in Marrakech, Morocco, 20-25 October 2002. 2004 ISBN 1-4020-1861-4
115. H. Kitagawa and Y. Shibutani (eds.): *IUTAM Symposium on Mesoscopic Dynamics of Fracture Process and Materials Strength.* Proceedings of the IUTAM Symposium held in Osaka, Japan, 6-11 July 2003. Volume in celebration of Professor Kitagawa's retirement. 2004 ISBN 1-4020-2037-6

Kluwer Academic Publishers – Dordrecht / Boston / London

9 781402 022326